APPLIED GENETICS

RECENT TRENDS AND TECHNIQUES

APPLIED GENETICS
RECENT TRENDS AND TECHNIQUES

C EMMANUEL
Director and Senior Scientist
International Centre for Biomedical Sciences and Technology
A unit of Frontier Lifeline and Dr. K. M. Cherian Heart Foundation
Chennai

Rev. Fr. S IGNACIMUTHU s.j.
Director
Entomology Research Institute
Loyola College
Chennai

S VINCENT
Reader
P.G. & Research Department of Advanced Zoology and Biotechnology
Loyola College
Chennai

MJP PUBLISHERS
Chennai 600 005

Cataloguing-in-Publication Data

Emmanuel, C. (1969–   ).
    Applied genetics: recent trends
and techniques / by C. Emmanuel,
S. Ignacimuthu and S. Vincent. –
Chennai : MJP Publishers, 2006.
    xxiv, 450p. ; 23 cm.
    Includes glossary and index.
    ISBN 81-8094-023-3 (pbk.)
    1. Genetics, applied. I. Ignacimuthu, S
II. Vincent, S III. Title
    576.5 dc22   EMM   MJP 018

ISBN 81-8094-023-3
Copyright © Publishers, 2006
All rights reserved
Printed and bound in India

MJP PUBLISHERS
A unit of Tamilnadu Book House
47, Nallathambi Street
Triplicane, Chennai 600 005

Publisher   : J.C. Pillai
Managing Editor   : C. Sajeesh Kumar
Project Coordinator : P. Parvath Radha

Edited and Typeset at ▨ Editorial Services, Chennai, eserve@rediffmail.com
Cover : R. Shankari   CIP : Prof K. Hariharan

PREFACE

Genetics is the science of heredity that involves the structure and function of genes and the way genes are passed from one generation to the next. Applied genetics refers to the structure and function of genes and their involvement in various practical fields, and reflects the dynamic nature of the field of genetics.

Gregor Mendel, the Austrian monk is considered the father of Genetics for his original contributions related to the theory of inheritance and the theory of transmissible factors in 1866. From 1900 onwards many discoveries were made to identify the transmissible factors, such as that of chromosome, nucleic acid, DNA, etc. Watson and Crick described the structure of DNA in 1953 and from then on genetics has become one of the most exciting and ground-breaking sciences.

Applied genetics has revolutionized the world today through the development of gene cloning in the 1970s, PCR in the 1980s, sequencing of genomes in the 1990s and stem cell research in the 2000s. As we move into the postgenomic era, our knowledge about genes and gene functions will increase enormously. Hence, it is very important that students learn the basics of genetics with its varied applications.

This book contains 30 chapters (which include the basic genetic principles, the molecular aspects and the latest applied areas). Every chapter has been presented in a simple style and sequential organization for effective learning. Questions are also provided for review. Care has been taken to make the text easy to read. We are hopeful that this book will cater to the immediate needs of students and the faculty members.

C Emmanuel
Rev. Fr. S Ignacimuthu s.j.
S Vincent

CONTENTS

THE CELL

1

1.1 INTRODUCTION

The **cell** is the basic structural and functional unit of all living organisms. It is actively delimited by a selectively permeable membrane and is capable of self-reproduction. The term "Cell" was coined by Robert Hooke in 1665. The structural unit (cell) is now known as the unit of life and the concept that the cell is basic unit of life is known as the **cell theory**. The **cell theory** was established by Schleiden (1838) and Schwann (1839). Figure 1.1 is the sectional view of a eukaryotic cell showing its components.

1.2 CHEMISTRY OF THE CELL

Cells of all living organisms are made up of chemicals, which obey the laws of chemistry and physics. Cells are composed of **water, inorganic ions** and **organic molecules**. Water is the most abundant molecule in cells accounting for more than 70% of the total wet cell mass. Being a polar molecule, its hydrogen atoms have a slight positive charge and the oxygen has a slight negative charge. They can form hydrogen bonds with each other or with other polar molecules as well as interact with positively or negatively charged ions. Due to this, many ions and polar molecules are readily soluble in water (hydrophilic). However the nonpolar molecules cannot interact with water (hydrophobic) and also in other aqueous environment. They minimize their contact with water by associating closely with each other. Interactions of macromolecules with water and with each other play important roles in the formation of biological structures.

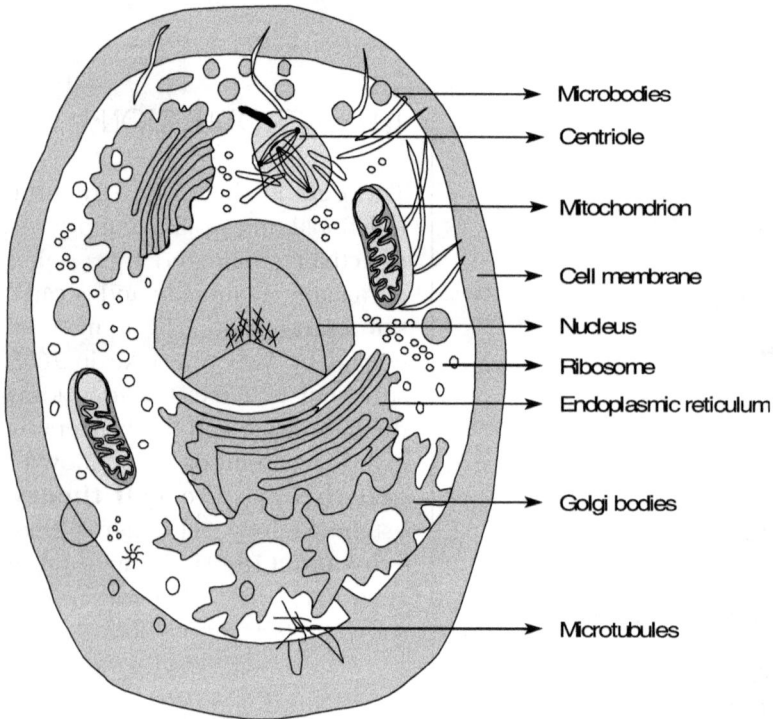

Figure 1.1   Eukaryotic cell and its components

The inorganic ions of the cell such as sodium (Na^+), potassium (K^+), magnesium (Mg^{2+}), calcium (Ca^{2+}), phosphate (HPO_4^{2-}), chloride (Cl^-), and bicarbonate (HCO_3^-) constitute less than 1% of the cell mass. These ions are unique constituents of cells and are involved in almost all cell metabolisms. Most of the cell organic compounds belong to one of four classes of molecules, viz. carbohydrates, lipids, proteins and nucleic acids. Proteins, nucleic acids and most carbohydrates (the polysaccharides) are macromolecules formed by the polymerization of hundreds of low molecular weight precursors such as amino acids, nucleotides and sugars respectively.

Such macromolecules constitute 80 to 90% of the dry weight of most cells. Lipids are the other major macromolecular components of the cells.

1.2.1   Carbohydrates

The carbohydrates include **simple sugars** as well as **polysaccharides**. The simple sugars such as glucose and galactose, are the major nutrients of cells. Their breakdown yields energy for cellular activities and substrate for the synthesis of other cellular molecules. Polysaccharides are storage forms of sugars and form structural components of the cell. They also act as the markers for a variety of cell recognition processes, including the adhesion of cells to their neighbours and the transport of proteins to appropriate intracellular destinations.

1.2.2   Lipids

Lipid has three major roles in cells. It

- provides an important form of energy storage
- forms the major component of cell membranes
- plays crucial role in cell signalling

The simplest lipids are **fatty acids**, which consist of long hydrocarbon chains, containing 16 or 18 carbon atoms, with a carboxyl group (COO^-) at one end. Unsaturated fatty acids contain one or more double bonds between carbon atoms whereas in saturated fatty acids the carbon atoms are mostly bonded to the hydrogen atoms. The long hydrocarbon chains of fatty acids contain more C—H bonds, which are nonpolar and hydrophobic in nature and which are responsible for the formation of biological membranes. Fatty acids are stored in the form of **triacylglycerols.** Triacylglycerols are insoluble in water. They accumulate as fat droplets in the cytoplasm of cells of the adipose tissue. During emergency conditions, these fat droplets can be broken down for yielding energy.

1.2.3   Nucleic Acids

The nucleic acids—deoxyribonucleic acid (DNA) and ribonucleic acid (RNA) are the principal informational molecules of the cell. DNA has a unique role as the genetic material and is located within the nucleus. There are different types of ribonucleic acids and they participate in a number of cellular activities. Messenger RNA (mRNA) carries information from DNA and binds to the ribosomes, where it serves as a template for protein synthesis. Two other types of RNA such as ribosomal RNA (rRNA) and transfer RNA (tRNA) are also involved in decoding the genetic information.

1.2.4   Amino Acids and Proteins

Amino acids are the building blocks of proteins. All amino acids contain a carboxylic group and an amino group, both linked to a single carbon atom called the 2-carbon. Although, there can be many different kinds of amino acids, nearly 100, only 20 of them are commonly found in proteins in all classes of organisms ranging from bacteria to higher plants and animals. Proteins are polymers of 20 different amino acids. The primary responsibility of proteins is to execute the tasks directed by the genetic information. The proteins serve as structural components of cells and tissues, act in the transport and storage of molecules (e.g. the transport of oxygen by haemoglobin), transmit signals within and in between cells (e.g. hormones), provide defence against invading pathogens (e.g. immunoglobulins) and act as enzymes (e.g. pepsin, renin).

1.3   CELLULAR COMPONENTS AND MEMBRANES

All living forms carry on processes such as assimilation, metabolism, excretion, growth, maturation, reproduction, differentiation, movement and responsiveness, which distinguish them from non-living things. Each eukaryotic cell has a variety of subcellular organelles with specific functions.

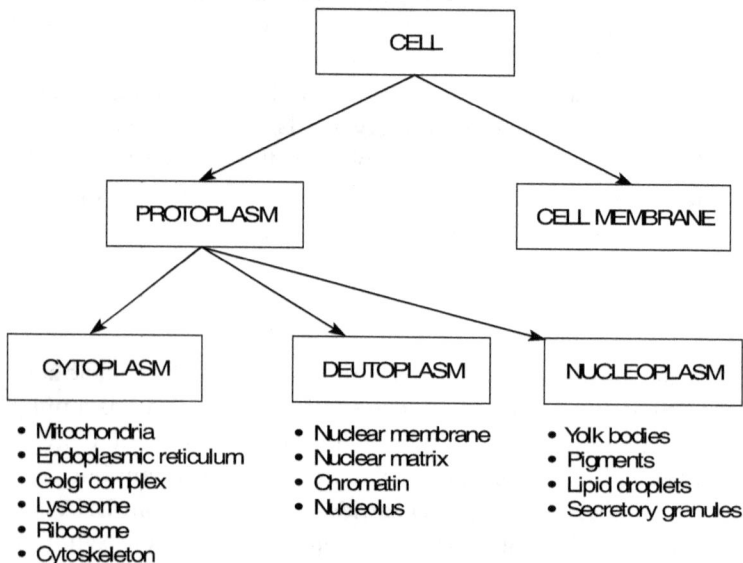

CELL

PROTOPLASM   CELL MEMBRANE

CYTOPLASM   DEUTOPLASM   NUCLEOPLASM

- Mitochondria
- Endoplasmic reticulum
- Golgi complex
- Lysosome
- Ribosome
- Cytoskeleton

- Nuclear membrane
- Nuclear matrix
- Chromatin
- Nucleolus

- Yolk bodies
- Pigments
- Lipid droplets
- Secretory granules

1.3.1   Nucleus

Brown in 1831 first discovered the nucleus. The presence of a definite nucleus is the main feature which differentiates the eukaryotic cell from the prokaryotic cell. Nucleus is present in almost all eukaryotic cells except mature erythrocytes and platelets. Nucleus contains most of the hereditary units of the cell, called **genes**, which are coiled as chromatin fibres. During cell division, chromatin fibres condense and pack into a compact structure known as chromosome. DNA replication, transcription and RNA processing take place within the nucleus. The nuclear envelope provides the structural framework of the nucleus and acts as a barrier that separates the contents of the nucleus from cytoplasm. The **nuclear envelop** consists of two **nuclear membranes**, (inner and outer nuclear membrane), **nuclear lamina** and **nuclear pore** complexes. The inner membrane carries unique proteins that are specific to the nucleus and the outer membrane is continuous with the endoplasmic reticulum. Underlying the inner nuclear membrane is the nuclear lamina, a fibrous meshwork that provides structural support to the nucleus. The nuclear lamina is composed of proteins called lamins. They exist in dimeric form and associate with each other to the filaments that make up lamina. The nuclear pore complexes allow selective traffic of ions, proteins and RNAs. **Nucleolus** is the most prominent substructure within the nucleus. Nucleolus is the site of rRNA transcription and processing.

1.3.2   Mitochondria

Mitochondria are small globular bodies, known as the "**powerhouses**" of the cell. In response to the cellular need, mitochondria can multiply in

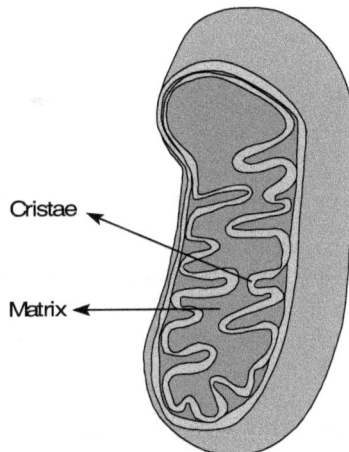

Figure 1.2   Mitochondrion

number. Figure 1.2 shows a mitochondrion consisting of two membranes, smooth outer membrane and inner membrane with lots of cristae to provide surface area for cellular respiration (to provide cell's energy). Mitochondrial matrix is the central cavity enclosed by the inner membrane. Mitochondrion is involved in the oxidative phosphorylation reaction. It also functions as a site for oxidation of carbohydrates, lipids and amino acids. The hepatocytes and myocytes have a large number of mitochondria.

1.3.3   Endoplasmic Reticulum

Endoplasmic Reticulum (ER) is a system of membranous structure (Figure 1.3). There are two types of ER, viz. **smooth endoplasmic reticulum** (SER) and **rough endoplasmic reticulum** (RER) studded with ribosomes. The ER may assume the form of cisternae, tubules or vesicles. ER gives mechanical support for the colloidal structure of cytoplasm. Glycogen, proteins, lipids, cholesterol and steroid hormones are synthesized and partially stored here. In the stomach, the chloride ions are secreted from the ER of oxyntic cells. Enzymes within the SER can inactivate or detoxify a variety of chemicals.

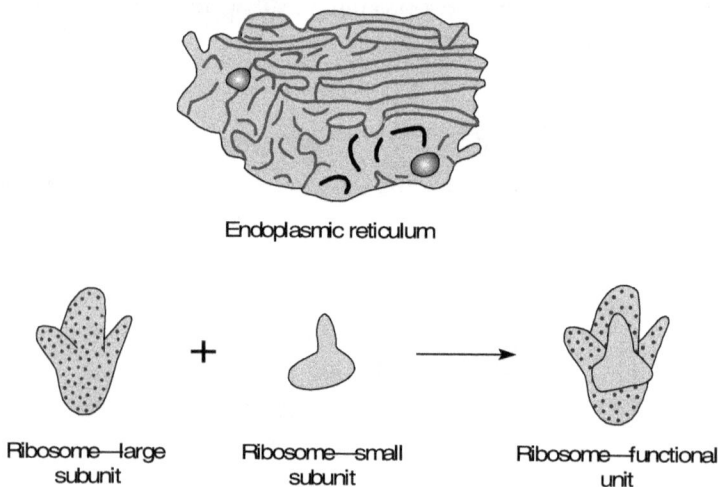

Endoplasmic reticulum

Ribosome—large
subunit
+
Ribosome—small
subunit
→
Ribosome—functional
unit

Figure 1.3   Ribosomes attached to endoplasmic reticulum

1.3.4   Golgi Complex

Golgi complex (GC) consists of a series of flattened membranous sacs called **cisternae**. GC is formed either from plasma membrane or endoplasmic reticulum and has the chemical composition in between them. It processes, sorts, packages and delivers the polypeptides and lipids to the cell membrane. It also forms lysosomes and secretory vesicles. Thiamine pyrophosphatase and four types of glycosyl transferase are found in higher concentration. Synthesis of glycosphingolipids and glycoproteins is the major role of golgi complex.

1.3.5 Lysosomes

Lysosomes are membrane-enclosed vesicles with different kinds of hydrolytic enzymes and are capable of breaking down a wide variety of cellular materials. Lysosomal enzymes mostly function best in acidic pH. Lysosomes show **polymorphism** (primary lysosome, secondary lysosome, residual bodies and autophagosome). Lysosomal enzymes help to recycle the cell's own structure (intracellular digestion) and also function in extracellular digestion. The lysosomal enzyme **hyaluronidase** helps in the penetration of sperm into the egg envelope.

> pH is negative logarithm of hydrogen ion concentration. $pH = -\log H^+$.
>
> Neutral pH is 7, Acidic 1 to 6 and Alkaline 7.2 to 12.

1.3.6 Ribosomes

Ribosomes are granular bodies; they contain **ribosomal RNA** (rRNA) and many **ribosomal proteins**. Ribosomes are found in the cytoplasm and also in the mitochondria. They were so named because of the high content of ribonucleic acid. Functionally, ribosomes act as the sites of protein synthesis. Structurally a ribosome consists of two subunits, viz. large subunit (60S) and small subunit (40S). The mRNA which passes between the two subunits of the ribosome is protected from nuclease action.

1.3.7 Peroxisomes

Peroxisomes are a group of organelles like lysosomes but smaller in nature. They use molecular oxygen to oxidize various organic substances and produce **hydrogen peroxide** (H_2O_2). Further, H_2O_2 is also used by their catalase enzymes to oxidize a variety of substances including phenol, formic acid, formaldehyde and alcohol. This kind of oxidation reaction is found especially in liver and kidney cells, where peroxisomes detoxify potentially harmful substances.

1.3.8   Cell Membrane

Each cell has a different internal milieu from that of its external environment, which is maintained by the cell membrane. The red blood cell membrane contains 52% protein, 40% lipid and 8% carbohydrate. Cell membrane gives a definite shape to the cell. It also acts as a semipermeable or selectively permeable membrane to various substrates, such as proteins, nuclear materials and ions. It determines the antigen specificity and also responds to a variety of chemical stimuli.

The structure and function of cells are critically dependent on membranes. All cell membranes share a common structural organization: bilayers of phospholipids with associated proteins. These membrane proteins are responsible for many specialized functions; some act as receptors that allow the cell to respond to external signals, some are responsible for the selective transport of molecules across the membrane, and others participate in electron transport and oxidative phosphorylation. In addition, membrane proteins control the interactions between cells of multicellular organisms.

The fundamental building blocks of all cell membranes are phospholipids, which are amphipathic molecules, consisting of two hydrophobic fatty acid chains linked to a phosphate-containing hydrophilic head group. Because their fatty acid tails are poorly soluble in water, phospholipids spontaneously form bilayers in aqueous solutions, with the hydrophobic tails buried in the interior of the membrane and the polar head groups exposed on both sides, in contact with water. Such phospholipid bilayers form a stable barrier between two aqueous compartments and represent the basic structure of all biological membranes.

Mammalian plasma membranes are more complex, containing four major phospholipids—phosphatidylcholine, phosphatidylserine, phosphatidylethanolamine and sphingomyelin—which together constitute 50 to 60% of total membrane lipid. In addition to the phospholipids, the plasma membranes of animal cells contain glycolipids and cholesterol, which generally correspond to about 40% of the total lipid molecules. An important property of lipid bilayers is that they behave as two-dimensional fluids in which individual molecules (both lipids and proteins) are free to rotate and move in lateral directions. Such fluidity is a critical property of membranes and is determined by both temperature and lipid composition.

Proteins are the other major components of the cell membrane, constituting 25 to 75% of the mass of various membranes of the cell. The widely accepted model of membrane structure was proposed by Jonathan Singer and Garth Nicolson in 1972. They viewed membranes as a fluid mosaic in which proteins are inserted into a lipid bilayer (Figure 1.4). The proteins have been floating like the icebergs in a sea of phospholipid bilayer. While phospholipids provide the basic structural organization of membranes, membrane proteins carry out the specific functions of the different membranes of the cell. These proteins are divided into two general classes based on the nature of their association with the membrane.

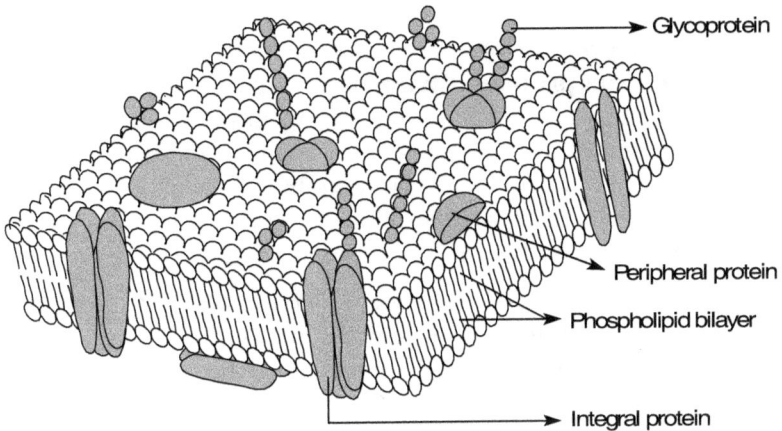

Figure 1.4　Cell membrane—fluid mosaic model

Integral membrane proteins are embedded directly within the lipid bilayer. Peripheral membrane proteins are not inserted into the lipid bilayer but are associated with the membrane by interactions with integral membrane proteins.

While the Danielli–Davson model postulates hydrophilic bonding among lipids and proteins, the Singer–Nicolson model considers the association to be hydrophobic. The fluidity of the membrane is the result of this hydrophobic interaction. It is also clear that the phospholipids and many intrinsic proteins of the membrane are amphipatic molecules, i.e., having both hydrophilic and hydrophobic groups.

Transport across cell membranes takes place in many ways (Table 1.1). The selective permeability of a membrane to different substances depends on several factors such as solubility, size, charge and presence of channels/ transporters that are related to the membrane structure. To support the cellular metabolic reactions, certain substances move into the cell, while nitrogenous as well as other waste materials move out. The mechanism that moves substances across a plasma membrane without using cellular energy is a passive process. In an active process, the cell uses some of its own energy to transport the substance.

The selective permeability of biological membranes to small molecules allows the cell to control and maintain its internal make-up. Small nonpolar molecules such as O_2 and CO_2 are soluble in the lipid bilayer and therefore can readily cross cell membranes. Similarly small uncharged polar molecules such as H_2O also can diffuse through membranes, but larger uncharged polar molecules such as glucose, cannot. Charged molecules such as ions are unable to diffuse through a phospholipid bilayer regardless of size; even H^+ ions cannot cross a lipid bilayer by free diffusion.

There are two general classes of membrane transport proteins—**channel proteins** and **carrier proteins**. **Channel proteins** form open pores through the membrane, allowing the free passage of any molecule of the appropriate size. For example, ion channels allow the passage of inorganic ions such as Na^+, K^+, Ca^{2+} and Cl^- across the plasma membrane. They open and close in response to cellular signals and also allow the cell to control the movement of ions across the membrane.

The **carrier proteins** bind specific molecules and then undergo conformational changes that open channels through which the molecule to be transported can pass across the membrane and be released on the other side.

Table 1.1   Movement of materials across cell membranes

Transport	Substances that are transported
Passive mode	
Simple diffusion	Oxygen, carbon dioxide, water, nitrogen, steroids, fat-soluble vitamins, glycerol, small alcohols and ammonia.
Osmosis	A solvent, usually water in living systems.
Bulk flow	Air flowing into and out of lungs; fluid flowing through blood capillary walls.
Facilitated diffusion	Different kinds of ions, glucose, fructose, galactose, urea and certain vitamins.
Active mode	
Primary active transport	Sodium, potassium, calcium, hydrogen ion, iodide ion, chloride ion.
Secondary active transport	Symport—glucose, into cells lining the small intestine and the kidney tubules; amino acids into most body cells.
	Antiport—calcium, hydrogen ion out of many cells.

1.3.9   Cell Organelles and Their Markers

Each cell organelle has a marker to make it unique from others. These markers are listed below.

Organelle	Marker
Cell membrane	Na^{2+}/K^+ ATPase
Endoplasmic reticulum	Glucose 6-phosphatase
Golgi complex	Galactosyl transferase
Lysosome	Acid phosphatase
Mitochondria	Succinate dehydrogenase
Nucleus	DNA
Cytoplasm	Lactate dehydrogenase

1.3.10   Cytoskeleton

Cellular shape and cellular movement depend on a complex internal network called **cytoskeleton**. Three main types of protein filaments such as microfilaments, microtubules and intermediate filaments comprise the cytoskeleton. The **microfilaments** are rod like structures and are formed from the actin protein (and myosin in muscle tissue). Microfilaments help in cell movement such as phagocytosis and secretion. **Microtubules** are larger than microfilaments. They consist of a protein called tubulin. Microtubules function like a conveyor belt to move various substances and organelles through the cytosol. **Intermediate filaments** are so named because their size is in between those of microfilaments and microtubules. They provide structural reinforcement inside cells, hold organelles in place and are associated with microtubules to give shape to the cell.

1.3.11   Cell Addition

Cell includes a large and diverse group of chemical substances produced by cells. For example, **Melanin** is a pigment synthesized and stored in cells of the skin and hair. It protects the body from ultraviolet light. **Glycogen** is a polysaccharide stored in the liver and in the inner lining of the uterus and vagina. When the body requires quick energy, glycogen is broken down to glucose and released into the blood. **Triglycerides** are stored in adipocytes (fat cells) and are metabolized to produce adenosine triphosphate (ATP).

1.3.12   Extracellular Matrix (ECM)

Eukaryotic cells secrete a complex network of proteins and carbohydrates, called the **Extracellular Matrix** (ECM). The matrix helps to fix the cells in tissues together and is a reservoir for various hormones which are controlling cell growth and differentiation. During early stages of differentiation, the matrix provides a lattice through which cells undergo morphogenetic movement.

The ECM has three major protein components:

1. collagen fibres for strength and resilience
2. proteoglycans for cushioning and
3. multiadhesive matrix proteins (fibronectin, laminin, etc.) that bind to receptors on the cell surface for interaction of matrix substances.

Collagens are the major fibrous proteins present in connective tissue. They are insoluble and comprise the ECM. There are at least 19 types of collagen. Type I, II and III are the most abundant (80–90%) in human body tissues. Type IV collagen forms a two-dimensional reticulum and is a major component of the basal lamina. Collagens are predominantly synthesized by fibroblasts but epithelial cells also synthesize these proteins. Collagens are synthesized as longer precursor proteins called **procollagens**. These pro-domains are globular and form multiple intrachain disulphide bonds. The disulphides stabilize the proprotein, allowing the triple helical section to form.

Collagen fibres start to assemble in the ER and golgi complexes. The signal sequence is removed and numerous modifications take place in the collagen chains. Specific proline residues are hydroxylated by prolyl 4-hydroxylase and prolyl 3-hydroxylase. Specific lysine residues also are hydroxylated by lysyl hydroxylase. Both prolyl hydroxylases are absolutely dependent upon vitamin C as cofactor.

Glycosaminoglycans are linear chains of 20–100 sulphated disaccharides. The most common disaccharides are chondroitin sulphate, heparin and heparan sulphate, and dermatan sulphate. Proteoglycans consist of multiple glycosaminoglycan chains that branch from a linear protein core. Extracellular proteoglycans are large, highly hydrated molecules that help cushion cells.

In cartilage, a proteoglycan called aggrecan binds at regular intervals to a central hyaluronan molecule, forming a very large aggregate. Smaller proteoglycans are attached to cell surfaces, where they facilitate cell–matrix interactions and help present certain hormones to their cell-surface receptors. Hyaluronan forms viscous hydrated gels that resist compression forces. When bound to specific receptors on certain cells, hyaluronan inhibits cell–cell adhesion and facilitates cell migration.

The function of ECM varies according to cell types and the combinations of the above three major protein components. In tendons, ECM gives strength whereas in cartilage it provides cushioning. The matrix also helps in intracellular communication directly or indirectly through signalling pathways. The ability of hepatocytes to express liver-specific proteins depends on its association with a matrix of appropriate composition.

Laminin is a multiadhesive protein in the basal lamina that binds heparan sulphate, type IV collagen, and specific cell-surface receptor proteins. **Fibronectins** are multiadhesive proteins that link collagen and other matrix proteins to integrins in the plasma membrane, thereby attaching cells to the matrix.

1.3.13   Gap Junctions

Gap junctions serve as direct connections between the cytoplasms of adjacent cells. They provide open channels through the plasma membrane, allowing ions and small molecules (less than approximately a thousand daltons) to diffuse freely between neighbouring cells, but preventing the passage of proteins and nucleic acids. Consequently, gap junctions couple both the metabolic activities and the electric responses of the cells they connect. Epithelial cells, endothelial cells, and the cells of cardiac and smooth muscle communicate by gap junctions.

In electrically excitable cells such as heart muscle cells, the direct passage of ions through gap junctions couples and synchronizes the contractions of neighbouring cells. Gap junctions also allow the passage of some intracellular signalling molecules such as cAMP and Ca^{2+}, between adjacent cells, potentially coordinating the responses of cells in tissues.

Gap junctions are constructed of transmembrane proteins called connexins. Six connexins assemble to form a cylinder with an open aqueous pore at its centre. Such an assembly of connexins in the plasma membrane of one cell then aligns with the connexins of an adjacent cell, forming an open channel between the two cytoplasms. The plasma membranes of the two cells are separated by a gap corresponding to the space occupied by the connexin extracellular domains.

1.4   CELL DIVISION

Cell division is a process by which cells reproduce themselves. There are two kinds of cell division, **somatic cell division** and **reproductive cell division**.

1.4.1   Somatic Cell Division

In a complete cycle of a cell division (Figure 1.5), a parent cell divides into two identical daughter cells. The cell passes through **interphase**, **mitosis** and **cytokinesis** stages during the cell cycle.

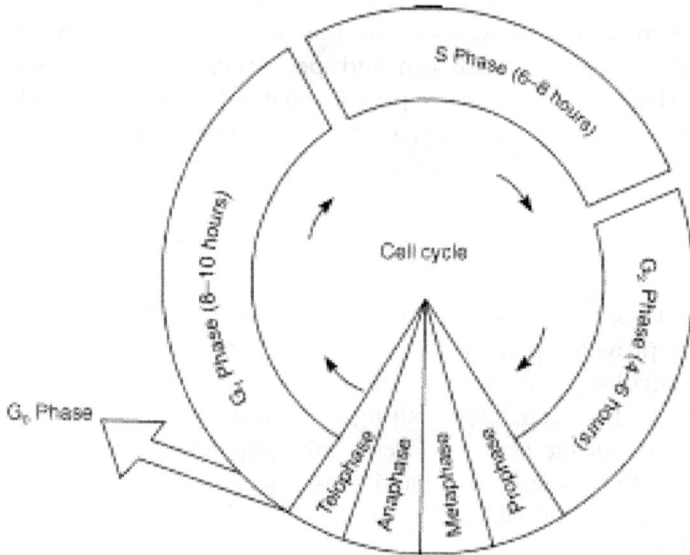

Figure 1.5   The cell cycle

During **interphase**, a cell prepares for its division in three phases.

1. G_1 (Gap 1) Phase   Rapid growth and metabolic activity.

2. S (Synthesis) Phase   DNA replicates, increase in chromosome copies, and duplication of centriole.

3. G_2 (Gap 2) Phase   DNA proofreading, additional protein synthesis to prepare for mitosis, new membrane synthesis.

Many times a cell will leave the cell cycle, temporarily or permanently. It exists in the cycle at G_1 and enters a stage designated **G_0 phase** (G zero). A G_0 cell is often called "**quiescent**". Such cells are busy carrying out their functions in the organism till they die. These functions include secretion, conducting nerve impulses and attacking pathogens. Most of the lymphocytes in human blood are in G_0. However, with proper stimulation (pathogen entry), they can be stimulated to re-enter the cell cycle (at G_1) and proceed on to mitotic stages.

The mitosis consists of four phases (Figure 1.6)

- Prophase
- Metaphase
- Anaphase
- Telophase

In the Prophase

- Chromatic fibres shorten and coil into visible chromosomes (paired chromatids)
- Nucleoli and nuclear envelope disappear
- Centrosome moves with its centrioles to opposite poles of the cell and forms the mitotic spindle
- Occurs in longest time duration

In the Metaphase

- The centromeres of the chromatid pairs line up at the exact centre of the spindle (metaphase plate)

In the Anaphase

- Centromeres split and separate
- Two sister chromatids of each pair move towards opposite poles of the cell. The separated sister chromatids are referred to as daughter chromosomes
- Occurs in shortest time duration

In the Telophase

- Nuclear envelope reappears and encloses chromosomes (two daughter nuclei formed)
- Chromosomes resume chromatin fibre form
- Nucleoli reappear
- Mitotic spindle breaks up

During Cytokinesis

- The cleavage furrow forms at the centre, progresses inward and separates cytoplasm into two separate equal portions, so that two daughter cells are formed.

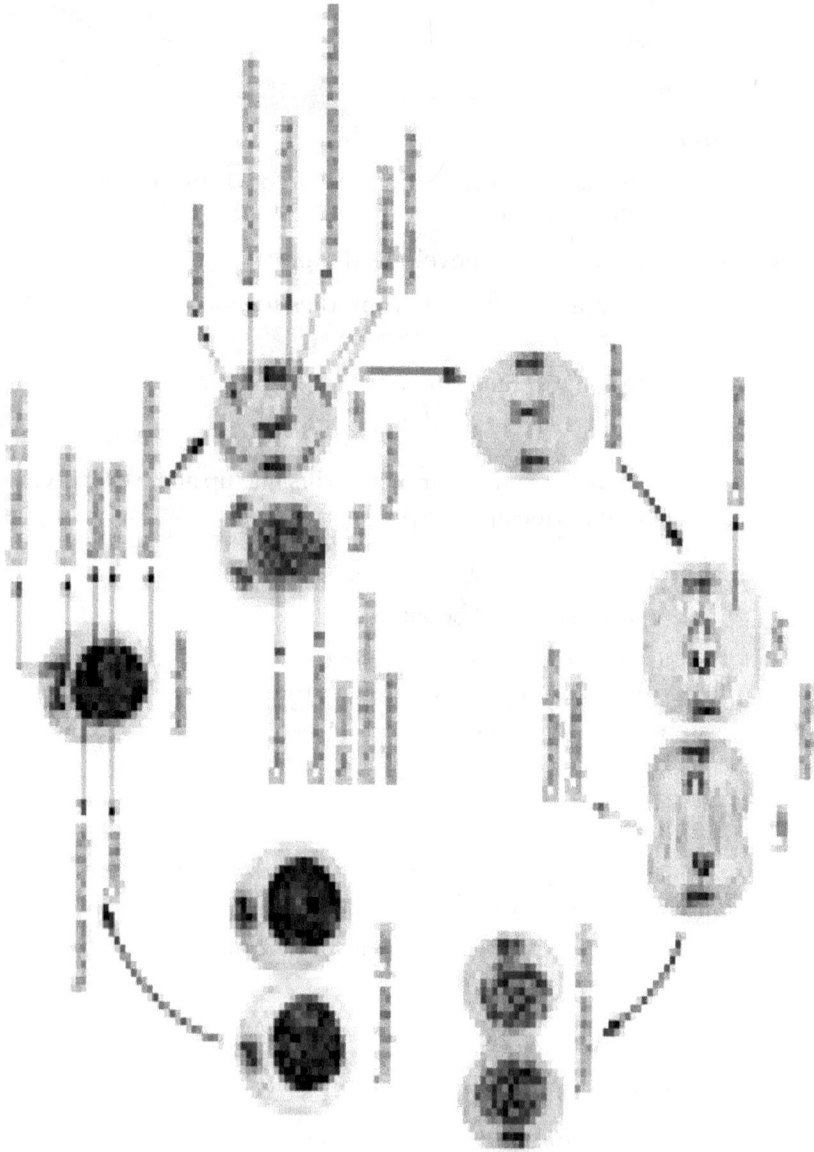

Figure 1.6   Mitosis

1.4.2   Length of the Cell Cycle

The time required for one cell cycle varies according to the type of cell, its location and other factors such as temperature, pH, signals, etc. Peripheral blood lymphocytes studied in laboratory cultures often have the following time intervals during a cell cycle.

G_1 phase	6–10 hours
S phase	6–8 hours
G_2 phase	4–6 hours
Mitosis and cytokinesis	1–2 hours

Within mitosis, prophase is the longest and anaphase the shortest. Together, the various phases of the cell cycle require 18–24 hours in many cultured mammalian cells.

1.4.3   Fate of a Cell

A cell has three possible destinies:

- to remain alive and function without dividing (G_0 phase),
- to grow and divide or
- to die

Homeostasis is maintained when there is a balance between cell proliferation and cell death. A key signal that induces cell division (both mitosis and meiosis) is called **maturation promoting factor (MPF)**. One component of MPF is a group of enzymes called **cdc2 proteins** because they participate in the cell division cycle (cdc). Another component of MPF is a protein called **cyclin**. It is so named because its level rises and falls during the cell cycle. Cyclin builds up in the cell during interphase and activates cdc2 proteins. As a result the cell undergoes division, mitosis (or meiosis).

1.4.4   Control of the Cycle

The passage of a cell through the cell cycle is controlled by proteins, viz. cyclins, cyclin-dependent kinases and anaphase-promoting complex in the cytoplasm.

Cyclins: There are 3 kinds of cyclins, they are

- G_1 cyclins—rise in G_1 phase
- S phase cyclins—rise in S phase
- M phase cyclins—rise in G_2 phase

There are three groups of Cyclin-dependent kinases (CDKs). They are

- G_1 CDKs
- S phase CDKs
- M phase CDKs

The three CDKs levels in the cell remain stable and they bind to the respective cyclin to activate. They add phosphate groups to a variety of proteins that control cell cycle.

The **anaphase-promoting complex (APC)** triggers the events leading to sister chromatids to separate and degrade the mitotic (M-phase) cyclins.

1.4.5   Control Steps in the Cycle

1. **G_1 cyclin** gives signal to the cell to prepare the chromosomes for duplication.

2. **S-phase promoting factor (SPF)** prepares the cell to enter S phase and to duplicate DNA and centriole.

3. **M-phase promoting factor** (the complex of mitotic cyclins with M-phase CDK) initiates assembly of the mitotic spindle, disintegration of the nuclear envelope and condensation of the chromosomes.

4. **Anaphase promoting complex (APC)**
   - allows the sister chromatids to separate and move to the poles
   - destroys the M-phase cyclins. It does this by conjugating them with the protein **ubiquitin** which targets them for destruction by **proteosomes**
   - turns on synthesis of G_1 cyclins for the next turn of the cycle when there is a signal
   - degrades geminin protein which preserves the freshly synthesized DNA in S phase from being re-replicated before mitosis

   The cell has several systems for interrupting the cell cycle if there is an error.

5. **A check on completion of S phase**   The cell seems to oversee the presence of the **Okazaki fragments** on the lagging strand during DNA duplication. The cell is not permitted to proceed in the cell cycle until these have disappeared.

6. **DNA damage checkpoints**   Sense DNA damage/mutation during duplication.

7. **Spindle checkpoints** detect any failure of spindle fibres to attach to **kinetochores** and arrest the cell in metaphase.

8. **Mitotic checkpoints** detect misalignment of the spindle itself and block **cytokinesis** and trigger **apoptosis** if the damage is irreparable.

The p53 protein senses DNA damage and halts progression of the cell cycle in both G_1 and G_2 phase. The p53 protein is also a key player in **apoptosis**, forcing abnormal cells to commit suicide. So if the cell has a copy of the p53 mutant gene and their mutant versions of the protein, then it can live on to develop into a tumour. More than half of all human tumours harbour p53 mutations and have no functioning p53 protein.

1.4.6 Reproductive Cell Division

A single diploid parent cell (primordial human cell) undergoes reduction and equatorial division to produce four haploid gametes that are genetically different from the parent cell.

This reduction in chromosome number is accomplished by two sequential rounds of nuclear and cell division (meiosis I and meiosis II) with a single round of DNA replication. Meiosis I resembles mitosis in that the sister chromatids separate and segregate to different daughter cells and this is followed by meiosis II. Meiosis II results in the production of four haploid daughter cells, each of which contains only one copy of each chromosome (Figure 1.7).

Prophase of meiosis I is divided into five stages (leptotene, zygotene, pachytene, diplotene, and diakinesis) on the basis of chromosome morphology.

Leptotene	The chromosomes become visible as thin threads within the nucleus.
Zygotene	The homologous pairs of chromosomes become closely associated called synapsis, to form a bivalent structure.
Pachytene	In a bivalent condition, the chromosome becomes short and thick.
Diplotene	Chromosomes continue to condense and the synaptonemal complex breaks down. The homologous chromosomes are held together at chiasmata.
Diakinesis	Nuclear membrane breaks down. Chiasmata move to the ends of the chromosome arms.

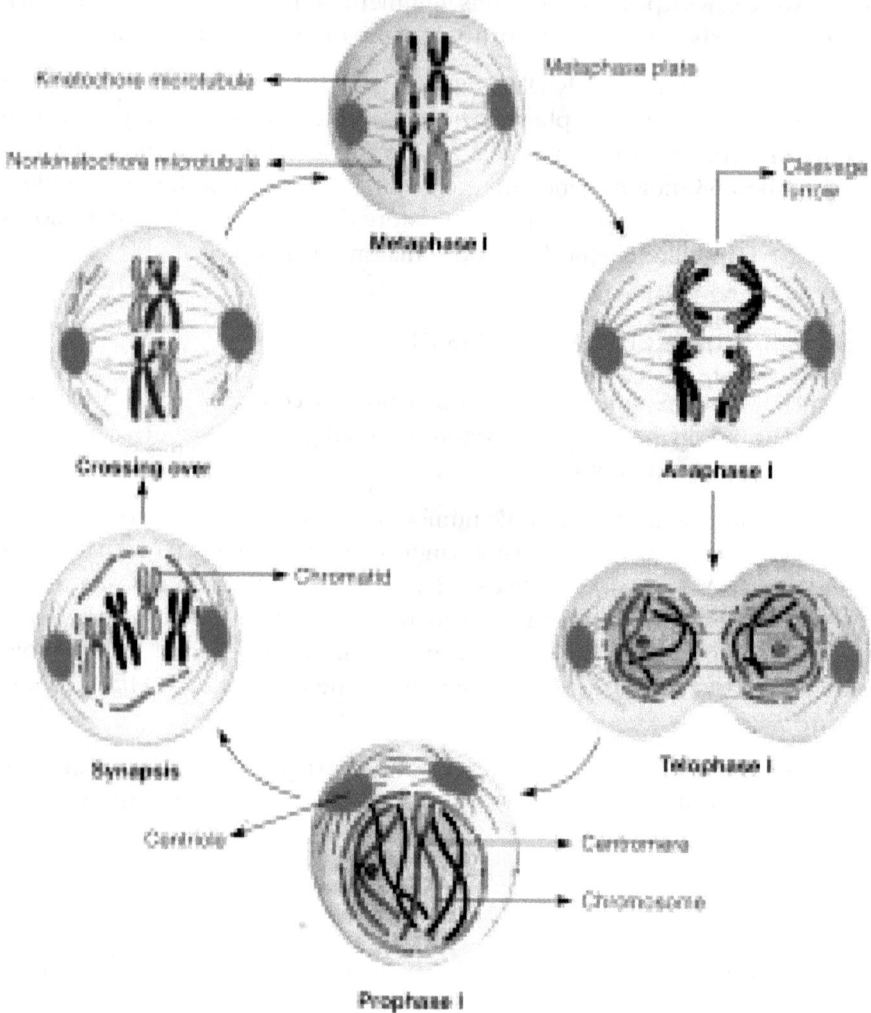

Figure 1.7   Meiosis stage I

Metaphase I is characterized by the completion of spindle formation and the alignment of the bivalents. The two centromeres of each bivalent now lie on opposite sides of the equatorial plane.

During **anaphase I**, the chiasmata disappear and the homologous chromosomes are pulled by the spindle fibres towards opposite poles of the cell. The key feature is that, the centromeres do not duplicate and divide, so that only half of the original number of chromosomes migrate towards each pole. The chromosomes migrating towards each pole consist of one member of each pair of autosomes and one of the sex chromosomes.

In **telophase I**, chromosomes reach opposite sides of the cell and uncoil. A new nuclear membrane begins to form. The two daughter cells contain haploid number of chromosomes, each having two sister chromatids. In humans, cytokinesis also occurs during this phase. The cytoplasm is divided approximately equally between the two daughter cells in the gametes formed in males. In females, nearly all of the cytoplasm goes into one daughter cell, which will later form the egg. The other daughter cell becomes the polar body.

Meiosis II then begins with a brief interphase II but there is no DNA replication. **Prophase II** is quite similar to mitotic prophase, except that the cell nucleus contains only the haploid set of chromosomes. During prophase II the chromosomes thicken, the nuclear membrane disappears, and new spindle fibres are formed. Following this **metaphase II**, the spindle fibres pull the chromosomes towards the equatorial plane.

In **anaphase II** the centromeres split and each carries a single chromatid towards a pole of the cell. The chromatids have now separated, but chiasma formation and crossing over produce newly separated sister chromatids that may not be identical.

Telophase II begins when the chromosomes reach opposite poles of the cell. New nuclear membranes are formed around each group of chromosomes and cytokinesis occurs. In male gametogenesis, at the end of meiosis, four functional daughter cells (sperms) are formed and each has an equal amount of cytoplasm. In female gametes, unequal division of the cytoplasm occurs, forming one egg cell (ovum) and three polar bodies.

FOR ADDITIONAL READING

1. Alberts B., Alexander, J., Lewis, J., Raff, M. *et al*. (2002). *Molecular Biology of the Cell*, 4th edn. Garland Publishing Inc., New York.

2. Bainton, D. (1981). "The discovery of lysosomes." *J. Cell Biol.* 91:66s–76s.

3. Baltimore. (2000). *Molecular Cell Biology*, 4th edn. W.H. Freeman Company, New York.

4. Becker, W.M. and Deamer, D.W. (1991). *The World of the Cell*, 2nd edn. The Benjamin Cummings Publishing Company, CA.

5. Cooper, G.M. (2000). *The Cell – A Molecular Approach*, 2nd edn. Sinauer Associates Inc., Sunderland, Massachusetts.

6. Dai, J. and Sheetz, M.P. (1998). "Cell membrane mechanics." *Meth. Cell Biol.* 55:157–171.

7. De Rijk, P., Van de peer, and De Wachter, R. (1997). "Database on the structure of large ribosomal subunit RNA." *Nucl. Acids. Res.* 25:117–122.

8. Elledge, S.J. (1996). "Cell cycle checkpoints: preventing an identity crisis." *Science.* 274:1664–1672.

9. Farquahr, M. and Palade, G. (1998). "The Golgi apparatus: 100 years of progress and controversy." *Trends Cell. Biol.* 8:2–10.

10. Ford, H.L. and Pardee, A.B. (1998). "The S Phase: beginning, middle, and end: a perspective." *J. Cell Biochem. Suppl.* 30–31:1–7.

11. Fowler, V.M. and Vale, R. (1996). "Cytoskeleton, a collection of reviews." *Curr. Opin. Cell Biol.* Vol 8.

12. Frank, J. (1998). "How the ribosome works." *Scientist.* 86:428–439.

13. Gerald Karp. (2001). *Cell and Molecular Biology.* Wiley, John & Sons Inc., New York.

14. Hartwell, L.H. and Weinert, T.A. (1989). "Checkpoints: controls that ensure the order of cell cycle events." *Science.* 246:629–634.

15. Lamond, A. and Earnshaw, W. (1998). "Structure and function in the nucleus." *Science.* 280:547–553.

16. Malandro, M. and Kilberg, M. (1996). "Molecular biology of mammalian amino acid transporters." *Ann. Rev. Biochem.* 66:305–336.

17. Masters, C. and Crane, D. (1996). "Recent developments in peroxisome biology." *Endeavour.* 20:68–73.

18. McIntosh, J.R and Koonce, M.P. (1989). "Mitosis." *Science.* 246:622–628.

19. Mueckler, M. (1994). "Facilitative glucose transporters." *Eur. J. Biochem.* 219:713–725.

20. Murray, A. and Hunt, T. (1993). *The Cell Cycle: An Introduction.* W.H. Freeman and Company, New York.

21. Murray, A.W. and Szostak, J.W. (1985). "Chromosome segregation in mitosis and meiosis." *Ann. Rev. Cell Biol.* 1:289–315.

22. Nasmyth, K. (1996). "Viewpoint: putting the cell cycle in order." *Science.* 274:1643–1645.

23. Norbury, C. and Nurse, P. (1991). "Animal cell cycles and their control." *Ann. Rev. Biochem.* 61:441–470.

24. Nurse, P. (1990). "Universal control mechanism regulating onset of M phase." *Nature.* 344:503–508.

25. Nurse, P. (1994). "Ordering S phase and M phase in the cell cycle." *Cell.* 79:547–550.

26. Prescott, D.M. (1988). *Cells.* Jones and Bartlett Publishers, Boston, MA.

27. Van de Peer, Y., Jansen, J., De Rijk, P. and De Wachter, R. (1997). "Database on the structure of small ribosomal subunit." *Nucl. Acids Res.* 25:111–116.

28. Glick, B. and Malhotra, V. (1988). "The curious status of the Golgi apparatus." *Cell.* 95:883–889.

29. Gant, T.M. and Wilson, K.L. (1997). "Nuclear assembly." *Ann. Rev. Cell. Biol.* 13:669–695.

30. Weis, K. (1998). "Importins and exportins: How to get in and out of the nucleus." *Trends Biochem. Sci.* 23:185–189.

31. Maxwell, E.S. and Fournier, M.J. (1995). "The small nucleolar RNAs." *Ann. Rev. Biochem.* 35:897–934.

32. Sommerville, J. (1986). "Nucleolar structure and ribosome biogenesis." *Trends Biochem. Sci.* 11:438–442.

33. Von Figura, K. and Hasilik, A. (1986). "Lysosomal enzymes and their receptors." *Ann. Rev. Biochem.* 55:167–193.

34. Pfeffer, S.R. and Rothman, J.E. (1987). "Biosynthetic protein transport and sorting by the endoplasmic reticulum and Golgi." *Ann. Rev. Biochem.* 56:829–852.

35. Rose, J.K. and Doms, R.W. (1988). "Regulation of protein export from the endoplasmic reticulum." *Ann. Rev. Cell Biol.* 4:257–288.

36. Hurtley, S.M. and Helenius, A. (1989). "Protein oligomerization in the endoplasmic reticulum." *Ann. Rev. Cell Biol.* 5:277–307.

37. Kornfeld, S. and Mellman, I. (1989). "The biogenesis of lysosomes." *Ann. Rev. Cell Biol.* 5:483–525.

38. Spirin, A.S. (2000). *Ribosomes.* Kluwer Academic Publishers, The Netherlands.

MODEL QUESTIONS

1. Define a cell.

2. Describe the structures and functions of the following.
   i.   Nucleus
   ii. Ribosome
   iii. Endoplasmic reticulum
   iv. Lysosome
   v.   Golgi complex

3. What is cell addition or inclusion? Give examples.

4. Comment on peroxisomes.

5. The lipid bilayer of the cell membrane exists as
   a. solid   b. fibre   c. gel   d. semi-solid

6. The main function of mitochondrion is
   a.   electron transfer   b.   fat synthesis
   c.   protein synthesis   d.   oxidation

7. Marker enzyme for lysosome is
   a.   acid phosphatase   b.   endonuclease
   c.   exonuclease   d.   lipase

8. Which of the following is associated with the endoplasmic reticulum?
   a.   hexokinase   b.   pyruvate kinase
   c.   cytochrome P-450   d.   none

9. Cells with decreased oxidative phosphorylation have
   a.   less ribosome   b.   less mitochondria
   c.   no motility   d.   low protein

10. What is the main purpose of mitosis?

11. How many chromosomes does each cell have at the end of mitosis? Are the two cells identical? Are they smaller or bigger than the parent cell?

12. Draw how a few of the chromosomes would line up along the metaphase plate during metaphase of mitosis.

13. What is the role of the centriole and what does it attach to?

14. What happens during S phase of interphase?

15. What happens during prophase of mitosis?

16. Contrast mitosis with meiosis stages.

17. Discuss in detail the various control steps in the cell cycle.

18. How does male gamete formation differ from female gamete formation?

19. What is extracellular matrix? List various components of ECM and their functions.

20. Discuss the role of proteins that control the cell cycle.

21. Define interphase of a cell cycle.

22. Describe the principle events that occur in the prophase I of meiosis.

23. How is cell density regulated?

24. List and describe the function of nucleolus.

25. Draw and discuss the following.
    i.   Cell   ii.   Cell membrane
    iii. Phenocytosis   iv.   Phagocytosis

MENDELIAN PRINCIPLES

2

2.1 INTRODUCTION

The basic idea of this chapter is to learn how characters (genes) are transmitted from parent generation to the offspring generation. This was first studied by Gregor Johann Mendel (1822–1884), an Augustinian monk, who is also considered to be the Father of Genetics. He began a series of experiments since 1854, but the scientific world failed to recognize or to appropriate it until 1900, after its rediscovery almost simultaneously by three different investigators.

Before we enter into details let us understand certain basic terminology that are used in this chapter. The character of an individual or a plant is termed **trait**. Traits are transmitted from one generation to another; this is called **heredity**. The traits are under the control of **genes**. The alternative form of a gene is an **allele** or **allelomorph**. A gene may exist in dominant or recessive condition. When both alleles are identical, then that organism is said to be **homozygous** (true breeding). If the alleles are different, the organism is

heterozygous (or hybrid). In a heterozygous organism, when one allele is dominant and the other is recessive, normally the phenotype of the organism is determined by the dominant allele.

Inherited characters (traits) are defined by their ability to be passed from one generation to the next in a predictable manner. Visible or measurable property is called **phenotype**, while the genetic factor (genetic make-up) responsible for creating the phenotype is called the **genotype**. A gene can be subject to alteration in its sequence, called **mutation**. The organism which is carrying an altered gene, is called a **mutant** and an organism carrying normal gene is called **wild type**.

The parental generation is otherwise called **P generation**. The generation produced by breeding (mating) of parental generation is called **first filial generation or F1 generation**. The progeny of the F1 generation is called the **F2 generation** or **second filial generation**.

Mendel (1854) began a series of experiments with the garden pea plant (*Pisum sativum*) to learn about inheritance of factors (now called genes). In 1900, Hugo de Vries, Von Tschermak and Carl Correns were able to confirm the result in pea and other plants. Mendel's selection of garden pea was a good choice because of its suitability for genetic experiments; it is easily cultivable, it bears flowers and fruits within a year, and it produces large seeds. The other advantage is that, it can reproduce by self fertilization, that is, the anthers of stamens produce pollen which falls on the pistil of the same flower and fertilizes. This process is also termed **selfing**. Fortunately selfing can be prevented by removing stamens of the flower before mating and pollen production. Cross-fertilization also takes place through the fusion of male gametes (pollen) of one pea plant and female gametes (eggs) of another pea plant. Once it is over the zygote develops into seeds, which are then planted. The external features of the plants that grow from the seeds are analysed.

Mendel selected pure breeding or true breeding strains (plant) for his experiments. That is, the traits (character) remained unchanged from parent to offspring for many generations. Some of the important traits (character) studied by Mendel are (Figure 2.1).

1. Colour of the flower (purple or white)
2. Coat colour of the seed (grey or white)
3. Seed colour (yellow or green)
4. Shape of the seed (round or wrinkled)
5. Colour and shape of the pod (inflated or pinched)
6. Height of the stem (tall or dwarf)

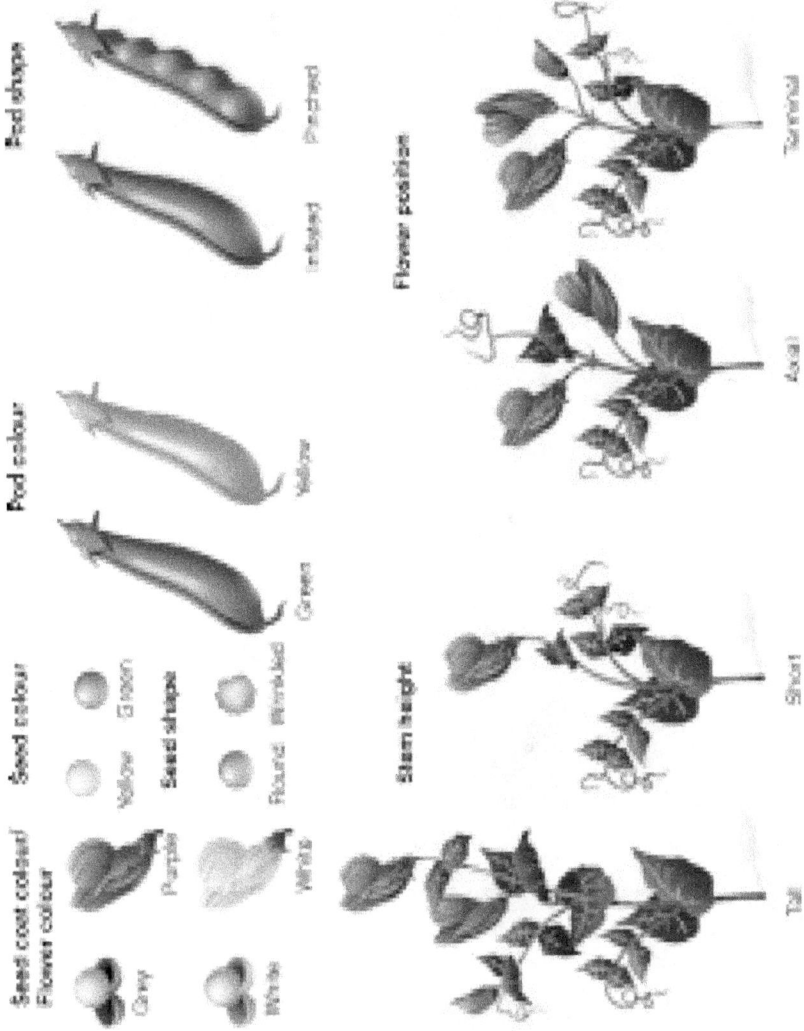

Figure 2.1   Mendel's pure breeding or true breeding strain

2.2   PRINCIPLES OF GENETIC TRANSMISSION

Mendel first made a cross of tall and dwarf pea plants of true breeding nature. Such cross is called **monohybrid cross**, because he took only one character or trait into consideration. Figure 2.2 (a and b) represents a simple

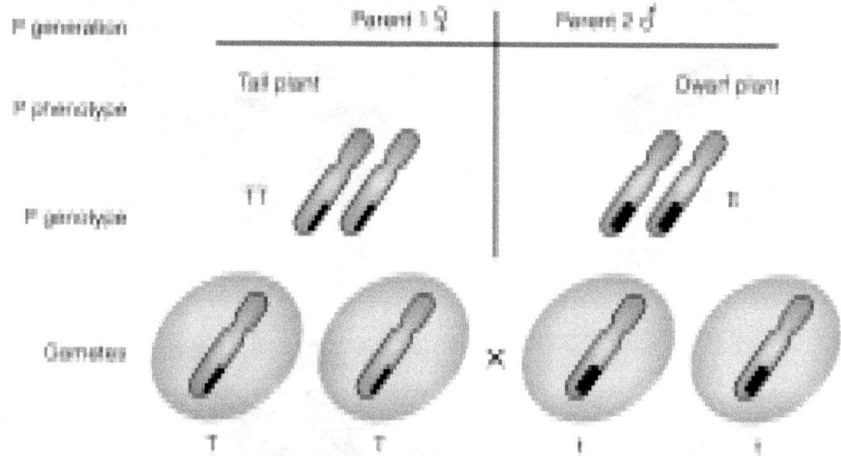

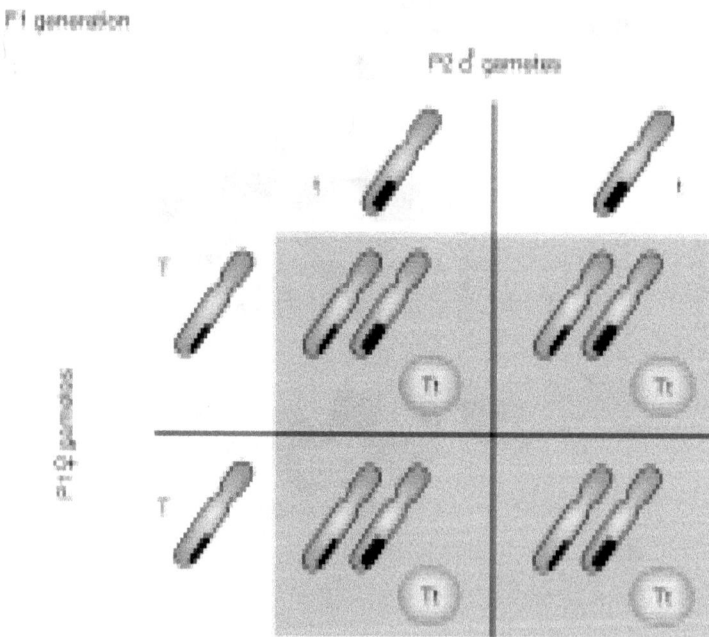

F1 genotype : all Tt

F1 phenotype : All tall plants (tall plant is dominant over dwarf plant)

Figure 2.2a   Monohybrid cross (F1 progeny)

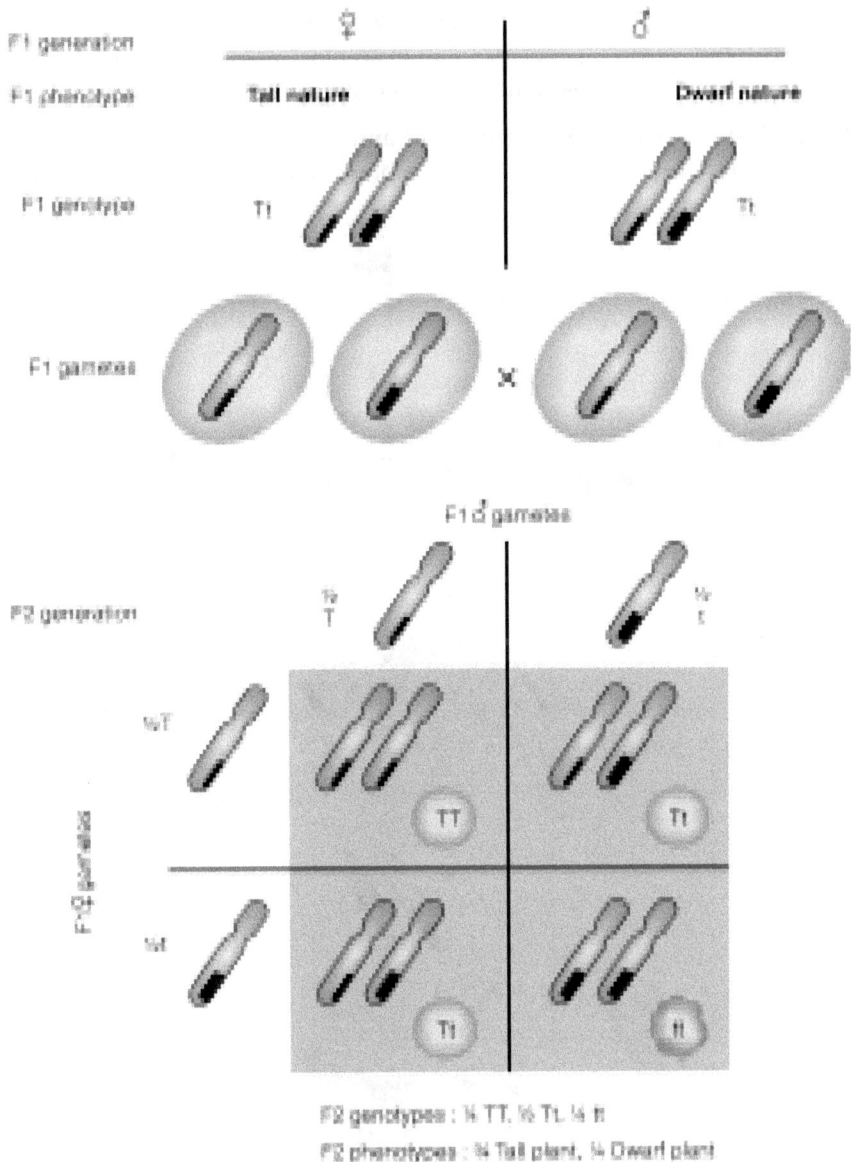

F1 generation ♀ ♂

F1 phenotype **Tall nature** **Dwarf nature**

F1 genotype Tt Tt

F1 gametes ×

F1 ♂ gametes

F2 generation ½T ½t

F1 ♀ gametes ½T ½t

TT Tt

Tt tt

F2 genotypes : ¼ TT, ½ Tt, ¼ tt
F2 phenotypes : ¾ Tall plant, ¼ Dwarf plant

Figure 2.2b   Monohybrid cross (F2 progeny)

way of visualizing the crosses of the pea plant. Two laws of heredity could represent Mendel's hypotheses. They are **law of segregation** (law of purity of gametes) and law **of independent assortment** (or law of free recombination).

Law of segregation states that though alternative alleles (dominant and recessive) remain together in the heterozygous condition, they segregate from each other during gamete formation so that each gamete receives only one allele. The phenotypic ratio is 3:1 and the genotypic ratio is 1:2:1.

Mechanism As shown in Figure 2.2a and 2.2b the dominant alleles are 'TT' (tall) and recessive alleles are represented by 'tt'. During gamete

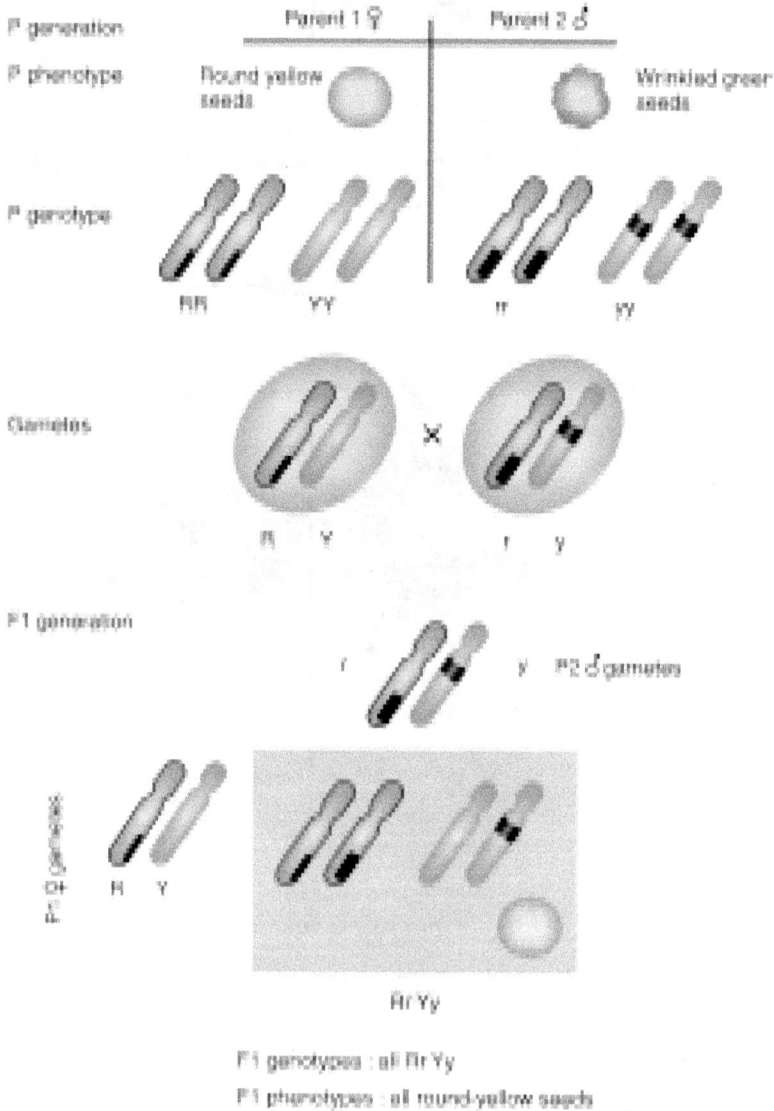

Figure 2.3a   Dihybrid cross (F1 progeny)

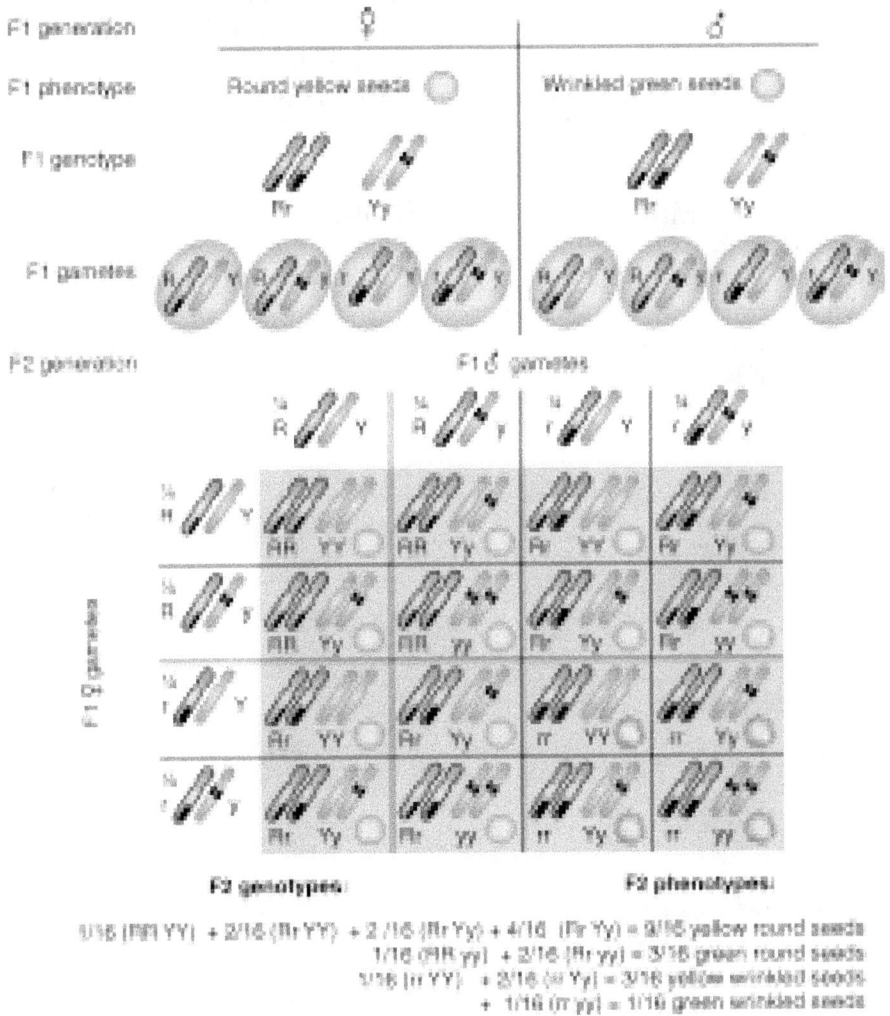

Figure 2.3b Dihybrid cross (F2 progeny)

formation, each gamete receives one allele either 'T' or 't'. During fertilization they fuse to form 'Tt' (heterozygous dominant) in F1 generation. Though 'Tt' (dominant and recessive alleles) remain together, they segregate at the time of F1 gamete formation.

The **Law of independent assortment** states that during gamete formation the members of the different pairs of characters (genes) segregate quite independently in all possible combinations. Mendel analysed two pairs of characters for this. The phenotypic ratio is 9:3:3:1 (Figure 2.3b).

Mechanism   The dominant alleles are 'YR'; 'Y' represents yellow colour and 'R' represents round nature of seed. The recessive alleles are 'yr', where 'y' is green colour and 'r' is wrinkled seeds. Even though 'YR' or 'yr' remain together they assort independently during gamete formation, so that various combinations are produced.

2.3   CONFIRMING THE LAW OF SEGREGATION

Mendel did a number of tests to check the validity of his outcome. He made selfing of the plants up to the F6 generation and found that every generation had dominant and recessive allele. His hypothesis came true, that is, the principle of segregation was valid no matter how many generations were involved.

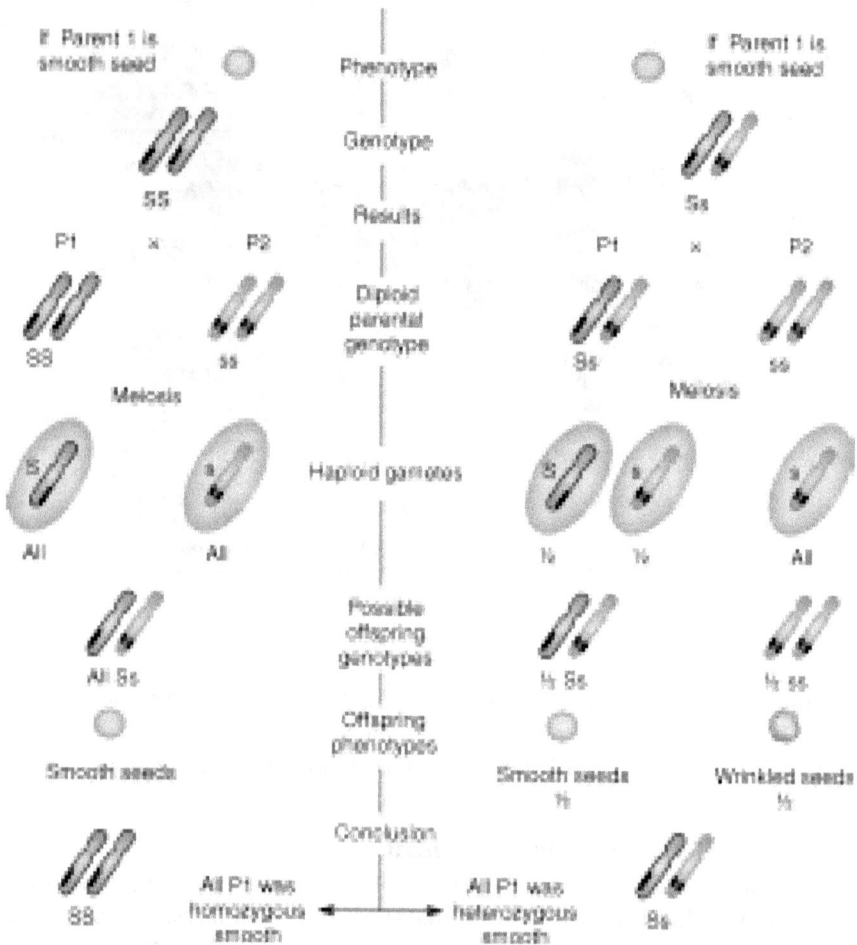

Figure 2.4   Test cross

Another important test to ascertain the genotype of the progeny with a given phenotype is to perform test cross. It is a cross between unknown genotype with a homozygous recessive plant. As shown in Figure 2.4 if the F2 individuals are homozygous SS, then the result of test cross with ss plant (P2 plants) will give only smooth seeds (Ss).

As in Figure 2.4, if the F2 individuals are heterozygous Ss, the result of test cross with 'ss' plant is 1:1, that is one heterozygous smooth seeds (Ss) and one homozygous wrinkled seeds (recessive) (ss).

This monohybrid cross can also be explained by using other characters of pea plant as listed earlier in this chapter, red versus white eyes in *Drosophila* and unattached versus attached ear lobes in humans.

2.4 MULTIPLE ALLELES

Though Mendel's principles form the foundation of heredity in all eukaryotic organisms including humans, there are a lot of exceptions and extensions. Some of these cases are detailed in this chapter. Let us get a broader knowledge of genetic analysis with reference to the phenotype of an organism. The alleles of a gene that usually appear in a population are of two types: Wild type or mutant (or alternative type), more generally dominant or recessive type. Sometimes a particular gene may have several alleles (more than two alleles). Such genes are said to have multiple alleles. But a single diploid individual can carry only any of these two alleles.

Human ABO blood grouping is the best example of multiple alleles of a gene. As shown in Table 2.1, the three ABO blood group alleles such as I^A, I^B, I^O respectively, give rise to the four phenotypes. Individuals homozygous for the recessive i allele, have O blood group. Both I^A and I^B are dominant to i allele. The genetics of blood group system follows mendelian principles. An individual with O blood group (i/i genotype) inherits from parents with the following genotypes (i/i × i/i), (I^A/i × I^A/i) (I^B/i × I^B/i) (I^A/i × I^B/i).

Another example of multiple alleles of a gene is the coat colours in rabbit. There are five different kinds of coat colours in rabbits such as wild type, chinchilla, light grey, Himalayan and albino. Combination of any of these five alleles can produce different coat colours. In *Drosophila* the multiple alleles of a gene are represented using eye colour. Though we have limited the number of allele forms of a gene to five, many hundreds of alleles are known for some genes.

When we look at multiple alleles in modern perspective, if there is any nucleotide change in a gene, it can result in an amino acid change at one or many places in a protein. This affects the function of a particular protein depending on the position and type of amino acid loss. In fact, if you analyse some of the human genetic diseases such as sickle cell anaemia and breast cancer, many alleles are often identified. The disease symptoms also vary depending on the mutation in the allele.

2.5   MODIFICATION OF DOMINANT RELATIONSHIPS

In studies of Mendel, all the allelic pairs showed complete dominance (that is when one allele is dominant to another, the phenotype of heterozygote is similar to homozygote dominant) or complete recessive relationships. But experiments by other scientists showed that many allelic pairs do not have such relationships. These allelic pairs have incomplete (partial) dominance or co-dominance relationships.

In incomplete dominance one allele is not completely dominant over another allele. The heterozygote phenotype is in between or intermediate to those of the two homozygotes.

The best example is plumage colour in chicken. As shown in Figure 2.5, mating between a true breeding black strain (C^B/C^B) and a true breeding white strain (C^W/C^W) gives F1 birds with bluish grey plumage (C^B/C^W). This new colour is called Andalusian blue by chicken breeders. When two F1 Andalusian blue birds are crossed the two alleles segregate in the offspring and produce one black strain, two Andalusian blue strain and one white fowl. The genotypic as well as phenotypic ratio is 1:2:1. Here two C^B alleles are required to produce black colour, whereas in the heterozygote, C^W allele dilutes black to bluish grey.

Another example of incomplete dominance is the Palomino horse. It has a golden yellow body colour and a mane and tail that are almost white. Palominos do not breed true. When they interbreed the result is one cremellos, two palominos and one chestnut in the ratio of 1:2:1. This is characteristic of incomplete dominance.

The other example of incomplete dominance in plants is flower colour in Snapdragon. In this plant, the red colour C^R is dominant and white colour C^W is recessive. When the red is crossed (Figure 2.6) with white, the F1 progeny has pink flowers ($C^R C^W$).

The other type of modification of the dominance relationship is **co-dominance.** In co-dominance the heterozygote exhibits the phenotypes of both homozygotes. The human ABO blood series is a good example of co-dominance. The blood group AB has alleles I^A and I^B respectively. A antigen and B antigen are produced. Thus I^A and I^B are co-dominant.

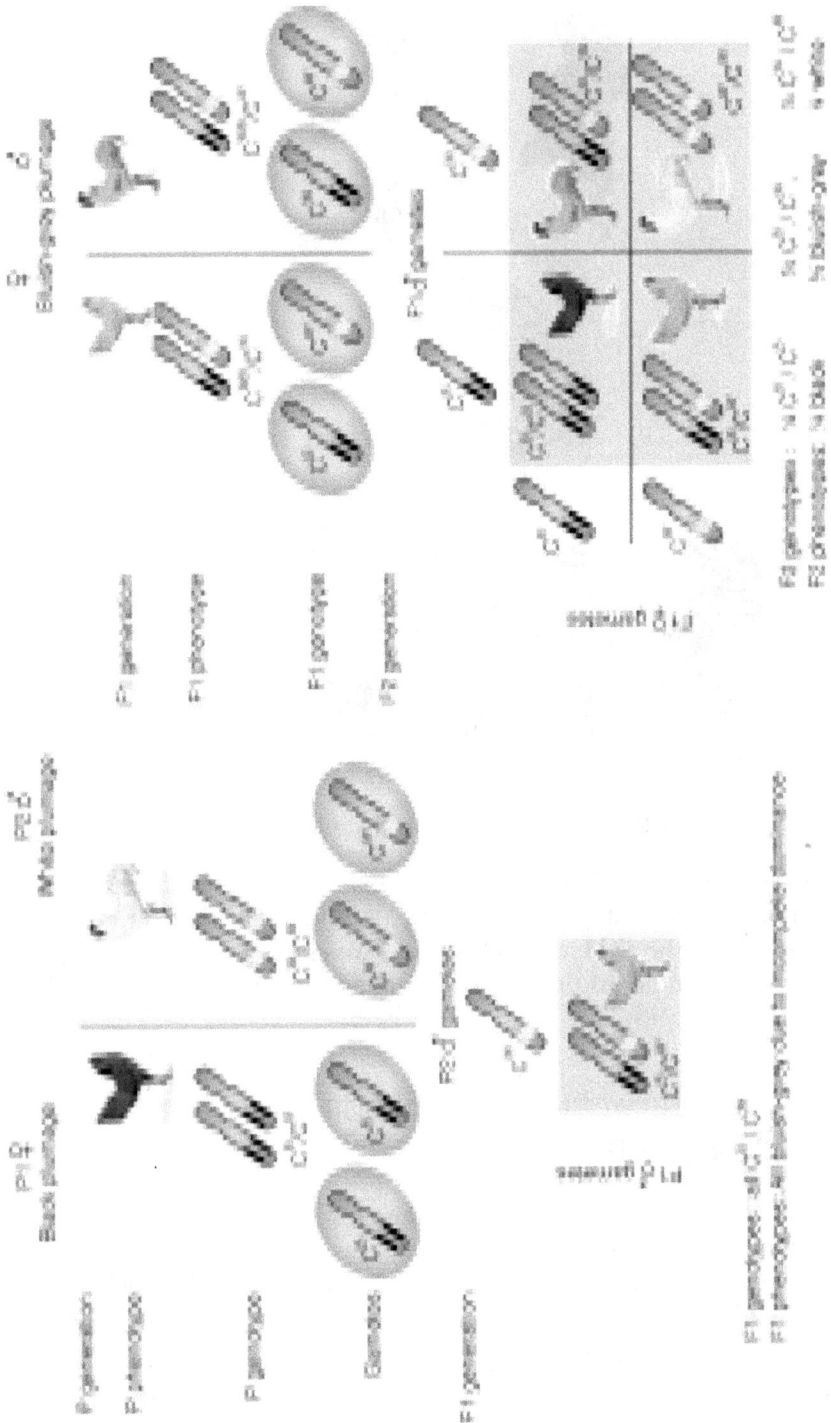

Figure 2.5 Incomplete dominance in chicken

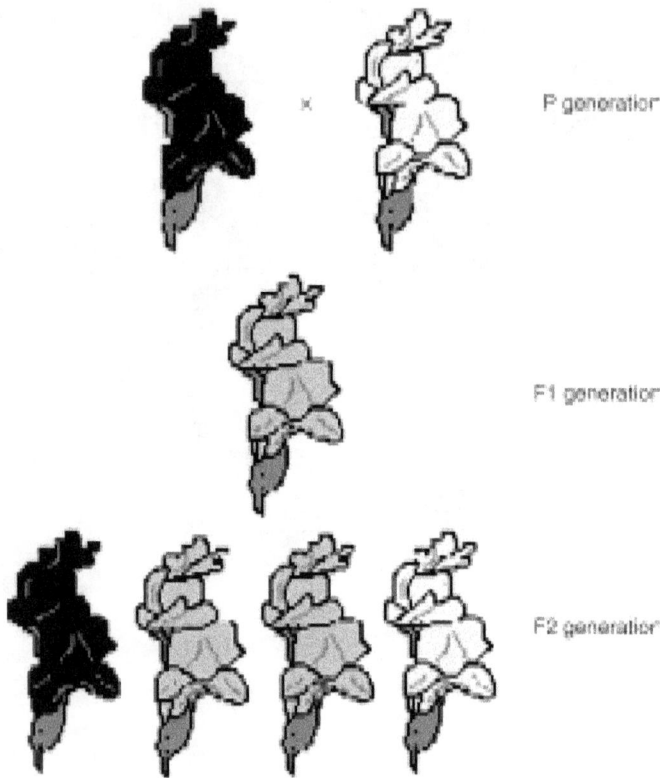

Figure 2.6   Incomplete dominance in Snapdragon

MN blood group system in humans is another example of co-dominance. There are three kinds of MN systems—M, MN and N. These M and N alleles form antigens on the red blood cell surface. A person with MN blood group (heterozygous condition) expresses both M and N antigens on the surface of the red blood cells.

2.6   GENE INTERACTIONS AND MODIFIED MENDELIAN RATIOS

Genes never act independently in determining a phenotype. Phenotype is the result of complex interactions at molecular level. Sometimes the interaction between genes results in new phenotypes. Comb shape in chickens is a good example.

In chickens four different comb shape phenotypes are obtained as a result of the interaction between the alleles of two gene loci. Mating between true breeding rose-combed and single-combed birds show F1 progeny with rose comb. When the F1 rose-combed birds are interbred, a ratio of 3 rose-combed to 1 single-combed birds result in F2. Rose is completely dominant over single-combed.

Similarly when true breeding rose and pea varieties are crossed, the F1 progeny showed Walnut comb. This result is completely different and interesting. As shown in Figures 2.7 and 2.8, when F1 walnut-combed birds are interbred, all four comb types appeared in a ratio of 9:3:3:1. Nine walnut, 3 rose, 3 pea and one single-combed birds were obtained.

Figure 2.7 Comb shape in chickens

The reason may be due to the following:

- The walnut comb depends on interaction of two dominant alleles R- P-, which are located on two different loci.
- The rose comb depends only on R- pp alleles
- The pea comb depends only on rr P- alleles
- The single comb depends on rr pp (both recessive alleles). These birds produce no functional gene product to influence the comb phenotype. When at least one dominant allele either R or P, or both

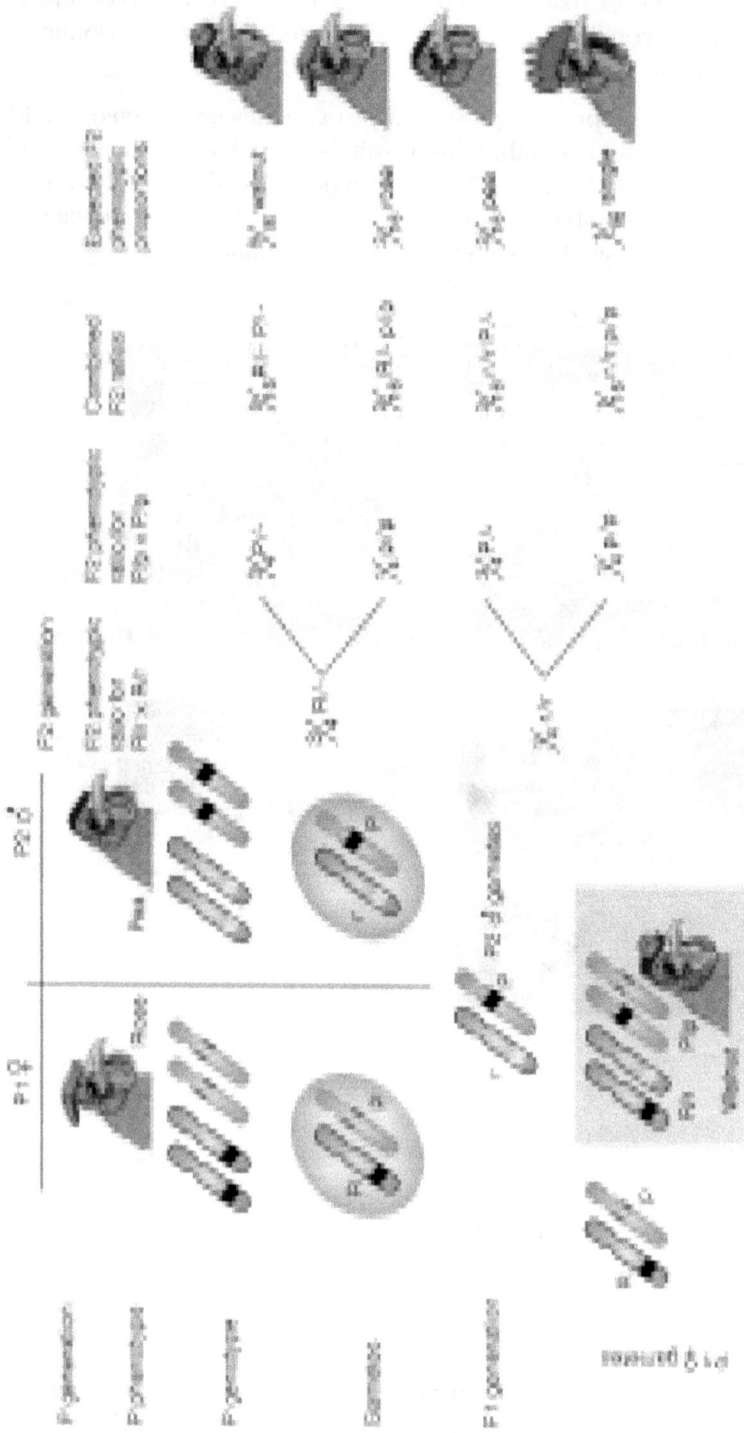

Figure 2.8   Comb shape in chickens (F1 and F2 progeny)

dominant alleles are found, that influences the comb phenotype to rose or pea or walnut respectively.

Another example of gene interaction is fruit shape in summer squash. The interaction of two dominant genes results in a new phenotype and modifies the ratio to 9:6:1. When two different varieties of true breeding sphere-shaped plants are crossed, the F1 fruit is disc-shaped. These F1 plants are interbred and F2 progeny results in nine disc-shaped fruit, 6 sphere-shaped fruit and one long-shaped fruit bearing plants.

As shown in Figure 2.9 when two dominant alleles are present they interact together to produce a new phenotype, disc shaped fruit. The homozygous

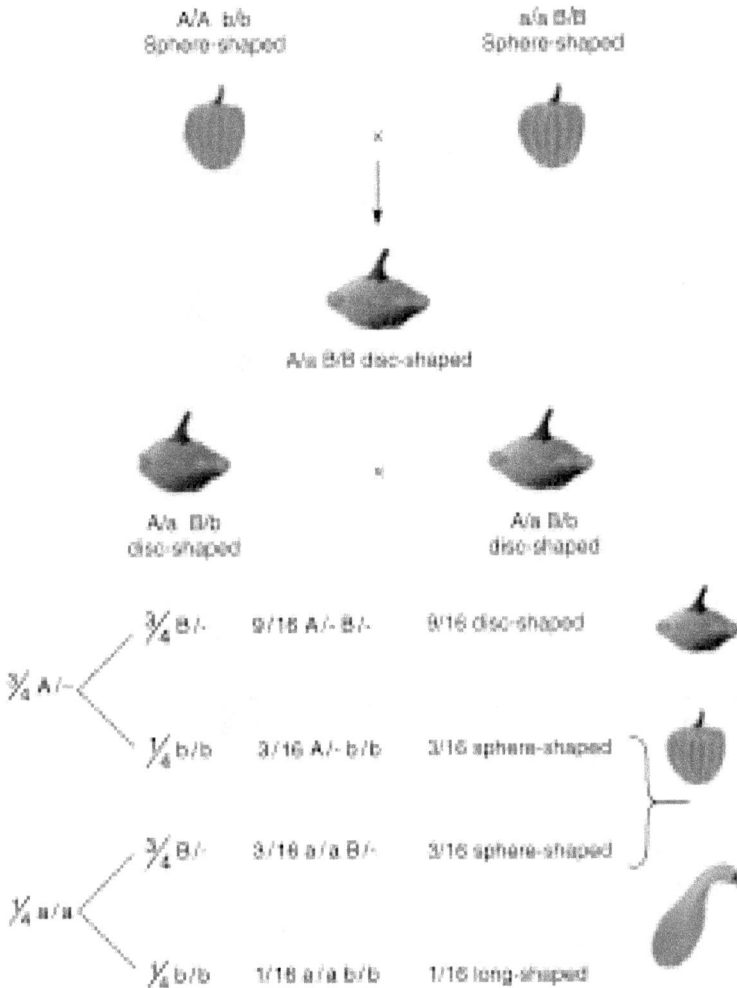

Figure 2.9 Fruit shape in summer squash

recessive (aa bb) gives long shaped fruit and existence of single dominant allele (either A or B) produces sphere shaped fruit, giving a ratio of 9:6:1.

Epistasis is another form of gene interaction. In this one gene masks the phenotypic expression of the other gene. A gene that suppresses another gene's expression is called **epistatic gene** (phenomenon is called epistasis) and the gene whose expression is suppressed is said to be **hypostatic gene** (phenomenon is called hypostasis). In this type of gene interaction, no new phenotypes are produced. Epistasis can occur in both directions between two gene pairs. An example is coat colour in rodents. The phenotypic ratio in the F2 is 9:3:4 rather than 9:3:3:1 as shown in Figure 2.10.

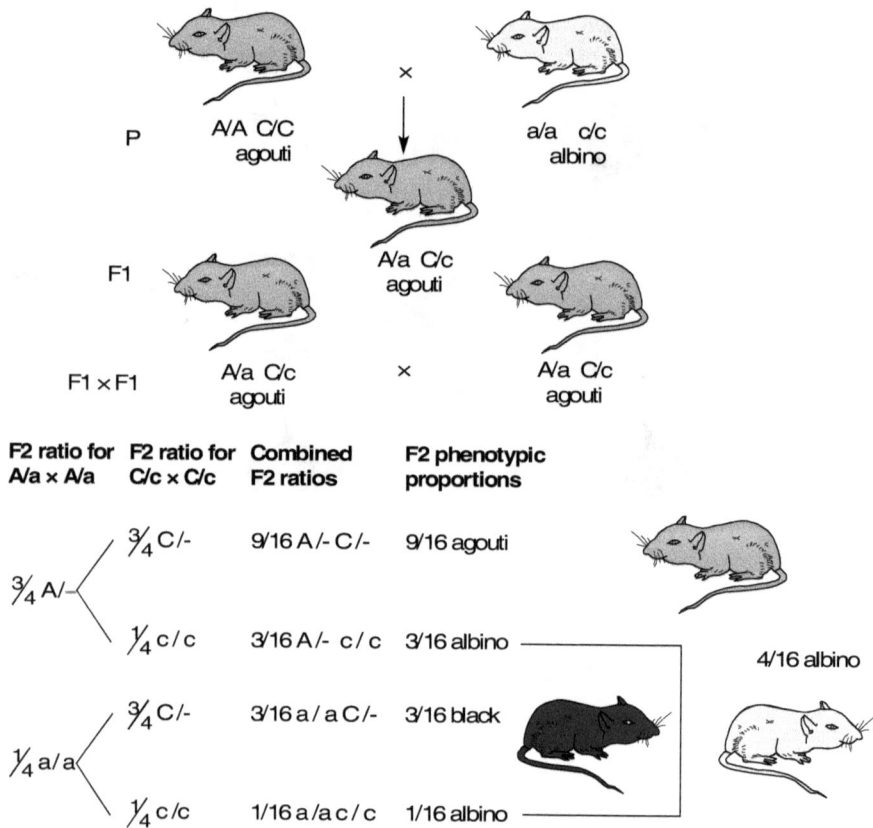

Figure 2.10   Recessive epistasis and coat colour in rodents

In this example, the epistatic gene is cc (homozygotic recessive allele) and the hypostatic gene is AA or Aa. The C allele in homozygous condition prevents pigment formation but dominant C allele specifies production of any pigment in the coat. The dominant allele 'A' is responsible for agouti pattern and recessive allele 'a' is responsible for nonagouti pattern in mice (Figure 2.10). When true breeding agouti mice (AA CC) was crossed with albino mice (aa cc), the F1 progeny was having heterozygotic agouti (Aa Cc) pattern. The F1 progeny was inbred and F2 generation resulted in phenotypic ratio of 9:3:4 (9 agouti A/C/, 3 black aaC/ and 4 albino A/CC and aacc).

Another example is flower colour in sweet peas. The ratio is modified to 9:7 instead of 9:3:3:1. Here the interaction of two genes gives rise to a specific product in the form of epistasis called **duplicate recessive epistasis of complementary gene action**.

In this a colourless compound is converted into compound 2 by C gene product. The compound 2 is converted into purple coloured pigment by P gene product. Recessive allele of C and P genes (homozygosity) leads to accumulation of white pigment in the flower; the presence of dominant alleles of C and P genes will result in purple flower. When F1 hybrids are self fertilized the F2 progeny consists of 9 purple flowered and 7 white flowered sweet peas (Figure 2.11).

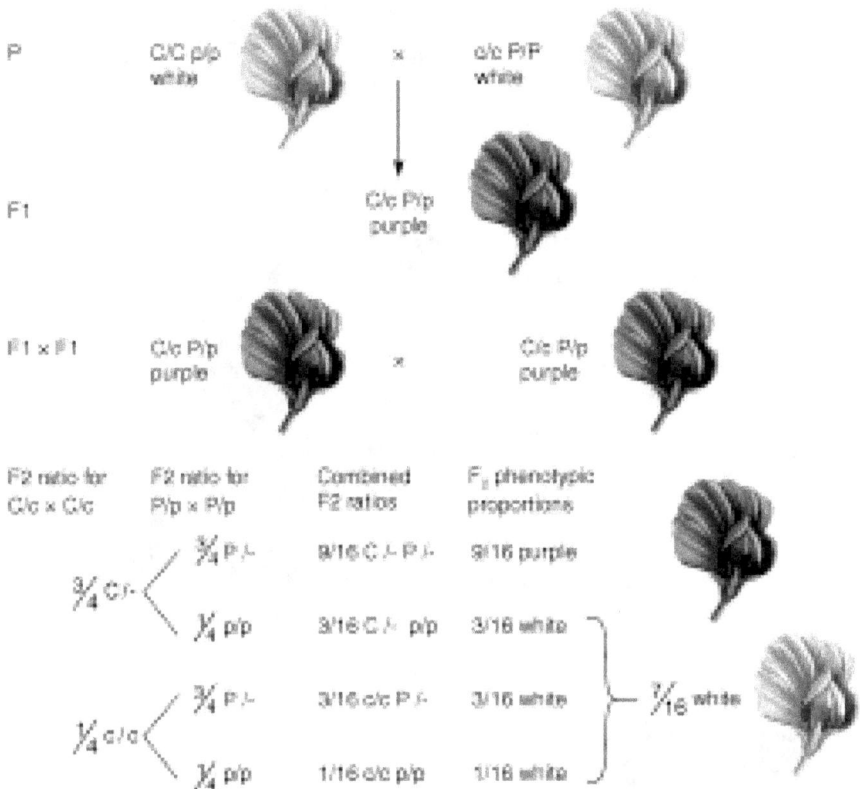

Figure 2.11 Duplicate recessive epistasis. Generation of an F2 9:7 ratio for flower colour in sweet pea.

2.7   ESSENTIAL GENES AND LETHAL GENES

Mutation in the allele not only changes the colour of mice coat or flower, but could cause death also. The allele that causes death of an organism is called a lethal allele. The lethal alleles are variants of essential genes. The allele can be lethal to an organism when it appears as a recessive (homozygotes) or dominant (homozygotes and heterozygotes).

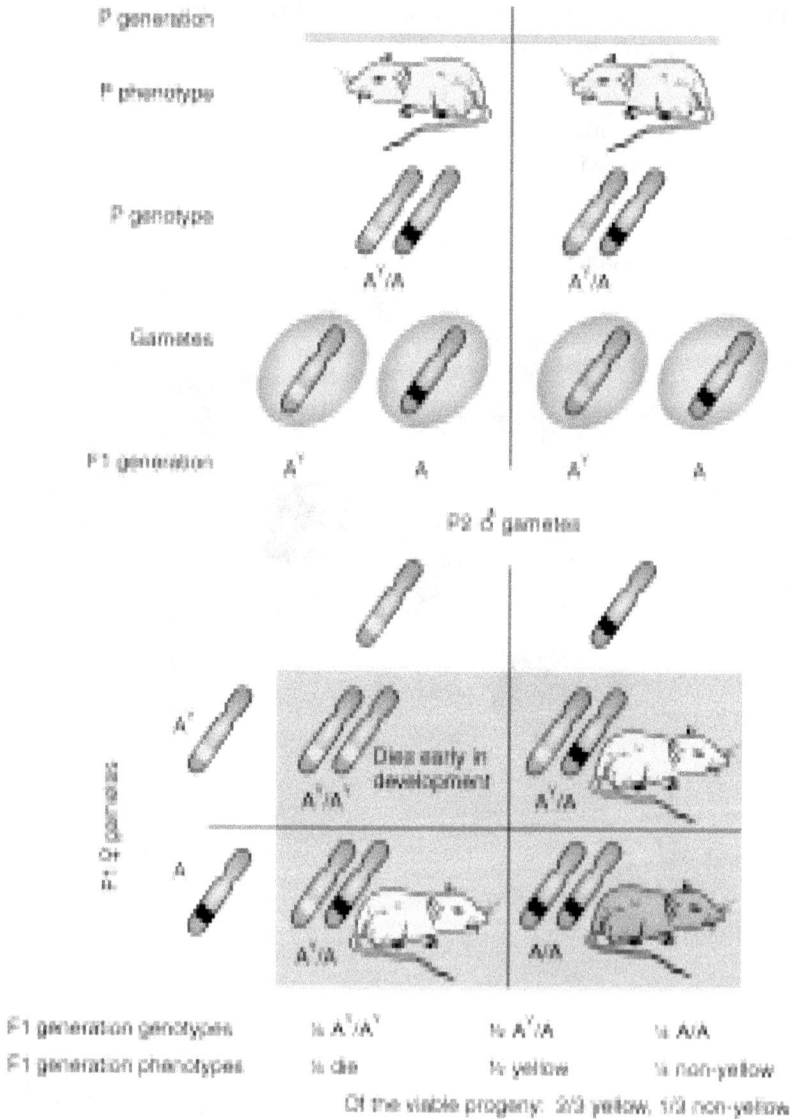

Figure 2.12   Lethal genes in mice

In 1905, Lucien Cvento studied the role of essential gene for yellow body colour in mice. When a yellow mouse is bred to non-yellow mouse, the progeny was similar to parental type. The phenotypic ratio was 1 : 1 (yellow : non yellow). When the yellow heterozygotes are interbred, a phenotypic ratio of 2 : 1 (yellow :non yellow respectively) was observed rather than 3 : 1 mendelian ratio. Later in 1910, Chenot's work was correctly interpreted by W. Castle and C. Little. As shown in Figure 2.12, when the yellow mice (A^YA) was crossed with another yellow mice (A^YA), the progeny was 1 A^YA^Y : 2 A^YA (yellow) : 1 AA (non-yellow). The A^YA^Y yellow mice died during early developmental stage.

Similarly in humans, mutations in essential genes sometimes lead to lethality. Mutations in HEXA gene cause Tay–Sachs disease (OMIM272800). Haemophilia and Huntington disease (OMIM 143100) are caused by lethal mutations in essential genes.

2.8   GENE EXPRESSION AND THE ENVIRONMENT

The development of a zygote into a complete organism is regulated by growth and differentiation of cellular system. For this, the interaction of genome with internal cellular environment and external environment is mandatory. This internal or external environment helps in the development of products, which control the complex pathways. But abnormal internal or external environment can influence the gene expression by controlling the pathways. Now it is important to find the extent of gene expression at varying environmental conditions. This can be adhered by using the concepts of **penetrance** and **expressivity** in the phenotypic expression of a genotype (Figure 2.13).

Penetrance depends on genotype of an individual as well as the effect of environmental condition. The frequency with which a dominant or recessive (homozygous) allele or gene manifests itself in individuals in a population is called penetrance. Penetrance is complete or 100 percent when recessive (homozygous) shows one kind of phenotype, dominant (homozygous) shows another kind of phenotype and all heterozygous are alike. If the yield of identical genotype drops to less than 100 percent of the expected phenotype, then it is incomplete penetrance. Further, if you see all individuals carrying mutant gene showing mutant phenotype, then the gene is completely penetrant. Alleles in the human ABO blood group system are a good example.

The degree of phenotypic expression varies according to the influence of gene. **Expressivity** is the degree of phenotypic expression to any level

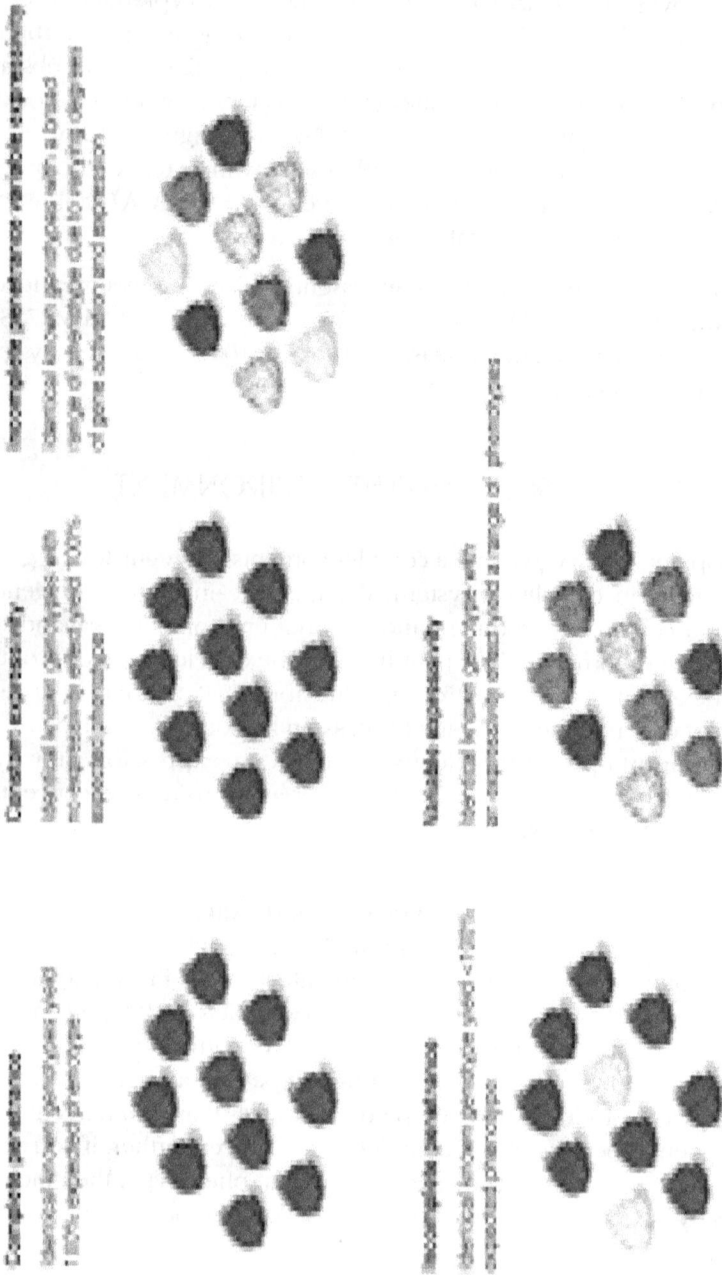

Complete penetrance
Identical known genotypes yield 100% expected phenotype

Constant expressivity
Identical known genotypes with representation of expected phenotype

Incomplete penetrance variable expressivity
Identical known genotypes with a broad range of phenotype due to varying degrees of gene activation and expression

Incomplete penetrance
Identical known genotypes yield < 100% expected phenotype

Variable expressivity
Identical known genotype with an expression of a range of phenotypes

Figure 2.13   Penetrance and expressivity

by a penetrant allele, gene or genotype in an individual. Expressivity may be constant when there is 100 percent similarity in expected phenotype. But in variable expressivity a range of phenotypes appear for a known genotype. The best example of variable expressivity is *Osteogenesis impercta* in humans. It is an autosomal dominant disorder with complete penetrance. Blueness of the sclera, fragile bone and deafness are three salient features of the disease. A person may have one or a combination of the above features.

In neurofibormatosis condition the gene exhibits both incomplete penetrance and variable expressivity. Few brown-pigmented areas of the skin are mildest form of the disease. However, in severe condition neurofibromas, tumours on the eye, brain or spinal cord, high blood pressure and large head are found. The age of onset of the disease can also affect the function of the gene.

FOR ADDITIONAL READING

1. Griffiths, A.J.F., Gelbert, W.M., Lewontin, R.C. and Miller, J.H. (2002). *Modern Genetic Analysis*, W.H. Freeman and Company, New York.

2. Griffiths, A.J.F., Gelbert, W.M., Lewontin, R.C., Miller, J.H. and Suzuki, D.T. (2002). *An Introduction to Genetic Analysis*. W.H. Freeman and Company, New York.

3. McKusick, V.A. (1998). *Mendelian Inheritance in Man*, 12th edn., Johns Hopkins University Press, Baltimore.

4. Olby, R.C. (1966). *Origins of Mendelism*. Constables and Company Ltd., London.

5. Orel, V. (1996). *Gregor Mendel: The first Geneticist.* Oxford University Press, Oxford, England.

6. Orel, V. and Hartl, D.L. (1994). "Controversies in the interpretation of Mendel's discovery." *History and Philosophy of the Life Sciences.* 16:423.

7. Snustad, D.P. and Simmons, M.J. (2002). *Principles of Genetics.* John Wiley & Sons Inc., New York.

8. Stern, C. and Sherwood, E. (1966). *The origins of Genetics: A Mendel Source Book.* W.H. Freeman and Company, New York.

9. Sturtvant, A.H. (1965). *A short history of Genetics*, Harper & Row, New York.

10. Tom Strachan *et al.* (1996). *Human Molecular Genetics*. BIOS Scientific Publishers Ltd., UK.

11. Vogel, F. and Motulsky, A.G. (1986). *Human Genetics: Problems and Approaches,* 2nd edn. Springer, Berlin.

12. Sturtevant, A.H. (2001). *A History of Genetics.* Cold Spring Harbor Laboratory and Electronic Scholarly Publishing, New York.

13. Hartl, D.J. and Jones, E.W. (1998). *Genetics: Principles and Analysis*, 4th edn. Jones and Bartlett Publishers, London.

MODEL QUESTIONS

1. The term "X-linked" is used for genes carried on the Y chromosome.
   a. true   b. false

2. Mendel most likely would have observed incomplete dominance on occasion if he had chosen different characteristics than the ones he chose.
   a. true   b. false

3. Mutations increase the amount of variations among offspring.
   a. true   b. false

4. Genes rarely interact to produce the phenotype.
   a. true   b. false

5. All chromosome pairs are the same in males and females.
   a. true   b. false

6. Who determines the sex of the child?
   a. male   b. female   c. it varies

7. The ABO blood type in humans is controlled by how many alleles.
   a. two   b. three   c. four

8. Which genes are more likely to cross over?
   a. closer together genes   b. farther apart genes
   c. there is no difference.

9. The sex chromosomes in human females are
   a. X   b. Y   c. XX   d. XY

10. Each gene in polygenic inheritance has the following allele.
    a. contributing   b. noncontributing
    c. both a and b   d. neither a nor b

11. Which one of the following affects the phenotype.
    a. genotype   b. environment
    c. both a and b   d. neither a or

12. An individual with the blood type AB is exhibiting the following characteristics.
    a. complete dominant   b. incomplete dominant
    c. codominant

13. Define the following:

    i. hybrid   ii. gene   iii. penetrance   iv. expressivity

14. Define and explain with an example Mendel's law of segregation?

15. Define and explain with an example the law of recombination (or law of independent assortment)?

16. What are genes?

17. How are genes organized?

19. Which are the genes that segregate and how?

20. Comment on various patterns of gene interaction with examples.

21. Draw and explain the degree of relationships.

22. Describe in detail the role of environment in gene interaction.

23. What is meant by the term "pleiotrophy" and what is a pleiotropic effect?

24. Define the following terms:

    i.   Genotype   ii. Phenotype
    iii. Allele   iv. Dominant
    v.   Recessive

25. Consider a human family with 4 children and remember that each birth is likely to result in either a boy or a girl.

    i.   What proportion of sibships will have at least one boy?
    ii. What fraction of sibships will have the gender order GBGB?

26. In the cross AaBb × AaBb, what will be the fraction of the progeny that will be homozygous for one gene and heterozygous for the others?

CHROMOSOME

3

3.1 INTRODUCTION

Chromosomes are dense bodies found within the nucleus of the cell, containing deoxyribonucleic acid (DNA) and associated proteins. The DNA is long but highly coiled. DNA sequence for a character (trait) is called a **gene**. Each chromosome contains a few thousand genes. During interphase stage of the cell cycle, chromosomes are less condensed and are not visible under light microscope. However they become highly condensed and are visible during metaphase stage of mitosis.

W. Waldeyer coined the term "chromosome" in 1888. Human somatic cell contains 46 chromosomes, out of which 44 are autosomes (somatic characters) and 2 are sex chromosomes (XX in female and XY in male). Table 3.1 shows some common species and their diploid set of chromosomes.

The chromosomes transfer genetic information to next generation (daughter cell) or reduce genetic information to a haploid set (through meiosis) which enables male and female gametes to join to form a diploid zygote. This function of the chromosomes is of great biological significance.

3.2 MOLECULAR ORGANIZATION OF THE CHROMOSOME

The basic material of eukaryotic chromosome is called **chromatin**. There are two kinds of chromatin, viz. euchromatin and heterochromatin. During interphase, most of the chromatin fibres are less densely packed, dispersed throughout the nucleus and stain diffusely. Such material is called **euchromatin**. The **heterochromatins** are highly condensed,

Table 3.1 Diploid set of chromosomes found in some species

Species	Common name	Diploid chromosome number
Myrmecia pilosula	Ant	2
Drosophila melanogaster	Fruit fly	8
Allium cepa	Onion	16
Zea mays	Corn or Maize	20
Saccharomyces cerevisiae	Budding yeast	32
Xenopus laevis	South African frog	36
Felis catus	Cat	38
Sus scrofa	Pig	38
Mus musculus	House mouse	40
Homo sapiens	Human	46
Ovis aries	Sheep	54
Bos taurus	Cattle	60
Capra hircus	Goat	60
Equus asinus	Donkey	62
Equus caballus	Horse	64
Canis familiarus	Dog	78
Gallus domesticus	Chicken	78
Cambarus clarkii	Cray fish	200
Equisetum arvense	Field horsetail plant	216

transcriptionally inert and stain dark. There are two classes of heterochromatin—facultative heterochromatin and constitutive heterochromatin. **Facultative heterochromatin** has a dual role: it is either genetically active or inactive (X-chromosome inactivation) whereas **constitutive heterochromatin** is always in inactive state (certain repetitive DNA sequences near centromeres).

Table 3.2   Histone proteins

Histones	Molecular weight (dalton)
H1	23,000
H2A	13,960
H2B	13,774
H3	15,342
H4	11,282

The most fundamental subunit of chromosome packaging is the nucleosome that consists of basic proteins called histones (H2A, H2B, H3 and H4) and 146 bp of double-stranded DNA. The histone proteins are rich in lysine and arginine residues and are positively charged. They bind tightly to the negatively charged phosphates in DNA. Each nucleosome is linked with linker DNA that in turn has H1 proteins. By treating chromatin with the enzyme micrococcal nuclease, the nucleosomes are separated into individual units. Different types of histone proteins along with their molecular weights are given in Table 3.2.

Chromosomes are divided longitudinally into two sister chromatids (p-short arm and q-long arm) and they are held together by the centromere (primary constriction) as shown in Figure 3.1.

Based on the position of centromere the chromosomes are classified as: **metacentric** chromosome (centromere exactly at the central region), **submetacentric** (centromere closer towards one end), **acrocentric** (centromere very close to one end) and **telocentric** (centromere at the tip) (Figure 3.2).

The chromosomal function is dependent on three kinds of elements—centromeres, telomeres and replication origin. Centromeres are *cis*-acting DNA elements responsible for separation of chromosomes at mitosis and meiosis. The telomeres seal the ends of chromosomes using a specialized enzyme called telomerase and confer stability to the chromosome.

The replication origin is required to initiate DNA replication and maintain chromosome copy number.

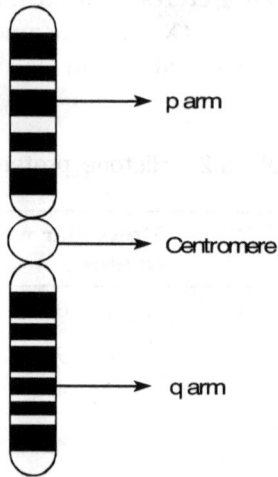

Figure 3.1   Chromosome

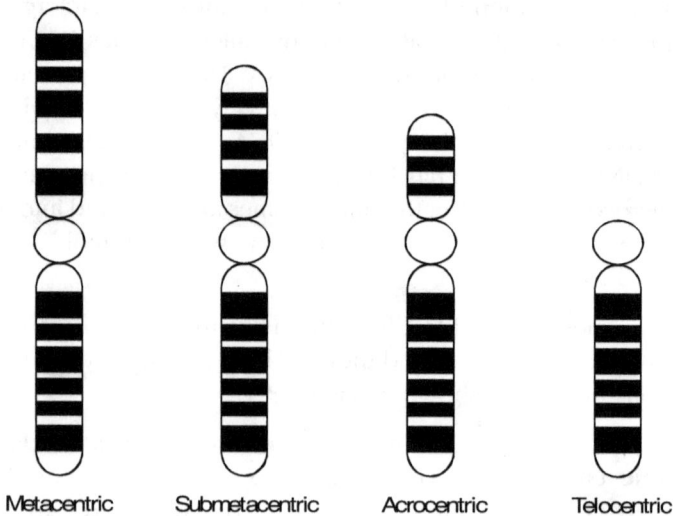

Metacentric   Submetacentric   Acrocentric   Telocentric

Figure 3.2   Chromosome types based on position of centromere

3.3   TYPES OF CHROMOSOMAL ANOMALIES

Normal chromosome constitutions in humans are described by a karyotype that states the total number of autosomes and sex chromosomes (The representation of the entire chromosome set as a series of banded chromosomes is called karyogram).

The loss or gain of a whole chromosome or a microchromosomal segment is known as chromosomal anomaly. It may be present throughout the body cells (constitutional anomalies) or present in an isolated region of the body cells (somatic anomalies or acquired anomalies).

Chromosomal anomalies are of two categories namely **numerical anomalies and structural anomalies**. The numerical chromosomal anomalies involve a change (increase or decrease) in the number of chromosomes. There are three classes of numerical chromosomal anomalies—polyploidy, aneuploidy and mixoploidy.

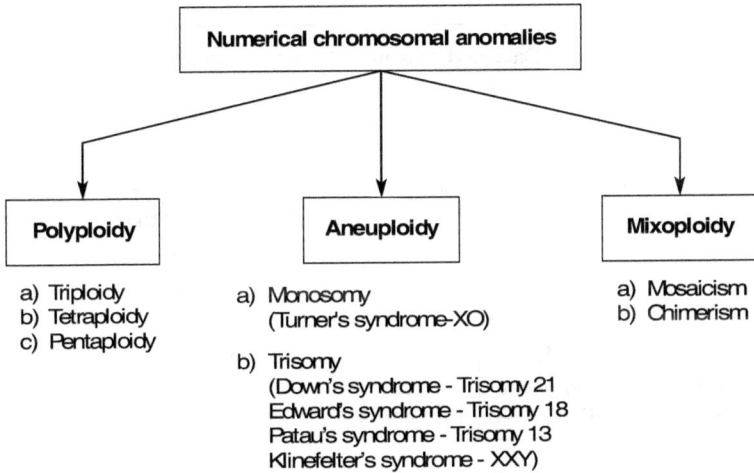

```
              ┌────────────────────────────────────┐
              │  Numerical chromosomal anomalies    │
              └────────────────────────────────────┘
```

Polyploidy	Aneuploidy	Mixoploidy
a) Triploidy b) Tetraploidy c) Pentaploidy	a) Monosomy (Turner's syndrome-XO) b) Trisomy (Down's syndrome - Trisomy 21 Edward's syndrome - Trisomy 18 Patau's syndrome - Trisomy 13 Klinefelter's syndrome - XXY)	a) Mosaicism b) Chimerism

3.3.1 Polyploidy

The most common polyploidy is triploidy which is caused by two sperms fertilizing a single egg (dispermy) or by fertilization involving an abnormal diploid gamete. Tetraploidy is usually due to failure to complete the first zygotic division. It should be noted that in all individuals some cells are naturally polyploid, viz. cells of regenerating liver, megakaryocytes and bone marrow cells. Certain cells do lack a nucleus; they are called nulliploid, e.g. red blood cells, platelets, squamous epithelial cells.

3.3.2 Aneuploidy

Aneuploidy condition is due to the addition or deletion of a chromosome in its set (in addition to the normal two homologous chromosomes). Aneuploidy cells arise as a result of non-disjunction or anaphase lag. Trisomy is the condition where addition of an extra chromosome is found in its set, for example Down's syndrome (Trisomy 21), Edward's syndrome (Trisomy 18),

Patau's syndrome (Trisomy 13). In monosomy condition there is loss of a homologous chromosome in its set—Turner's syndrome (XO).

3.3.3   Mixoploidy

Rarely, an individual may possess two or more genetically different cell lines. This is called mixoploidy. It can occur as a result of mosaicism or chimerism. If an individual possesses two or more genetically different cell lines, which are derived as a result of a single zygote, it is called mosaicism and if it is as a result of different zygotes, it is called chimerism.

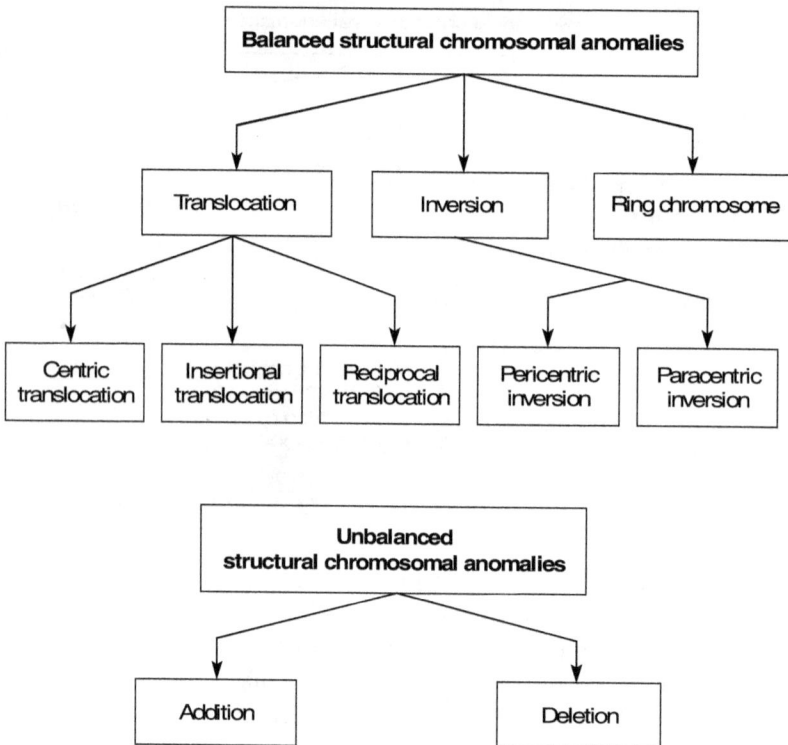

```
          Balanced structural chromosomal anomalies
         /                  |                    \
   Translocation        Inversion          Ring chromosome
   /      |      \         /      \
Centric Insertional Reciprocal Pericentric Paracentric
translocation translocation translocation inversion inversion
```

```
              Unbalanced
      structural chromosomal anomalies
         /                    \
     Addition              Deletion
```

The structural chromosomal anomalies are either due to balanced or unbalanced chromosomal breakage and reunion. Deletion, insertion, inversion and translocation of genetic material within or in between chromosomes are involved in structural chromosomal anomalies.

3.3.4   Balanced Structural Chromosomal Anomalies

Chromosomal inversion, chromosomal translocation or ring chromosomes occur due to balanced structural chromosomal anomalies. A chromosomal inversion occurs when the chromosomal segment between two break points is inverted. The inverted segment may be within one arm of the chromosome (paracentric inversion) or include the centromere (pericentric inversion). If breaks occur on more than one chromosome, hybrid chromosomes may be formed and this is called **chromosomal translocation**. Three types of chromosomal translocations are known, viz. **reciprocal translocation**, **robertsonian translocation** and **insertional translocation**. In reciprocal translocation, the material distal to the break points on two chromosomes is exchanged. The break points may occur on either the long or the short arms. In robertsonian translocation, breaks occur at or near the centromere of two acrocentric chromosomes and large fragments of the two chromosomes fuse (often results in dicentric chromosome). The acentric fragments will usually be lost in subsequent cell division. Robertsonian translocations are usually regarded as balanced because they cause no phenotypic change. An insertional translocation results from three chromosome breaks and involves excision of a DNA segment located between two break points on one chromosome or on a second chromosome at interstitial site. Offspring of carriers may be at a risk of partial monosomy or partial trisomy. A **ring chromosome** results when the two ends of the segment between the break points are joined. It carries a centromere and can pass through cell division.

3.3.5 Unbalanced Structural Chromosomal Anomalies

When there is a loss of chromatin from a chromosome, the unbalanced structural chromosomal anomalies appear. The terminal deletion arises from one break and the acentric fragments are lost at the next cell division. Interstitial deletion arises from two breaks; the acentric fragments are lost and fusion takes place at the break sites, e.g. DiGeorge syndrome 22q11 microdeletion.

Sometimes duplication of chromosome segments may arise from familial arrangements such as translocations or inversions. Duplicated segments may be arranged as a direct tandem repeat or inverted repeats of chromosome segments.

3.4 HUMAN KARYOTYPES

The highly condensed metaphase chromosomes (Figures 3.3 and 3.4) are visible under the light microscope. Individual chromosomes can be identified and arranged based on their size and location of the centromere. The chromosomes are stained with Giemsa and to obtain G-banding pattern, trypsin treatment is given.

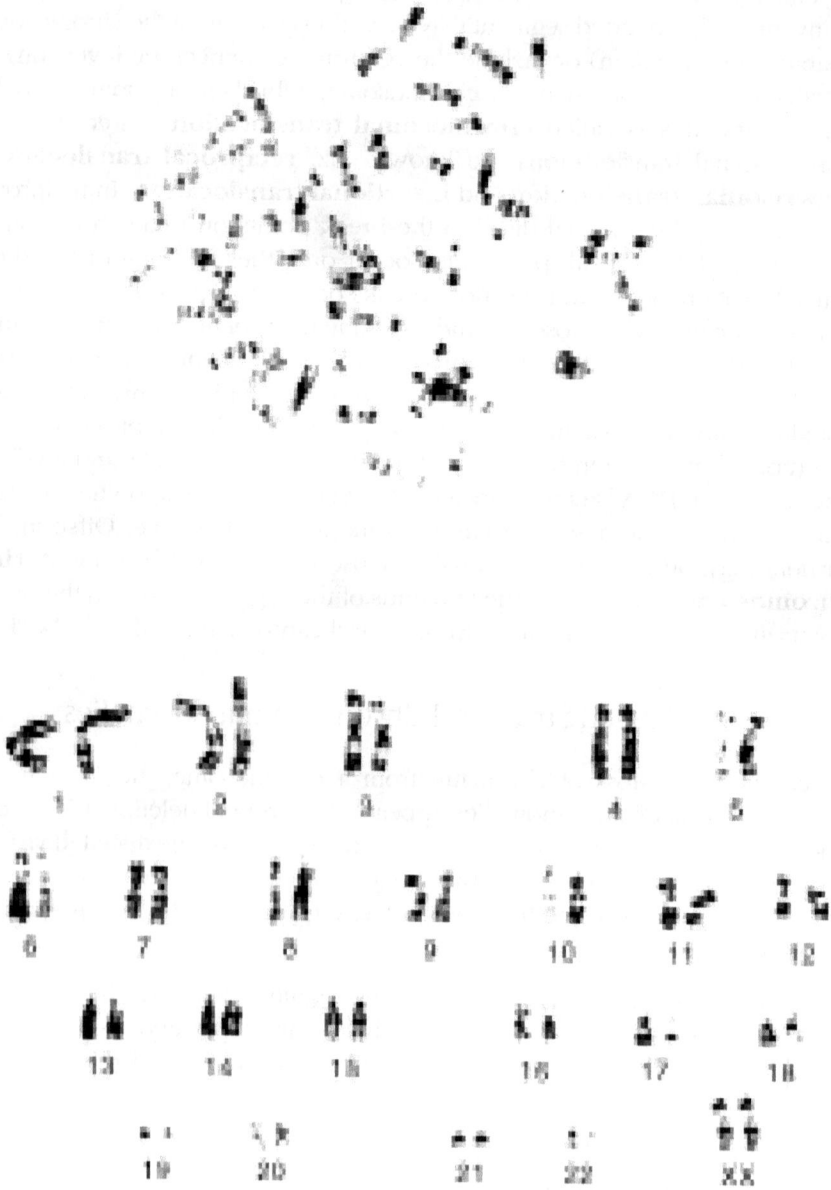

Figure 3.3   Normal female karyotype

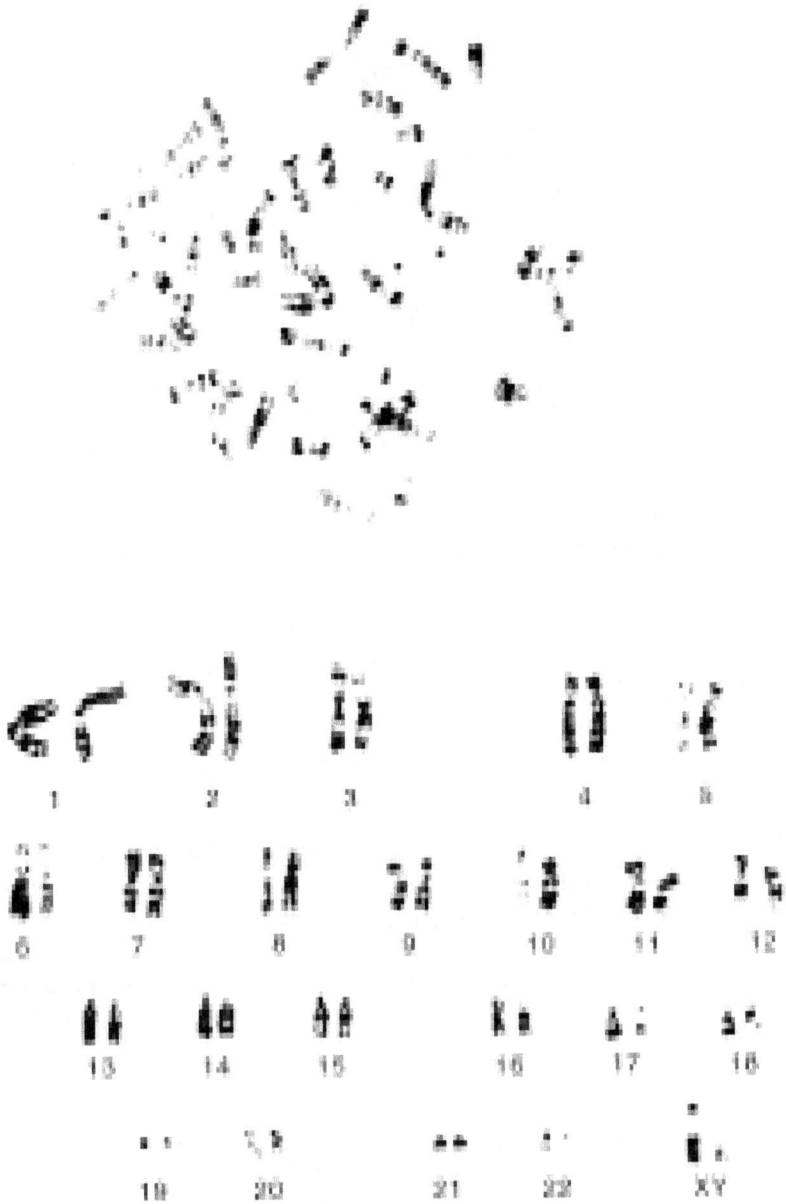

Figure 3.4 Normal male karyotype

3.5   CULTURE TECHNIQUE

3.5.1   Human Peripheral Blood Lymphocyte Culture

Inoculate 0.5 ml of blood into a sterile culture vial containing 5 ml of RPMI 1640 medium, 1 ml of foetal bovine serum, antibiotics (penicillin and streptomycin) and 0.2 ml (4%) of phytohaemagglutinin-M (PHA), aspetically. Incubate cultures for a period of 69½ hours at 37°C. Release carbon dioxide every 24 hours.

Add 0.05 ml of 0.01% colchicine to the culture vial after 69½ hours of incubation and mix well. Incubate further for 30 minutes at 37°C. Pour content into a centrifuge tube and spin at 1200 rpm for 5 minutes. Discard the supernatant leaving the cell button at the bottom. Resuspend cells in 6 ml of hypotonic solution (0.075M KCl) and incubate at 37°C for 7 minutes and then centrifuge at 1200 rpm for 5 minutes. Discard the supernatant and disturb the cells by tapping at the bottom of the tube. Fix the cells in freshly prepared fixative (3 parts of methanol and 1 part of glacial acetic acid) and leave undisturbed for 1 hour. Give two changes of fixative at 45-min. interval each.

Suspend the cells in a small quantity of fixative and place a drop of cell suspension on a clean pre-cooled slide. Examine the slide through a microscope (with the condenser lowered) to check for cell density and spread of chromosomes at metaphase. The rest of the slides should be prepared after suitable modifications.

Immerse one-day-old slides in 0.05% trypsin solution for 3 to 10 seconds. Wash slides in distilled water and stain in 4% buffered Giemsa solution for 5 min. Wash the slides in running tap water and air-dry. Analyse fifty well spread metaphase plates for the presence of chromosome abnormalities. Designate aberrations according to the standard nomenclature (ISCN, 1995). Take photographs of well-banded metaphases under an oil immersion lens.

Lymphocyte culture for micronucleus test   Stimulate lymphocyte culture by adding 0.5 ml of human peripheral blood (sodium heparinized) to 5 ml of RPMI-1640 (Himedia Lab Pvt. Ltd., Mumbai, India) medium supplemented with 20% gamma-irradiated foetal bovine serum, antibiotics (penicillin and streptomycin) and 4% of phytohaemagglutinin-M. Incubate the culture at 37°C and add cytochalasin B (3 mg/ml of culture) in the 44th hour after the initiation of cultures to block cytokinesis. After 72 hours, collect the cells by centrifugation, treat them with hypotonic solution (0.01 M KCl) and fix them in methanol/glacial acetic acid (3 : 1) and harvest them on slides. Air-dry the slides and stain them for 5 minutes in 4% buffered Giemsa solution. All slides must be coded and analysed microscopically at 100× magnification under oil immersion lens for micronuclei. A total of 1000 binucleated cells must be counted and present the values as percentage of micronucleus in 1000 binucleated cells.

Lymphocyte culture for sister chromatid exchange To stimulate lymphocytes, add half-a-millilitre (0.5 ml) of whole blood to 5 ml of culture medium RPMI-1640 supplemented with 20% gamma-irradiated foetal bovine serum and 4% phytohaemagglutinin-M. Incubate the culture at 37°C. After 24 hours of initiation, add 15 mg/ml of 5-bromo 2´-deoxyuridine (BrdU; Sigma Chemical Co. USA). Again incubate the culture at 37°C in the dark. After 71 hrs, add colchicine (0.1 mg/ml of culture) and an hour later harvest the cells with a hypotonic treatment (0.075 M KCl), and fix in methanol/ glacial acetic acid (3 : 1). Dry the slides and stain them with Giemsa, code and score for sister chromatid exchanges. Count one sister chromatid exchange when two adjacent segments of one of the chromatids in a chromosome is stained differently. The value of sister chromatid exchanges for each specimen should be taken as the mean rate of the sister chromatid exchanges per metaphase.

Lymphocyte culture for fragile sites In the first method, initiate lymphocyte culture in McCoy 5A negative folic acid medium supplemented with 20% foetal bovine serum and 10 mg/l of BrdU. Harvest cultures after 72 hours using standard methods. Fifty unbanded metaphases per individual are to be examined for fragile sites, and the location of the fragile site has to be confirmed in each individual, using trypsin-Giemsa banding.

In the second method, incubate peripheral blood lymphocyte cultures for 72 hours at 37°C in 5% CO_2 in RPM1 medium supplemented with 10% foetal calf serum and 2% phytohaemagglutinin. Dissolve Aphidicolin (0.12 mM) in dimethylsulphoxide and add in the culture vial for the last 24 hours. Add Colcemid (0.04 mg/ml) 2 hours before harvesting. Give 0.075 M KCl hypotonic shock and methanol–acetic acid (3 : 1) fixation followed by air-drying. Chromosome spreads are to be obtained according to the standard procedure.

3.6 MICRONUCLEUS

Micronucleus (MN) is the small nucleus, separate from and additional to the main nucleus (Figure 3.5). It is formed during the metaphase/anaphase transition of mitosis (or meiosis). It may arise from a whole lagging chromosome (aneugenic leads to chromosome loss) or an acentric chromosome fragment detaching from a chromosome after breakage (clastogenic), which does not integrate in the daughter nucleus.

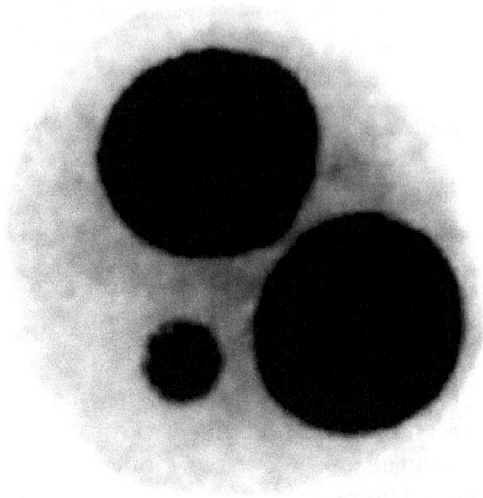

Figure 3.5   Micronucleus

3.6.1   Characteristics of the Micronucleus Test

- It is a biomarker for risk assessment.
- It is the endpoint in the identification of chromosomal and genomic mutations.
- MN contains either a whole chromosome or an acentric fragment.
- Micronucleus found within mononucleated cells depicts damage accumulated before cultivation of the cells.
- Micronucleus found within binucleated cells depicts damage accumulated before and after cultivation of the cells.

3.6.2   Selection of Binucleated Cells

- The diameter of the MN should be less than one-third of the main nucleus.
- MN should be separated from or marginally overlap with the main nucleus as long as there is clear identification of the nuclear boundary.
- MN should have similar staining as the main nucleus.
- Scoring cannot be done if there were trinucleated, quadranucleated, or multinucleated cells or cells whose nuclei have undergone apoptosis.

In the presence of cytochalasin B, mononucleated cells are recommended to be harvested at 24 hours post-PHA stimulation as there can be no doubt at this time-point that MN within such a cell is a result of *in vivo* rather than *ex vivo* division. Binucleated cells are recommended to be harvested at 72 hours post-PHA. Moreover, 24-hour post-PHA time-point may be the right time to count apoptotic/necrotic cells.

3.7 SISTER CHROMATID EXCHANGE

Sister chromatid exchange (SCE) involves breakage of two DNA strands, followed by an exchange of whole DNA duplexes. It occurs during 'S' phase and is efficiently induced by mutagens (carcinogens) that form DNA adducts or interfere with DNA replication. The formation of SCE has been correlated with recombinational repair and the induction of point mutations, gene amplification and cytotoxicity.

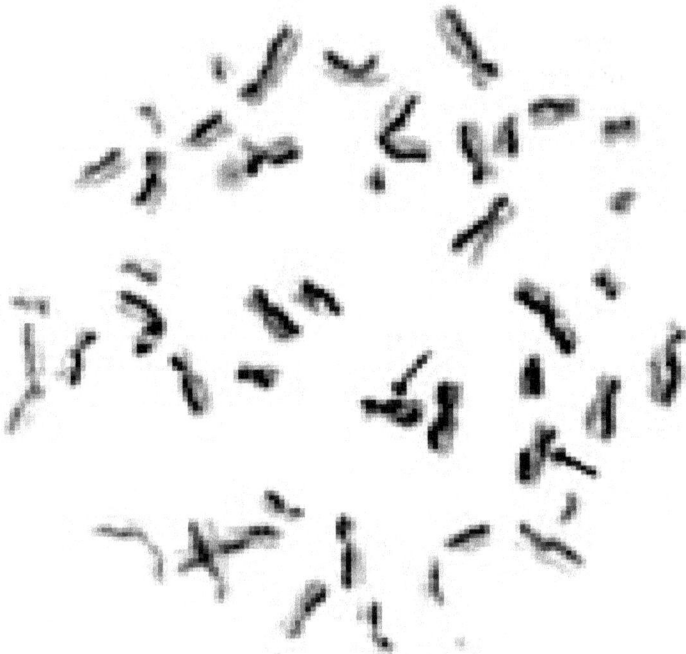

Figure 3.6 Sister chromatid exchange

The sister chromatid exchanges (SCE) assay provides a sensitive means for evaluating cytogenetic damage caused by chemical or physical agents. Chemical mutagens often induce SCE at lower concentrations and produce significant yields of chromosomal aberrations. The aberrations persist throughout the cell division. Also, exchanges occur between the two daughter chromatids during those mutagens exposures.

The sister chromatids are achieved by growing the cells in the presence of a thymidine analog, bromodeoxyuridine (BrdU). After two rounds of replication, a chromosome will have one chromatid in which the DNA in one strand is substituted with BrdU, and another in which both DNA strands are substituted. Metaphase chromosome spreads are prepared and differentially stained with Giemsa (Figure 3.6). Exchanges are evident as regions along the chromatid where the staining pattern changes.

3.8   FRAGILE SITES ON CHROMOSOMES

Fragile sites are specific points on chromosomes liable to be expressed as non-staining gaps of variable width and usually involve both chromatids (Figure 3.7). The production of acentric fragments, deleted chromosomes or triradials expresses the fragility. They are inherited in a dominant Mendelian fashion and are referred to as "heritable" or "rare" fragile sites. Majority of fragile sites are expressed in response to external culture conditions and are known as constitutive or "common" fragile sites. They have also been reported to be "hot spots" for the formation of sister chromatid exchanges.

Figure 3.7   Fragile sites

The chromatid gap formation is the result of despiralization of DNA due to failure of compact folding in the metaphase chromosome. Fragile sites could be the sites of viral DNA interaction, or may also be due to a section of DNA rich in thymidine. A positive correlation has also been found between bands where common fragile sites are located and bands particularly rich in CpG islands.

Fragile sites are broadly classified into **rare fragile sites** and **common fragile sites**. Further fragile sites are also classified according to external culture conditions. There are three groups of rare fragile sites. They are

- Folate-sensitive fragile sites
- Distamycin-A-inducible fragile sites
- Bromodeoxyuridine-requiring fragile sites
- Common fragile sites are classified into three groups; they are
- Aphidicolin-inducible fragile sites
- 5-Azacytidine-inducible fragile sites
- Bromodeoxyuridine-inducible fragile sites

Studies on fragile sites have gained great importance in human cytogenetics when two important discoveries took place. First, the fragile site on the X-chromosome was associated with a common form of X-linked mental retardation. Second it was shown that expression of certain fragile sites was possible only in the culture medium TC199. Expression of fragile sites is also enhanced by the addition of chemicals such as

- Aphidicolin
- 5-azacytidine
- Methotrexate
- Fluorodeoxyuridine
- Fluorodeoxycytidine
- Distamycin-A
- Bromodeoxyuridine
- Netropsin
- Hoechst 33258

The association of fragile sites with phenotypic abnormalities is well established for the fra (X) (q27.3). The fra (X) resulting from a mutation or DNA methylation was involved in X-inactivation.

Attention was also focused on the possible role of fragile sites as factors predisposing to chromosome breakage and rearrangements, specifically those characteristic of human cancers. A significant correlation was found between the cancer-specific break points and the location of fragile sites. The rare fragile sites at 8q22 and 16q22 correspond in location to the break points of rearrangements seen in acute non-lymphocytic leukemia. The common fragile site 3p14 is located in the bend that is rearranged in the small cell carcinoma of the lungs.

3.9   FLUORESCENCE *IN SITU* HYBRIDIZATION

Fluorescence *in situ* hybridization (FISH) is a method of investigation that covers a gap between classical cytogenetic and molecular genetic techniques. FISH is a technique that allows DNA sequences to be detected on metaphase chromosomes, in interphase nuclei or in a tissue section (Figure 3.8). FISH technique can be effectively applied to detect syndromic condition, prenatal diagnosis, postnatal diagnosis and also in pre-implantation diagnosis.

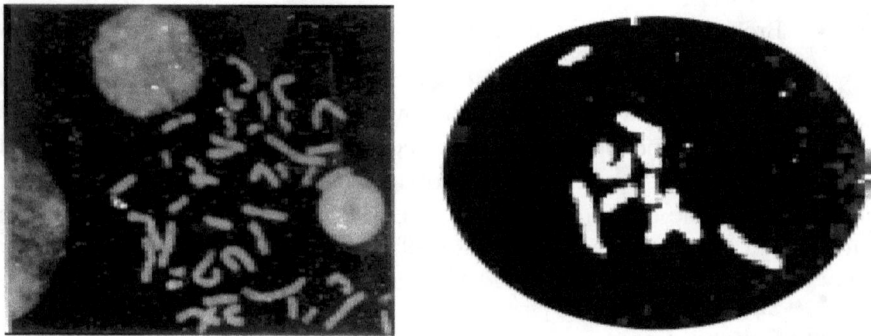

Figure 3.8   Fluorescence *in situ* hybridization

FOR ADDITIONAL READING

1. Alberts, B., Bray, D., Lewis, J., Raff, M., Roberts, K. and Watson, J.D. (1994). *Molecular Biology of the Cell.* Garland Publishing Inc., New York.

2. Black, L.W. (1989). "DNA packaging in dsDNA bacteriophages." *Ann. Rev. Microbiol.* 43:267–292.

3. Blackburn, E.H. (1991). "Structure and function of telomeres." *Nature.* 350: 569–573.

4. Blackburn, E.H. (1992). "Telomerases." *Ann. Rev. Biochem.* 61:113–129.

5. Blackburn, E.H. and Szostak, J.W. (1984). "The molecular structure of centromeres and telomeres." *Ann. Rev. Biochem.* 53:163–194.

6. Brown, S.W. (1966). "Heterochromatin." *Science.* 151:417–425.

7. Connor, J.M., Ferguson–Smith, M.A. (1994). *Essential Medical Genetics,* 4th edn. Blackwell Scientific Publications, Oxford.

8. Crow, J.F. (1983). *Genetics Notes*, 8th edn. Macmillan, New York.

9. Dupraw, E.J. (1970). *DNA and Chromosomes*. Holt, Rinehart and Winston, New York.

10. Gosden, J.R. (1994). *Chromosome Analysis Protocols*. Humana Press, New Jersey.

11. Hartl, D.J. and Jones, E.W. (1998). *Genetics: Principles and Analysis*, 4th edn. Jones and Bartlett Publishers, London.

12. ISCN. (1995). *An International System for Human Cytogenetic Nomenclature*. F. Mittelman, (ed.). Karger, Basel.

13. Lichter, P. (1997). "Multicolor FISHing: what's the catch?" *Trends Genet.* 13:475–479.

14. Lubs, H.A. (1969). "A marker X chromosome." *Am. J. Hum. Genet.* 21:231–244.

15. Magenis, R.E., Donlon, T.A. and Wyandt, H.E. (1978). "Giemsa-11 staining of chromosome 1: A newly described heteromorphism." *Science.* 202:64–65.

16. McKay, R.D.G. (1973). "The mechanism of G and C banding in mammalian metaphase chromosome." *Chromosoma.* 44:1–14.

17. Pasternak, J.J. (1999). *An introduction to Human Molecular Genetics: mechanisms of inherited diseases*. Fitzgerald Science Press, Maryland.

18. Strachan, T. and Read, A.P. (1996). *Human Molecular Genetics*. BIOS Scientific Publishers Ltd., UK.

19. Sutherland, G.R. and Hecht, F. (1985). *Fragile sites on human chromosomes*. Oxford University Press, Oxford.

20. Therman, E. and Sulsman, M. (1993). *Human Chromosomes: Structure Behaviour and Function*, 3rd edn. Springer, New York.

21. Tyler-Smith, C. and Willard, H.F. (1993). "Mammalian chromosome structure." *Curr. Opin. Genet. Dev.* 3, 390–397.

22. Verma, R.S. and Babu, A. (1989). *Human Chromosomes*. Pergamon Press, New York.

23. Verma, R.S. and Lubs, H.A. (1976). "A simple R banding technique." *Am. J. Hum. Genet.* 27:110–117.

24. Yunis, J.J. (1976). "High resolution of human chromosomes." *Science.* 191:1268–1270.

MODEL QUESTIONS

1. What are chromosomes?

2. What is a metacentric chromosome?

3. What is a submetacentric chromosome?

4. What is an acrocentric chromosome?

5. Define each of the following terms:
   - i.   Allopolyploid
   - ii. Aneuploid
   - iii. Duplication
   - iv. Inversion
   - v.   Translocation

6. What kind of rearrangement can make a metacentric chromosome into a submetacentric chromosome?

7. What is Lyoinization?

8. How many Bar bodies would be present in each of the following human chromosomal anomalies?
   - i.   Klinefelter's Syndrome
   - ii. Turner's Syndrome
   - iii. XXX
   - iv. XYY

9. What is crossing over?

10. What is sister chromatid exchange?

11. Sister chromatid exchange technique is also used to detect chromosomal aberration. Explain with examples.

12. Discuss the use of FISH technique in clinical diagnosis.

13. What is a fragile site? List various fragile sites associated with clinical features and diseases.

14. Discuss in detail the different kinds of fragile sites.

15. Write in detail on setting up of lymphocyte culture for micronucleus test and sister chromatid exchange.

16. Discuss various kinds of unbalanced and balanced chromosomal anomalies.

17. How are nucleosomes packed in a chromosome?

18. What is ideogram?

19. What are nucleosomes?

20. What are heterochromatins?

21. What is karyogram?

22. How are lymphocytes grown *in vitro*?

CHROMOSOME BASIS OF INHERITANCE

4

4.1 INTRODUCTION

It is nature's wonder, how chromosome number varies widely among species. Also interesting is the fact that, within a given species the total chromosomal number is constant in all cells. In the beginning of the twentieth century, Walter Sultan and Theodar Boveri (1902) recognized the transmission of chromosome generation after generation similar to Mendelian factor. They proposed a theory called chromosome theory of inheritance, which states that genes are located on chromosomes. They also gave some evidences, to support this theory.

Already in chapter 3, we have understood the features of chromosomes. Further, the chromosome which determines the sex of an individual is known as **sex chromosome** or **allosome** (also heterosome); the rest of the chromosomes are referred to as **somatic chromosomes** or **autosomes**. In humans and *Drosophila*, there are two kinds of sex chromosomes, X chromosome and Y chromosome. If there are two X chromosomes in a cell of an individual or organism (homogametic sex) then it is a female, but male (heterogametic sex) has one X and one Y chromosome. In fruit fly, the size of the X and Y-chromosomes are same but both differ in their shapes. In the case of humans, the size and shape of the X and Y chromosomes differ.

4.2   SEX LINKAGE

In 1910, Thomas Hunt Morgan presented evidence to support the chromosome theory of inheritance. He did genetic experiments using fruit flies (*Drosophila*). For this discovery, he received the Nobel Prize in physiology/medicine in the year 1933. He used true breeding wild type strains. Morgan saw male fly with white eye, showing true breeding. These mutant characteristics (white eyes) are produced due to mutant allele. This mutant allele is due to mutational changes of the wild type alleles. This mutant allele is either dominant or recessive to the wild type.

Figure 4.1. shows male and female fruit fly (*Drosophila*), which are used extensively in genetics experiment. Round abdomen, sex comb in first pair of legs and XY sex chromosome pairs are salient features of male fly. Pointed abdomen and XX sex chromosome pair are the features of female. Red-eyed fly is wild type strain and has dominant allele. **Sex linkage** is the linkage of genes with the sex (X) chromosome, also called X-linkage.

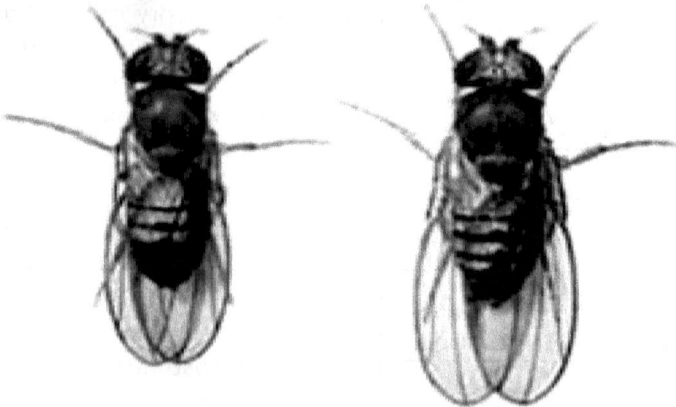

Figure 4.1   Fruit fly (*Drosophila*)

Morgan crossed true breeding white-eyed male with a red-eyed female. He found only red-eyed flies in the F1 generation. So he concluded that the white-eyed trait (character) was recessive. Then he interbred F1 red-eyed progeny. In the F2 generation, the recessive phenotype was very less in quantity, that is 3 :<1 (dominant : recessive). Another important observation was that all the white-eyed flies were males.

In the experiment conducted by Morgan, the proposed eye colour gene is found in the X chromosome. The condition of X-linked genes in female is said to be homozygous or heterozygous whereas in males it is said to be hemizygous. This hemizygous condition is due to unavailability of homologous gene on the Y chromosome.

4.3 CRISS-CROSS INHERITANCE

The red-eyed alleles are designated as w^+ (w plus) and the white-eyed alleles are **represented** as w. As given in the Figure 4.2 **homozygous (w^+w^+) female was crossed with hemizygous (w^-) male fly**. In the F1 progeny the males received their X chromosome from their mother (with w^+ allele) and Y chromosome from father. The males are red-eyed. The female received a dominant allele w^+ from mother (X chromosome) and recessive allele w^- from father (X chromosome), and they are also red-eyed.

When F1 progeny is interbred, the F2 progeny is unique. The male that receives an X chromosome with w allele from mother (F1 female) are white-eyed. But the males that receive an X chromosome with w^+ allele from father (F1 male) are red-eyed. This kind of uniqueness can also be explained as follows. The recessive allele or gene (in X chromosome, w allele for white-eyed) transferred from a male parent to his daughter and from his daughter to his grandson is called **criss-cross inheritance**.

In another experiment, Morgan crossed a true breeding white-eyed female with a red-eye male (Figure 4.2). The female was homozygous; both X chromosomes had 'w' allele for white-eye character. The male was hemizygous; the X chromosome had w^+ allele for red-eye character and Y chromosome did not carry any eye colour allele.

In the F1 progeny, the female receives an X chromosome from mother (w allele) and the other X chromosome with w^+ allele from father. The fly is heterozygous (w^+w) and has red eyes. But male receives X chromosome (w allele for white eye) from mother and Y chromosome from father. This hemizygous fly has white eyes. The F1 progeny of this experiment 1 : 1 (red-eyed female : white-eyed male) is different from previous experiment where all the F1 progeny were red-eyed.

When F1 flies were interbred the results were as follows;

- an approximately 3 : 1 red-eyed : white-eyed flies

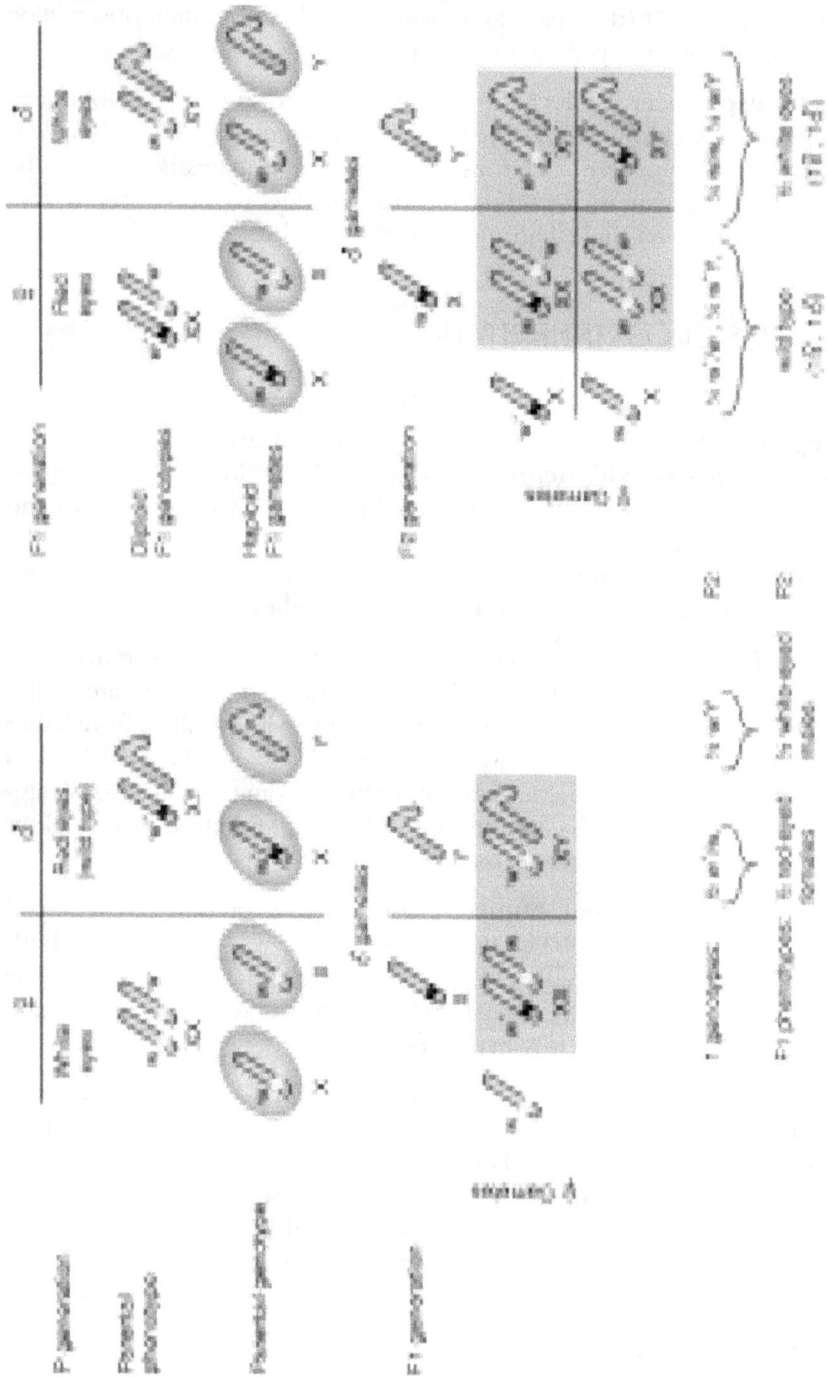

Figure 4.2   Reciprocal cross between red- and white-eyed *Drosophila*

- none of the females were white-eyed
- an approximately 1 : 1 red-eyed male : white-eyed male

This variation in the phenotypic ratio is due to the inheritance of sex chromosome, nature of gene they contain and combination of alleles. This work of Morgan strongly supported inheritance of sex-linked genes rather than proving chromosome theory of inheritance.

4.4   NON-DISJUNCTION

Morgan's finding supported that genes were located on chromosomes. Later, Morgan's student Calvin Bridges continued the experiments with *Drosophila*. Calvin Bridges paved the way for the chromosome theory of inheritance. Morgan found that a cross of a white-eyed female with a red-eyed male, yielded all males with white eye and females with red eye. Contrastingly Calvin Bridges found that white-eyed female or red-eyed male were present in about 1 in 2000 of the F1 progeny. Bridges explained that these exceptions are due to non-disjunction of sex chromosomes (Klinefelter's syndrome in human). Sometimes non-disjunction can occur in autosomes also (Patau's syndrome, Edward's syndrome and Down's syndrome in humans).

When there is a failure of movement (segregation) of homologous chromosome during anaphase to opposite poles, the chromosome non-disjunction occurs. As a result, cell with one extra chromosome is seen. When non-disjunction occurs in *Drosophila* germ cells, eggs were produced with either two X chromosomes or with no X chromosomes (Figure 4.3). This is called sex chromosome non-disjunction. If non-disjunction occurs for the first time in an organism when there is a normal set of chromosomes, then it is called primary non-disjunction.

Bridges explained the non-disjunction of X chromosome in *Drosophila*. When a female fly produces egg with two X chromosomes and another egg without any X chromosome due to non-disjunction, then there is much alteration in the genotypic and phenotypic ratio in all generations. When white-eyed female fly was crossed with red-eyed male fly, the egg produced due to non-disjunction was of two kinds; an egg with two X chromosomes carrying genes for white eye and another egg without X chromosome. In male, each gamete will have either X or Y chromosome. The result of F1 progeny will be as shown in Figure 4.4.

- XXX fly dies during early development due to overdosage of gene
- XXY female white-eyed fly survives

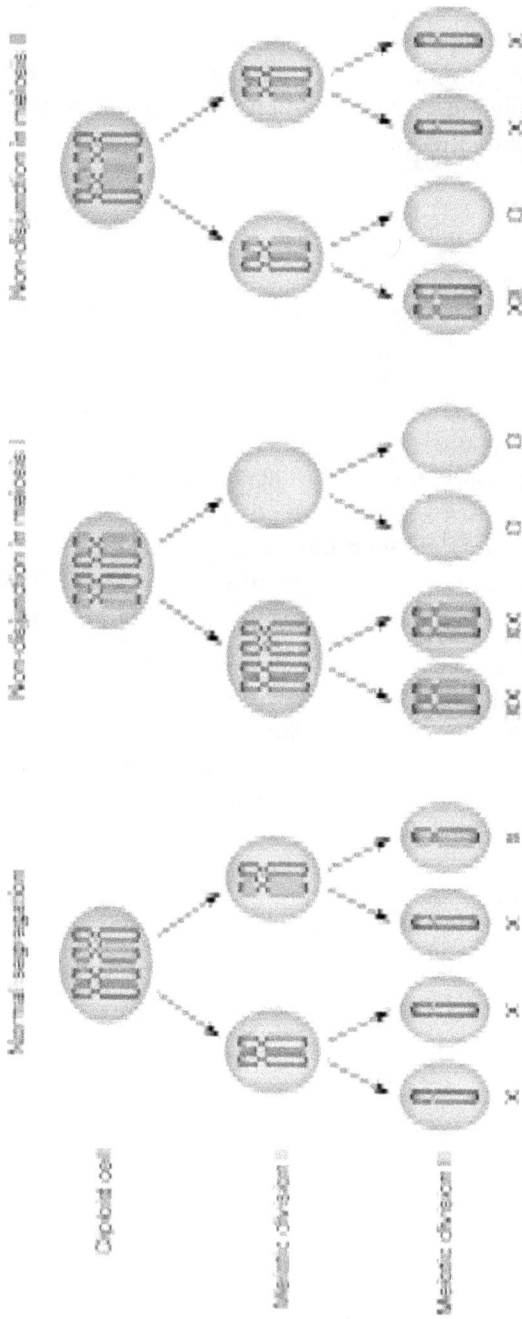

Figure 4.3   Non-disjunction of X chromosome

- X red-eyed sterile male
- Y fly dies during early development

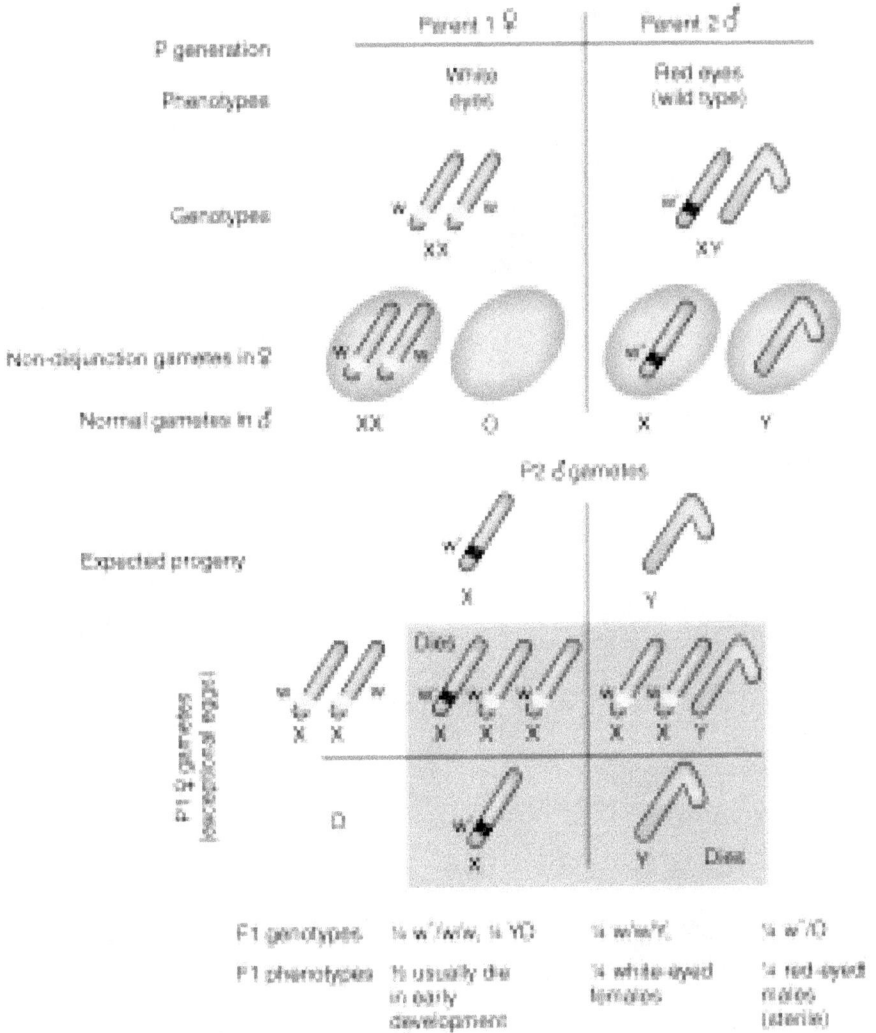

Figure 4.4 Non-disjunction in the cross of white-eyed female with a red-eyed male in *Drosophila*

Similarly the surviving XXY white-eyed fly was crossed with XY red-eyed male. There were two kinds of gamete formation which occurred in female fly. In the first kind, during meiosis, chromosome may segregate into X and XY type of egg formation and in the second kind, two X chromosomes remain together and an egg with Y chromosome can be formed (Figure 4.5).

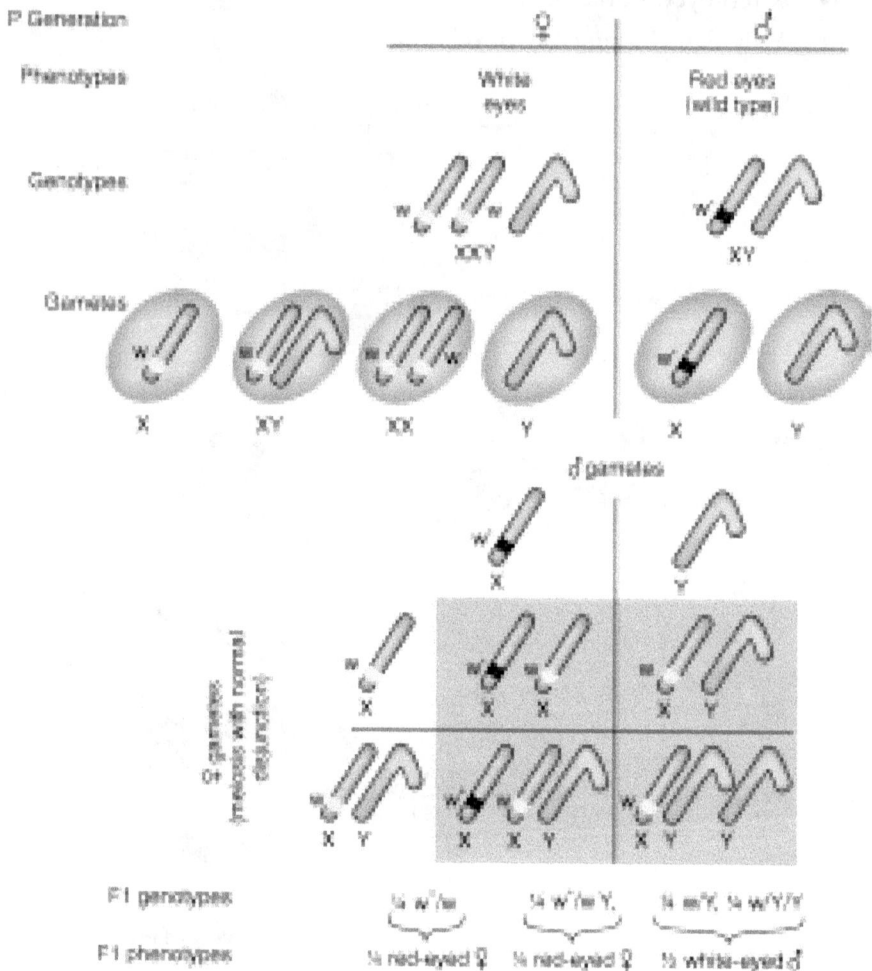

Figure 4.5 Non-disjunction in the cross of white-eyed female (XXY) with a red-eyed (wild type) male in *Drosophila*

When eggs from the first kind of fly were fused with normal red-eyed male, both eye colour were seen in four different genotypes (Figure 4.5). The $X^{w+} X^w$ (heterozygote) female with red eye, XXY red-eyed female, XY white-eyed male and XYY also a white-eyed male (with an extra dosage of Y chromosome) were obtained.

Contrastingly if the second kind of egg such as XX and Y was fused with X or Y male (Figure 4.6) gamete, the progenies were as follows;

- XXX fly dies during early development
- XXY white-eyed female fly

- XY red-eyed male fly
- YY fly dies during early development

These experiments of Bridges proved the correlation between gene segregation and chromosome movement in meiosis. This gave the proof for chromosome theory of inheritance.

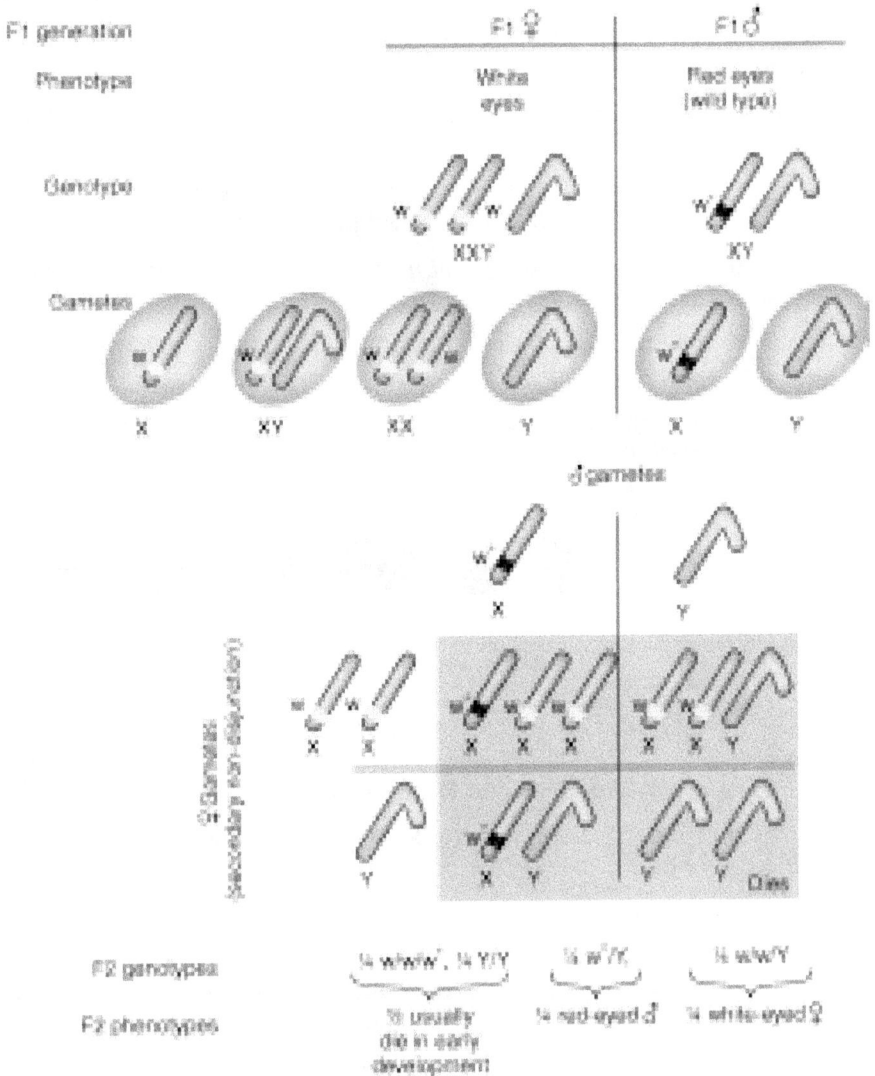

Figure 4.6   Non-disjunction in the cross of normal red-eyed male with white-eyed female in *Drosophila*

FOR ADDITIONAL READING

1. Griffiths, A.J.F., Gelbert, W.M., Lewontin, R.C. and Miller, J.H. (2002). *Modern Genetic Analysis.* W.H. Freeman and Company, New York.

2. Griffiths, A.J.F., Gelbert, W.M., Lewontin, R.C., Miller, J.H. and Suzuki, D.T. (2002). *An Introduction to Genetic Analysis.* W.H. Freeman and Company, New York.

3. McKusick, V.A. (1998). *Mendelian Inheritance in Man,* 12th edn. Johns Hopkins University Press, Baltimore.

4. Snustad, D.P. and Simmons, M.J. (2002). *Principles of Genetics.* John Wiley & Sons Inc., New York.

5. Stern, C. and Sherwood, E. (1966). *The Origins of Genetics: A Mendel Source Book.* W.H. Freeman and Company, New York.

6. Tom Strachan, *et al.* (1996). *Human Molecular Genetics.* BIOS Scientific Publishers Ltd., UK.

7. Vogel, F. and Motulsky, A.G. (1986). *Human Genetics: Problems and Approaches,* 2nd edn. Springer and Verlag, Berlin.

8. Hartl, D.J. and Jones, E.W. (1998). *Genetics: Principles and Analysis,* 4th edn. Jones and Bartlett Publishers, London.

MODEL QUESTIONS

1. How did the discovery of sex-linked genes support the chromosomal theory of inheritance?

2. What do we mean by linked genes?

3. How and why do the results of genetic crosses involving linked genes deviate from those expected according to Mendel's Law of Independent Assortment?

4. How is independent assortment responsible for genetic recombination in unlinked genes?

5. How is crossing over during meiosis responsible for genetic recombination in linked genes?

6. How are the patterns of inheritance of sex-linked (X–linked) traits different from traits determined by genes located on autosomes?

7. What are some common sex-linked disorders in humans?

8. What is X-inactivation or lionization?

9. How does X-inactivation result in a Barr Body?

10. Why can X-inactivation result in females that consist of a mosaic of different cell types?

11. How can non-disjunction of chromosomes result in aneuploidy?

12. What is the difference between a monosomic and a trisomic aneuploid cell?

13. How does a polyploid (triploid or tetraploid) cell result?

LINKAGE, RECOMBINATION AND MAPPING

5

5.1 LINKAGE

The concept of linkage is difficult to understand. Let us go through a stepwise approach to make it easy. Each chromosome of an organism contains many genes. These genes are held together, that is, they are **linked**. Linked genes belong to a linkage group. However at meiosis, during crossover they assort independently between loci. This crossing over between loci is used as a tool to determine how close one locus actually is to another locus on a chromosome. If the distances between any two loci on a chromosome are greater, greater is the probability of a meiotic recombination. That is, **recombination frequency** (RF) is proportional to the physical distance between two loci. If we measure the RF, we can create a linkage map. Presently, the focus of mapping studies is on constructing genetic maps of genome using gene markers and DNA markers. A marker or genetic marker is a visible phenotypic feature, which will be useful for investigating genes and their functions. Centimorgan is the term used as a unit for Linkage mapping. This is named in honour of Thomas Hunt Morgan.

Let us understand first a simple case. To start with, presume that we have a disease gene located on a chromosome with some structural chromosomal aberration (marker) that has nothing to do with the disease. Remember, the disease gene is located apart from this marker region. Most of the time, the marker region of this chromosome and the disease will always segregate together. Now we can predict that whoever inherits this marker will also inherit the disease gene. But many times this is not happening? Because, as long as the marker is some distance from the actual disease gene, there is a possibility for recombination during meiosis. So the disease gene and marker region get segregated.

The accuracy of linkage analysis is increased by using a probability technique called the **lod score method.** This method was developed by Newton E. Mortan and is most widely used. A value is calculated based on the ratio of the probability of genotypes in a family given a certain crossover frequency compared with the probability of those genotypes if the loci are unlinked. Any lod score (z) greater than zero favours recombination. A lod score less than zero suggests that recombination frequency has been underestimated. A lod of greater than 3.0 shows a strong likelihood of linkage.

The law of independent assortment states that individual genes assort independently and get transmitted to the next generation. However, the principle has limitation because the genes located near one another on the same chromosome are transmitted together and not independently. Such loci are said to be linked.

Now let us study another example. From Figure 5.1 you will understand that two loci A and B, are located close together on the same chromosome. A third locus, C, is located on another chromosome. Each of these loci has two alleles, designated 1 and 2. A and B are linked. So A and B are

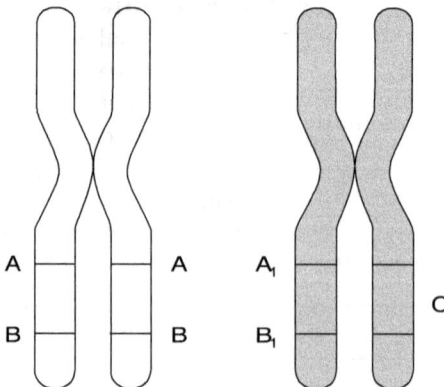

Figure 5.1   Gene loci and mapping

inherited together. Because A and C are on different chromosomes and thus unlinked, their alleles do follow the principle of independent assortment. Hence, in meiosis, if A is placed in a gamete, the chance of C and A being together is only 50%.

Homologous chromosomes sometimes exchange portions of their DNA during first prophase (crossing over). Each chromosome generally experiences one to three crossover events during meiosis. As a result of crossover, new combinations of alleles can be formed on a chromosome. Consider the linked loci, A and B, discussed previously. Alleles A_1 and B_1 are located together on one chromosome, and alleles A_2 and B_2 are located on the homologous chromosome (Figure 5.2). The combination of alleles on each chromosome is termed as haplotype (from "haploid genotype"). The two haplotypes of this individual are denoted as A_1B_1/A_2B_2. As Figure 5.3 shows, A_1B_1 is found in one gamete and A_2B_2 is found in the other gamete in the absence of crossover. But when there is a crossover, new allele combinations, A_1B_2 and A_2B_1 are found in each gamete. The process of forming such new arrangements of alleles is called **recombination**. Crossover does not necessarily lead to recombination every time; however, a double crossover can occur between two loci, resulting in no recombination.

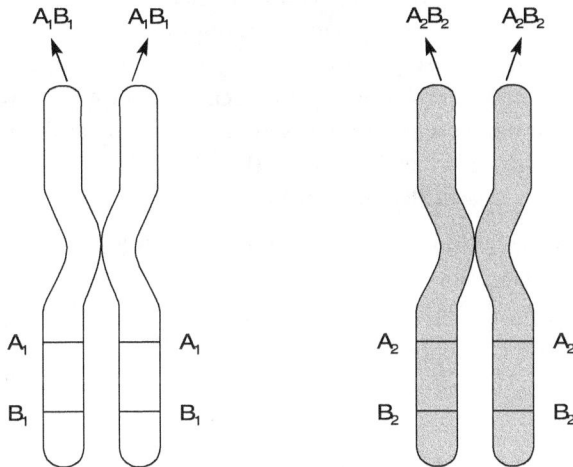

Figure 5.2 Homologous chromosomes

As shown in Figure 5.3 crossovers are more likely to occur between loci that are situated at a distance on a chromosome than between loci that are located close together. Thus, the distance between two loci can be inferred by estimating how frequently recombination occurs in families (this is called the **recombination frequency**). When a large series of meiosis is studied in families, the alleles of A and B undergo recombination 5% of the time. Now the recombination frequency for A and B would be 5%.

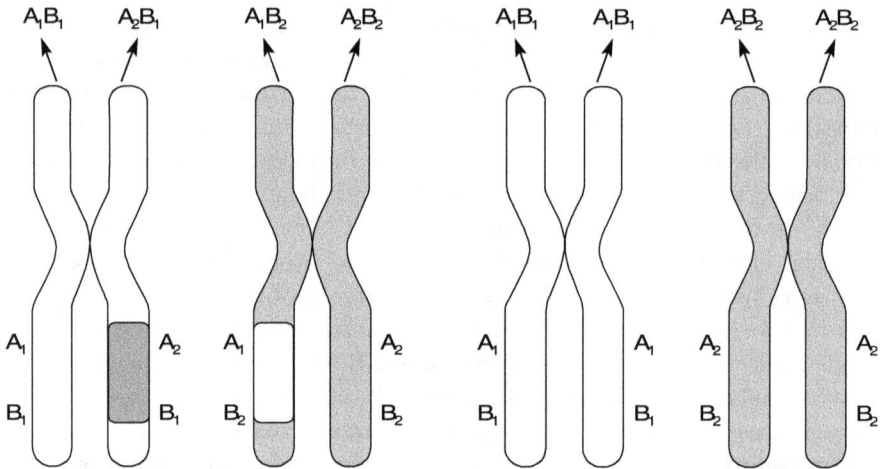

Figure 5.3   Outcome of crossover of homologous chromosomes

The genetic distance between two loci is measured in centiMorgans (cM). One cM is equal to a recombination frequency of 1%. The relationship between recombination frequency and cM is almost an approximate value because double crossovers produce no recombination. The recombination frequency thus underestimates map distance, especially as the recombination frequency increases above 5% and 10%. Mapping functions have been devised to correct this underestimation.

Recombination is the mechanism through which exchange of genetic materials takes place between chromosomes and DNA. This occurs mostly at the regions of homology. In prokaryotes the recombination is often non-reciprocal. But there are reciprocal exchanges in eukaryotes. The product after recombination is called a **recombinant**. Two different cellular processes produce the recombinants.

- Recombination by independent assortment
- Recombination by crossing over

In a test cross method, both the classes make up 50% of the progeny. If there is a recombinant frequency of 50% then the two genes assort independently. However, genes that are far apart on the same chromosome pair can act independently and bring out the same result.

Crossing over can produce recombinants when there is an exchange between two non-sister chromatids of the homologous chromosome. Generally during meiosis, there is no crossover between two specific genes. The crossover occurs only when 50% of the products are recombinants. For genes close together on the same chromosome pair, the physical linkage of parental allele combinations makes independent assortment impossible. So it produces recombinant frequency significantly lower than 50 percent.

5.2 MODELS OF RECOMBINATION

There are three models of recombination. They are Holliday model, Meselson–Radding model and Double-strand break-repair model.

5.2.1 Holliday Model

The main features of the **Holliday model** are as follows:

- the formation of heteroduplex DNA,
- creation of a cross bridge,
- advancement of branch migration,
- occurrence of mismatch repair and
- the production of recombinant molecules.

From Figure 5.4 you will understand that two homologous double helices are aligned. This alignment occurs after the rotation of both the strands to have same polarity ($5' \rightarrow 3'$). Then an endonuclease cleaves the two strands that have the same polarity. The free ends of the cut strands leave their original complementary strands and undergo hydrogen bonding with the complementary strands in the homologous double helix and get ligated.

5.2.2 Meselson–Radding Model

In Meselson and Radding model, a single-strand cut (nick) is made in one chromosome. After the nick, the displaced single strand invades the second duplex strand and grows. Now this generates a loop, which is later excised (Figure 5.5). The Meselson–Radding model has only one heteroduplex region at the bottom chromatid. This is in contrast to two chromatids having heteroduplex regions in the Holliday model.

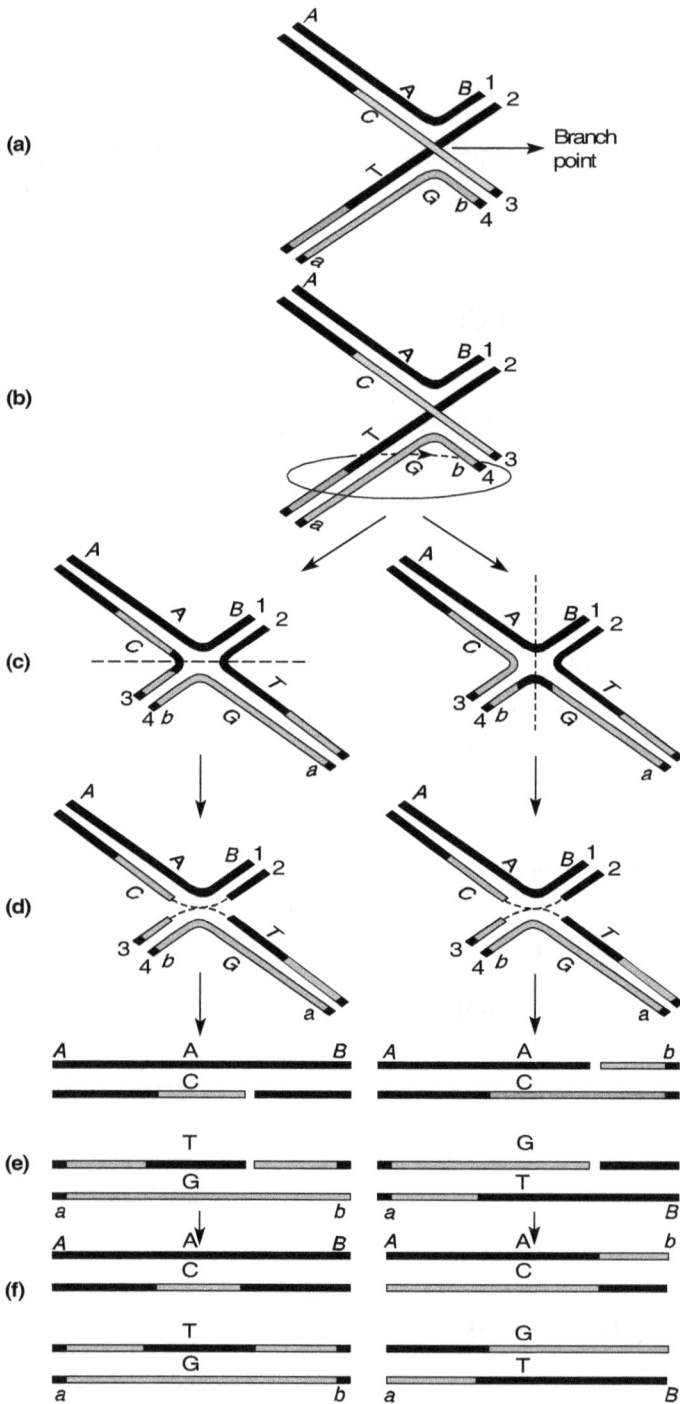

Figure 5.4   Holliday model

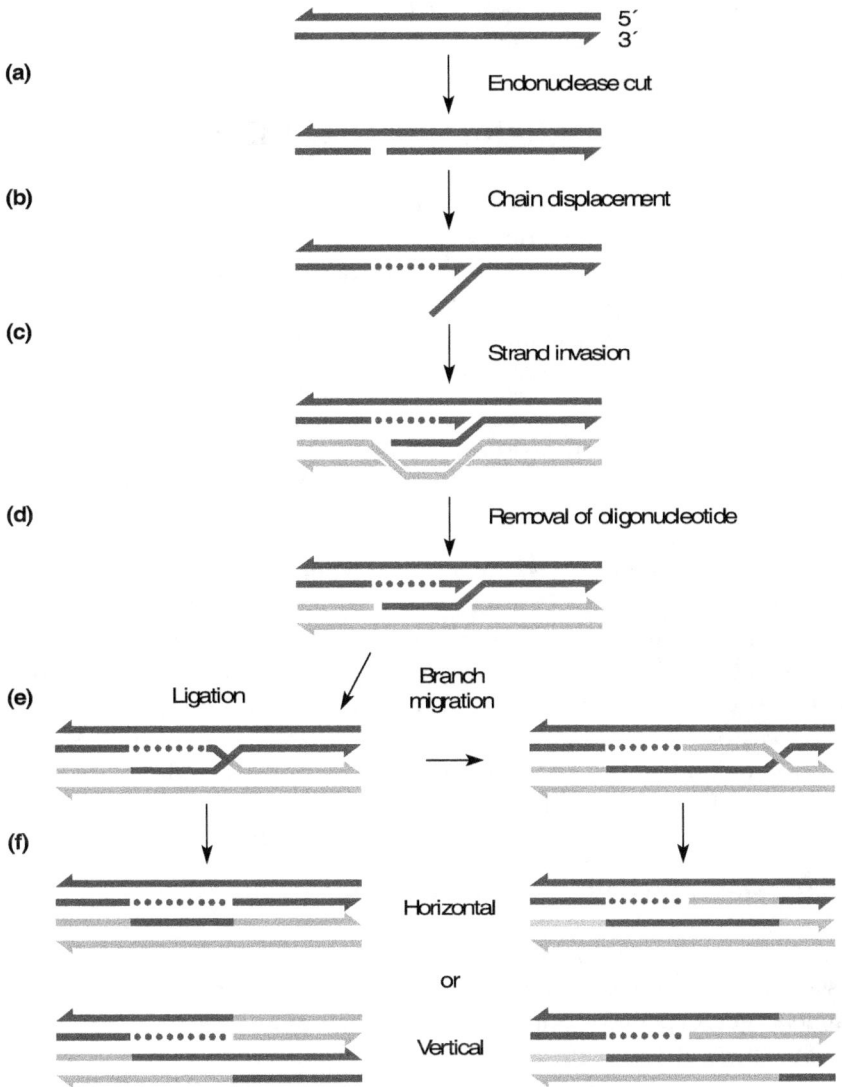

(a)

Endonuclease cut

(b)

Chain displacement

(c)

Strand invasion

(d)

Removal of oligonucleotide

(e) Ligation — Branch migration

(f) — Horizontal — or — Vertical

Figure 5.5   Meselson–Radding model

5.2.3   Double Strand Break-repair Model

In the Holliday and Meselson–Radding models of genetic recombination, single-strand nicks result in the generation of heteroduplex DNA. But in yeast, a double-strand break is introduced into a circular donor plasmid. This provides the impetus for an additional model to initiate recombination.

The breaks are enlarged to form a gap. The repair of the double-stranded gap results in gene conversion.

5.3   ENZYMATIC MECHANISM OF RECOMBINATION

In *E. coli*, the products of several genes such as *recA, recB, recC,* and *recD* are involved in recombination. Mutation in any of these genes will reduce the levels of recombination. There are three distinct recombination pathways: One major RecBCD pathway and two minor pathways. RecF and RecE are also activated in certain situations. In all these three pathways RecA proteins are used to form recombinant.

5.3.1   Production of Single-stranded DNA

In the initial step of *E. coli* the nicking and unwinding of a DNA duplex (Figure 5.6) is facilitated by a protein complex consisting of RecB, RecC and RecD proteins. This protein complex has both helicase and nuclease activities. This complex initially unwinds the DNA using energy through the hydrolysis of ATP. Then the enzyme moves and generates a single strand from a duplex DNA molecule. The nuclease activity recognizes an 8-bp sequence: **5´—G C TG G T G G—3´** called a *chi* site. These *chi* sites appear approximately at every 64 kb. As the complex unwinds the DNA, the free single-stranded DNA can be used to initiate recombination. The single-strand-binding protein binds to each single strand and stabilizes them.

5.3.2   RecA Protein-mediated Single-strand Exchange

The RecA protein plays an important role in the induction of the SOS repair system. It binds to single strands along their length and forms a nucleoprotein filament. RecA catalyses single strand invasion of a duplex and subsequent displacement of the corresponding strand from the duplex. The invasion and displacement take place by using energy through the hydrolysis of ATP. The displaced strand forms a D loop.

5.4   BRANCH MIGRATION

The movement of a Holliday junction increases the length of heteroduplex DNA. This is termed as branch migration. The RuvA and RuvB proteins catalyse branch migration. They get energy through the hydrolysis of ATP. The RuvA protein binds exactly at the crossover point. Then it is flanked by two RuvB ATPase hexameric rings.

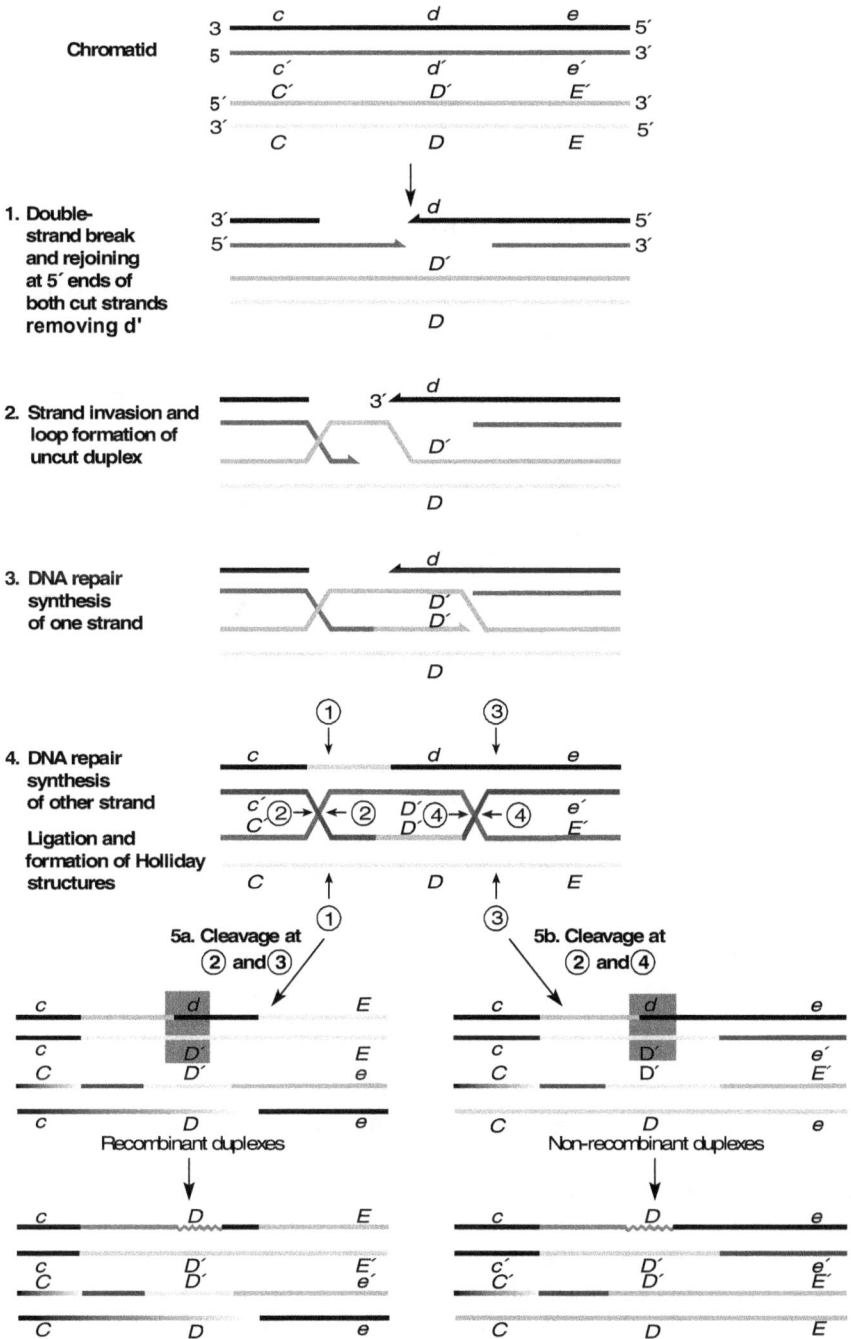

Figure 5.6 Enzymatic mechanism of recombination

The enzymatic cleavage of strands across the point of exchange to yield two duplexes is done either by RuvC protein (which is an endonuclease) or through RecG and Rus proteins.

5.5   GENE MAPPING

In humans, the genetic diseases have two modes of inheritance. They are simple and complex patterns of inheritance. The mapping of genes to specific chromosomal location (loci) is a very important step. This helps us in better understanding, diagnosis and treatment of genetic diseases. If the disease gene location in the chromosome is known correctly, then it is easy to provide an accurate prognosis for persons who are at risk of a genetic disease. For locating a disease gene, the first step is to clone the gene. Once a gene is cloned, by studying its sequences and protein nature we can understand the actual cause of the disease. There are two major types of gene mapping: **physical mapping and genetic mapping.** Physical mapping is to determine the actual locations of genes on chromosomes. Genetic mapping is to estimate the frequency of meiotic crossovers (recombination frequency) between loci.

5.6   LINKAGE ANALYSIS

Loci that are on the same chromosome are said to be **syntenic**. That means it is in same thread. If two syntenic loci are 50 cM apart, they are considered to be unlinked. This is because their recombination frequency is found to be large, as if they were on different chromosomes.

Recombination frequencies can be estimated by observing the transmission of genes in pedigrees. Let us consider for example that neurofibromatosis type 1 (NF1) is being transmitted in a family. The members of this pedigree have also been typed for a two-allele by RFLP method. They found that *1F10* like the *NF1* gene, is located on chromosome 17. The *1F10* genotypes are also found in each individual in the pedigree. Examination of generations 1 and 2 help us to determine that, under the hypothesis of linkage between *NF1* and *1F10*, the *NF1* gene must be on the same chromosome as allele 1 of the *1F10* locus in this family. This is because the individual I-2, who is homozygous for allele 2, does not have the disease. Here the affected father (I-1) is a heterozygote for the *1F10* locus and could have transmitted a chromosome containing both the disease allele and *1F10* allele 1 to the daughter (II-2). The arrangement of these alleles on each chromosome is referred to as **linkage phase**. Knowing the linkage phase, now the individual II-2's haplotypes would be *N1/n2*, where *N* indicates the allele causing NF1 and *n* indicates the normal allele. The 1 and 2 are referred to as two *1F10* alleles. The woman's husband (individual II-1) does not have the disease and is a homozygote for allele 2 at *1F10*. He must have the haplotypes *n2/n2*. If *NF1* and *1F10* loci are linked, the children of this union who are affected with *NF1* should usually have *1F10* allele 1. Those who are unaffected should have only allele 2. In seven of eight children in generation III, it is found to be true. In one case a recombination occurred (individual III-6). This gives a recombination frequency of 1/8, or 12.5%, supporting the hypothesis of linkage between the *NF1* and *1F10* loci. A recombination frequency of 50% would support the hypothesis that the two loci are not linked.

Polymorphisms such as *1F10*, which can be used to follow a disease gene through a family, are termed markers. These help us to "mark" the chromosome on which a disease allele is located. Since linked markers can be typed in an individual even from foetal stage, it is very useful for the early diagnosis of genetic disease. It should be noted that marker locus simply designates which chromosome is being transmitted but has nothing to do with the actual causation of the genetic disease.

In general, 1 cM corresponds to approximately 1 million base pairs (1 Mbp) of DNA. However, this relationship is only approximate, since several other factors are known to influence crossover rates. First, crossovers are roughly 1.5 times more common in females during oogenesis than in male (spermatogenesis). Also crossovers are found to be more common near the telomeres of chromosomes than around the centromeric region. Finally, some chromosome regions exhibit substantially elevated crossover rates. These regions are termed **recombination hot spots**. It is not clear still what causes recombination hot spots in humans, although specific DNA sequences (*Alu* sequences), are thought to be especially prone to crossover.

5.7 SIGNIFICANCE OF LINKAGE RESULTS

First we must be careful to ensure that the results obtained in a linkage study are not simply due to chance in statistical analyses. For example, consider a two-allele marker locus that has been typed in a pedigree. It is possible by chance for all affected offsprings to inherit one allele and all unaffected offsprings to inherit the other allele. This would happen even if the marker is not linked to the disease gene. This misleading result becomes less likely as we increase the number of subjects in our linkage study.

How do we find whether a linkage result is because of chance alone? In linkage analysis a standard method is used. We begin by comparing the likelihood that two loci are linked at a given recombination frequency (denoted θ) versus the likelihood that the two loci are not linked (that is recombination frequency = 50%, or $\theta = 0.5$). If you wish to test the hypothesis that two loci are linked at a recombination frequency of $\theta = 0.1$ versus the hypothesis that they are not linked. You use the pedigree data to form a likelihood ratio:

$$\frac{\text{Likelihood of observing pedigree data if } \theta = 0.1}{\text{Likelihood of observing pedigree data if } \theta = 0.5}$$

If pedigree data indicates that θ is more likely to be 0.1 than 0.5, then the likelihood ratio (or "odds") will be greater than one. For convenience, the common logarithm of the ratio is usually taken. This "logarithm of the odds" is termed as **lod score**. Conventionally, a lod score of 3.0 is accepted as evidence of linkage whereas a score of 3.0 indicates that the likelihood in favour of linkage is 1,000 times greater than the likelihood against linkage. But a lod score less than −2.0 is considered to be evidence that two loci are not linked.

5.8   LINKAGE ANALYSIS AND THE HUMAN GENE MAP

When we are studying a disease gene in a series of pedigrees, most of the time we end up with negative lod scores, indicating no linkage between the marker and the disease gene. This situation has changed dramatically over the past decade, as thousands of new polymorphic markers (RFLPs, VNTRs and microsatellite polymorphisms) have been generated. Since the human genome is large, one may have to type dozens, or even hundreds, of markers before finding linkage. Many important diseases have been localized using this approach, including cystic fibrosis, Huntington disease, Marfan syndrome, and NF1.

To be useful for gene mapping, marker loci should have three important properties.

- Firstly, they should be **codominant,** that is, the homozygotes should be distinguishable from heterozygotes. This makes it easier to find the linkage phase using RFLPs, VNTRs and microsatellite polymorphisms. The older type of markers such as the ABO and Rh blood groups will not help in finding.
- Secondly, marker **loci** should be **numerous** so that there is a chance for close linkage to the disease gene.
- Finally, marker loci are most useful when they are **highly polymorphic.** That is the locus has many different alleles in the

population. A high degree of polymorphism ensures that most parents will be heterozygous for the marker locus, making it easier to establish linkage phase in families.

Let us consider another example for the highly polymorphic condition. The affected man is a homozygote for a two-allele RFLP that is closely linked to the disease gene locus. His wife is having allele of heterozygotic condition. But their affected daughter is homozygous for the marker allele. Based on these genotypes, it is not possible to study the linkage phase in this generation. Now this leads to a condition where we cannot predict which children will be affected with the disease and who will be normal. Here the first generation mating is called an uninformative mating. In contrast, a microsatellite repeat polymorphism with six alleles has been typed in the same family. Because in the first generation the mother has two alleles that differ from those of the affected father, it is now possible to determine that the affected daughter (in second generation) has inherited the disease gene on the single chromosome that contains marker allele 1. Since she married a person who has alleles 4 and 5, we can predict that each offspring who receives allele 1 from her will be affected and each one who receives allele 2 will be normal. Exceptions will be due to the crossover and recombination. This example proves the value of highly polymorphic markers both for linkage analysis and for diagnosing genetic disease.

5.9 NONRANDOM ASSOCIATION OF ALLELES AT LINKED LOCI

Within families, most of the time, marker locus gets transmitted along with the disease allele if both the loci are linked. Let us assume that allele 1 of a linked two-allele marker could co-occur with the Huntington disease (HD) allele in a family. This association may be due to linkage. For linkage between HD and the marker locus, there can be heterozygosity when allele 1 will co-occur with the disease in some families, whereas allele 2 of the marker will co-occur with the disease in others. This may be due to the reason that disease-causing mutations have occurred multiple times, sometimes on marker allele 1 and other times on marker allele 2. Similarly, if the disease is the result of only one original mutation, crossovers occurring through time will eventually result in recombination of the marker and disease alleles. A disease allele and a linked marker allele will thus be associated within families, but not necessarily between families. If we study a large collection of families and find that there is no preferential association between the disease gene and a specific allele at a linked marker locus, the two loci are said to be in **linkage equilibrium**.

Sometimes, there is preferential association of a specific marker allele and the disease allele in a population. Because of this, the chromosome haplotype consisting of one marker allele and the disease allele is found more often than expected, based on frequencies of the two alleles in the population. For example, the disease allele has a frequency of 0.01 in the population and the frequencies of the two alleles (allele 1 and allele 2) of the marker are 0.4 and 0.6, respectively. Assuming independence between the two loci (linkage equilibrium), the multiplication rule would predict that the population frequency of the haplotype containing the disease allele and marker allele 1 would be $0.01 \times 0.4 = 0.004$. Based on family information, one can directly count the haplotypes in the population. If the actual frequency of this haplotype is 0.009 instead of the predicted 0.004, then the assumption of independence has been violated, indicating preferential association of marker allele 1 with the disease allele. This association of alleles at linked loci is termed **linkage disequilibrium**.

5.10   LINKAGE VERSUS ASSOCIATION IN POPULATIONS

The word linkage and association are sometimes confusing during analysis. Linkage refers to the positions of loci on chromosomes. When two loci are linked, specific combinations of alleles at these loci will be transmitted together. But at times, the specific combinations of alleles transmitted together can vary from one family to another. Association refers to a statistical relationship between two traits (characters) in the population. The two traits occur together in the same individual more often than expected by chance.

FOR ADDITIONAL READING

1. James, D., Watson, *et al.* (2001). *Recombinant DNA*, 2nd edn. W.H. Freeman and Company, New York.

2. Anthony, J.F., Griffiths, *et al.* (2002). *An introduction to Genetic Analysis*, 7th edn. W.H. Freeman and Company, New York.

3. Stahl, F.W. (1979). *Genetic Recombination: Thinking about it in Phage and Fungi*. W.H. Freeman and Company, New York.

4. Ruddle, F.H. and Kucherlapati, R.S. (1974). "Hybrid Cells and Human Genes." *Scientific American* (July).

5. Lynn, B., Jorde, John, C., Carey, *et al.* (1997). *Medical Genetics*, 2nd edn. Mosby Inc., Missouri.

6. James, D. Watson, *et al.* (2001). *Recombinant DNA*, 2nd edn. W.H. Freeman and Co., New York.

7. Griffiths, A.J.F., Miller, J.H., Suzuki, D.T., Lewontin, R.C. and Gelbart, W.M. (2000). *An introduction to Genetic Analysis*, 7th edn. W.H. Freeman and Company, New York.

8. Walter, M.A., Spillett, D.J., Thomas, P., Weissenbach, J. and Goodfellow, P.N. (1994). "A method for constructing radiation hybrid maps of whole genomes." *Nature Genet.* 7:22–28.

9. Dausset, J., Cann, H., Cohen, D., *et al.* (1990). "Program description: Centre d'Etude du polymorphisme Humaine—collaborative genetic mapping of the human genome." *Genomics.* 6:575–577.

10. Friefelder, D. (1987). *Microbial Genetics.* Jones and Bartlett, Boston.

11. Gerhold, D., Rushmore, T. and Caskey, C.T. (1999). "DNA chips: promising toys have become powerful tools." *Trends Biochem. Sci.* 24:168–173.

12. Heiskanen, M., Peltonen, L. and Palotie, A. (1996). "Visual mapping by high resolution FISH." *Trends Genet.* 12:379–382.

13. Jing, J.P., Lai, Z.W., Aston, C., *et al.* (1999). "Optical mapping of *Plasmodium falciparum* chromosome 2." *Genome Res.* 9:175–181.

14. Lichter, P., Tang, C.J., Call, K., *et al.* (1990). "High resolution mapping of human chromosome II by *in situ* hybridization with cosmid clones." *Science.* 247:64–69.

15. Lin, J., Qi, R., Aston, C., *et al.* (1999). "Whole-genome shotgun optical mapping of *Deinococcus radiodurans.*" *Science.* 285:1558–1562.

16. Marra MA, Hillier L and Waterson RH (1998). "Expressed sequence tags - ESTablishing bridges between genomes." *Trends Genet.* 14:4–7.

17. McCarthy, L. (1996). "Whole genome radiation hybrid mapping." *Trends Genet.* 12:491–493.

18. Michalet, X., Ekong, R., Fougerousse, F., *et al.* (1997). "Dynamic molecular combing: stretching the whole human genome for high-resolution studies." *Science.* 277:1518–1523.

19. Mir, K.U. and Southern, E.M. (2000). "Sequence variation in genes and genomic DNA: methods for large-scale analysis." *Ann. Rev. Genomics Hum. Genet.* 1:329–360.

20. Morton, N.E. (1955). "Sequential tests for the detection of linkage." *Am. J. Hum. Genet.* 7:277–318.

21. Oliver, S.G., van der Aart, Q.J.M., Agostoni–Carbone, M.L., *et al.* (1992). "The complete DNA sequence of yeast chromosome III." *Nature.* 357:38–46.

22. Orel, V. (1995). *Gregor Mendel: The First Geneticist.* Oxford University Press, Oxford.

23. Schwartz, D.C., Li, X., Hernandez, L.I., Ramnarain, S.P., Huff, E.J. and Wang, Y.K. (1993). "Ordered restriction maps of *Saccharomyces cerevisiae* chromosomes constructed by optical mapping." *Science*. 262:110–114.

24. SNP Group (The International SNP Map Working Group) (2001). "A map of human genome sequence variation containing 1.42 million single nucleotide polymorphisms." *Nature*. 409:928–933.

25. Sturtevant, A.H. (1913). "The linear arrangement of six sex-linked factors in *Drosophila* as shown by mode of association." *J. Exp. Zool.* 14:39–45.

26. Trask, B.J., Massa, H., Kenwrick, S. and Gitschier, J. (1991). "Mapping of human chromosome Xq28 by 2-color fluorescence *in situ* hybridization of DNA sequences to interphase cell nuclei." *Am. J. Hum.Genet.* 48:1–15.

27. Tyagi, S., Bratu, D.P. and Kramer, F.R. (1998). "Multicolor molecular beacons for allele discrimination." *Nature Biotechnol.* 16:49–53.

28. Wang, D.G., Fan, J.B., Siao, C.J., *et al.* (1998). "Large-scale identification, mapping and genotyping of single-nucleotide polymorphisms in the human genome." *Science*. 280:1077–1082.

29. Yamamoto, F., Clausen, H., White, T., Marken, J. and Hakamori, S. (1990). "Molecular genetic basis of the histo-blood group ABO system." *Nature*. 345:229–233.

30. Alberts, B., Bray, D., Lewis, J., Raff, M., Roberts, K. and Watson, J.D. (1989). *Molecular biology of the cell*, 4th edn. Garland Science, New York.

31. Stahl, F.W. (1994). "The Holliday Junction on its Thirtieth Anniversary." *Genetics*. 138:241–246.

32. Whitehouse, H.L.K. (1973). *Towards an Understanding of the Mechanism of Heredity*, 3rd edn. St. Martin's Press, New York.

33. Finchman, J.R.S., Day, P.R. and Radford, A. (1979). *Fungal Genetics*, 3rd edn. Blackwell, London.

34. Kemp, R. (1970). *Cell Division and Heredity*. St. Martin's Press, New York.

35. Murray, A.W. and Szostak, J.W. (1983). "Construction of Artificial Chromosomes in Yeast." *Nature*. 305:189–193.

36. Puck, T.T. and Kao, F.T. (1982). "Somatic Cell Genetics and Its Application to Medicine." *Annual Review of Genetics*. 16:225–272.

37. Stahl, F.W. (1969). *The Mechanics of Inheritance*, 2nd edn. Prentice Hall, Englewood Cliffs, NJ.

MODEL QUESTIONS

1. What is meant by the term recombination?

2. What is a recombinant?

3. What are the major types of recombination?

4. Describe various models of recombination.

5. What unusual types of tetrads provide evidence that a region of heteroduplex DNA is created in the process of recombination?

6. Define loss of heterozygosity.

7. What is meant by conservation of synteny?

8. Explain the difference between a chimeric animal and a mosaic animal using examples.

9. Explain the techniques for gene replacement by homologous recombination with an example.

10. Explain how recombinant inbred strains are constructed, and how they would be used to map genes for a quantitative trait for which they differ.

11. What is gene mapping? What are the two types of gene mapping?

12. Discuss in detail the mechanism of linkage analysis.

13. What are the significant points in linkage results?

14. Using linkage analysis, how is human gene mapping done?

15. How are linkage studies used for finding associations of disease in a population?

16. What is linkage disequilibrium and nonrandom association of alleles? Explain with examples.

SEX DETERMINATION

6

6.1 INTRODUCTION

We now know that the sex chromosome plays an important role in sex determination. Mammals have X and Y chromosomes. The Y chromosome is the sex-determining factor. Individuals with Y chromosome always develop into a male, no matter even if there are more than one X chromosome. Similarly individuals without Y chromosome develop into a female even if there is only one X chromosome. The sperm (with X or Y chromosome) that fertilizes the egg (with only X chromosome) determines the sex of the resulting zygote.

There are four different types of chromosome-based sex-determining mechanisms found in animals. They are XY, ZW, XO and compound chromosome mechanism. Each of this system is discussed in detail in this chapter. In XY condition, the females have homomorphic pairs of chromosomes (XX) as in human beings or fruit flies; the males are heteromorphic (XY). In the ZW cases such as birds, some fishes and moths, males are homomorphic (ZZ) and the females are heteromorphic (ZW). In grasshopper and beetles males have only one X (XO) chromosome and females are usually XX. But in *Ascaris* (a nematode) several X and Y chromosomes are involved (compound chromosome case) in sex determination.

6.2   THE XY SYSTEM

The Y chromosome has a crucial gene called *sry* (sex determining region of Y), which has the testis-determining function. In an experiment, when *sry* gene was introduced into XX mouse zygote, the transgenic embryo developed into male. But this transgenic mouse could not produce sperm due to the presence of two X chromosomes, which showed sperm suppressive activity.

Sry encodes a regulatory protein that activates the expression of other regulatory proteins including SOX 9 (*Sry*-related protein). *Sry* expression causes a subset of somatic cells to differentiate into sertoli cells. The sertoli cells govern the genital ridge (which produces primordial germ cell) in the following ways.

1. New primordial cells differentiate into sperm-producing cells.
2. Ani mullerian hormone is secreted, which suppresses female gonadial system.
3. Sertoli cells stimulate subset of somatic cells to develop into critical connective tissue that are required for normal sperm production.
4. They stimulate somatic cells to form Leydig cells, which secrete testosterone.

In the absence of *sry*, the primordial germ cells and genital ridge commit to develop into a female reproductive system.

6.2.1   Dosage Compensation

In human beings and other mammals, the necessary dosage compensation is achieved by inactivation of one of the X chromosomes in females. This leads to only one functional chromosome per cell in both males and females. A condensed body as a sex chromatin, now referred to as a Barr body, was first observed by M. Barr and E. Bertram. Mary Lyon then suggested that this Barr body represents an inactive X chromosome, which is tightly coiled into heterochromatin.

Various lines of evidence came in support of the Lyon hypothesis. XXY males have a Barr body, XO females have no Barr body, XXX females have two Barr bodies and XXXX females (super female) have three Barr bodies.

X chromosome inactivation is initiated from the X-inactivation centre (XIC). Portions of the X chromosome that are removed from the XIC and attached to the autosome, escape inactivation. But autosomes that are fused to the XIC region are transcriptionally silenced. An unusual RNA (XIST RNA) that is found within the XIC, is expressed solely from the inactive X chromosome. This XIST RNA remains in the nucleus instead of translating into a protein. This XIST RNA spreads over the XIC and coats them for inactivation.

6.2.2   Sex Determination in *Drosophila*

Male fly development is the default pathway in which *sxi* and *tra* genes are transcribed, but their RNAs are non-functional. In turn *dsx* (double sex gene) is another gene that turns off the genes that specify female characteristics. These genes determine the production of male flies. When X/A ratio is 1, it triggers the female differentiation pathway by activating a promoter within the *sxi* gene and translates into a functional Sxi protein. The Sxi protein not only helps in further Sxi protein production but also produces an active Tra regulatory protein through alternative splice of the Tra transcripts. Flies that lack *dsx* gene express both male and female-specific characters. The Tra protein in turn activates *dsx* gene to transcribe and produce female specific protein. The Tra protein also switches off the genes that specify male features.

6.2.3   Genic Balance in *Drosophila*

Calvin Bridges, in 1922, suggested that the sex in *Drosophila* can also be determined by the balance of autosomal alleles and X chromosome. The autosomal alleles favour maleness whereas the X chromosomes favour femaleness. He calculated the ratio of X chromosomes to autosome, to find the sex of a fly. The autosomal set (A) consists of 3 chromosomes (22 in human beings). Bridge's **genic balance theory** of sex determination shows that when X : A ratio is 1.00 or greater, the organism is a female. When the ratio is 0.50 or less, then the organism is a male (Table 6.1). If the ratio is 0.67 then the organism is an intersex. Below 0.33 ratio, the organisms are metamales and above 1.50, metafemales; these animals are usually very weak and sterile. Sometimes they do not even emerge from their pupal case.

A sex-switch gene is located on the X chromosome. Other genes located on the X chromosomes and the autosomes regulate this sex switch gene.

Table 6.1   Genic balance in *Drosophila*

Autosomal set (A)	Number of X chromosomes (X)	Total number of chromosomes	$\frac{X}{A}$ ratio	Sex of fly
2	3	9	1.50	Metafemale
3	4	13	1.33	Female
4	4	16	1.00	Female
3	3	12	1.00	Female
2	2	8	1.00	Female
1	1	4	1.00	Female
3	2	11	0.67	Intersex
2	1	7	0.50	Male
3	1	10	0.33	Metamale

6.3   THE XO SYSTEM

It is also referred as XO-XX system. In grasshopper, normal male produces gametes that contain either an X chromosome or no sex chromosome. The female grasshopper produces all the gametes with X chromosome. The result of this arrangement causes even number of chromosomes in females and odd number in males (Figure 6.1).

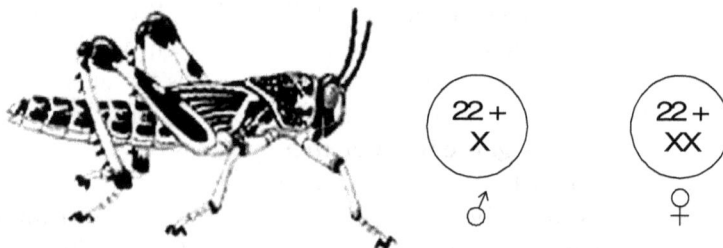

Figure 6.1   XO-XX system (Grasshopper)

Sex determination in the nematode *Caenorhabditis elegans* also occurs by an XO-XX system. *C. elegans* has two sexual types, they are hermaphrodites and males. Most individuals have both the sex organs, an ovary and two testes. Self-fertilization always produces more hermaphrodites. Sometimes males can mate with hermaphrodites and produce equal number of male and hermaphrodite individuals. Genetically hermaphrodites are XX and males are XO.

6.4   THE ZW SYSTEM

In birds, ZW system is found. This is identical to XY system of human beings but the major difference is that males are homogametic and females are heterogametic (Figure 6.2).

Figure 6.2   ZW system (Birds)

6.5   COMPOUND CHROMOSOME SYSTEM

The compound chromosome system is found to be more complex. In *Ascaris*, a nematode, there are twenty-six autosomes, eight X chromosomes and one Y chromosome. Males have thirty-five chromosomes out of which 26 are autosomes, 8 are X chromosomes and one is a Y chromosome (26A + 8X + 1Y). The females have forty-two chromosomes (26A + 16X). During gametic cell division (meiosis), the X chromosomes unite end to end and behave as one unit.

6.6   SEX-LIMITED TRAITS

Sex-limited traits are characters expressed in one sex (either male or female), although genes are present in both the sexes. Some examples are given below:

- Breast and ovary formation
- Facial hair and sperm production
- Plumage pattern in birds

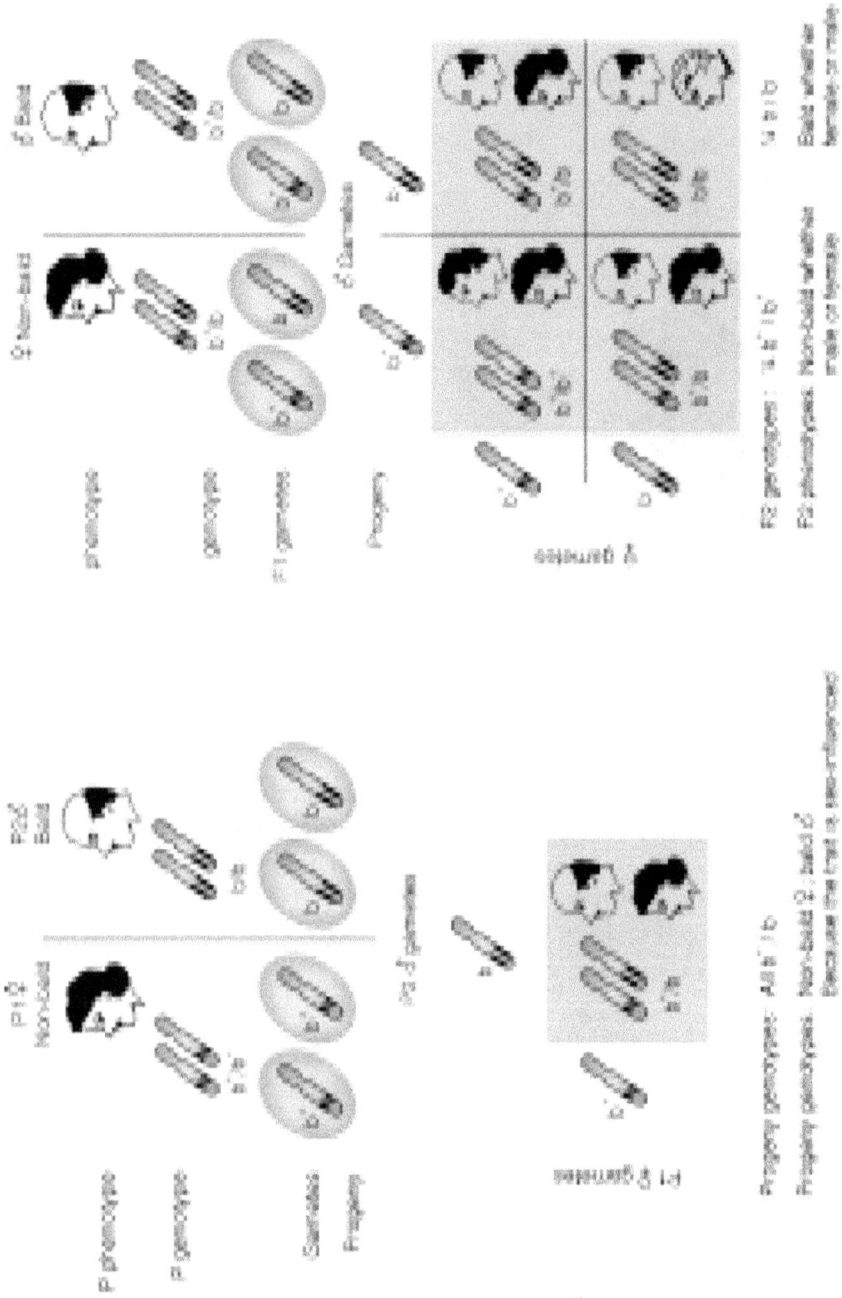

Figure 6.3   Sex-influenced traits

- Bright colouration in birds (male)
- Horns in sheep
- Milk production in female mammals

6.7   SEX-INFLUENCED TRAITS

Sex-influenced traits appear in both the sexes but occur in one sex more than the other (Figure 6.3). Hormones also play a role in influencing the trait. Some examples of influenced traits are

- Baldness in humans (more in male)
- Cleft lip and cleft palate (more in male)
- Club foot (more in male)
- Rheumatoid arthritis (more in females)
- Autoimmune disease (more in females)

FOR ADDITIONAL READING

1. Baker, B.S., Hoff, G., Kauman, T.C., Wolfner, M.F. and Hazerlrigg, T. (1991). "The doublesex locus of *Drosophila melanogaster* and its flanking regions: A cytogenetic analysis." *Genetics.* 127:125–38.

2. Charlesworth, B. (1991). "The evolution of sex chromosomes." *Science.* 251:1030–33.

3. Cherfas, J. (1991). "Sex and the single gene." *Science.* 252:782.

4. Corcos, A.F. (1983). "Pattern baldness: Its genetics revisited." *Amer. Biol. Teacher.* 45:371–75.

5. Elbrecht, A. and Gsmith, R. (1992). "Aromatase enzyme activity and sex determination in chickens." *Science.* 255:467–70.

6. Erickson, J.W. and Cline, T.W. (1991). "Molecular nature of the *Drosophila* sex determination signal and its link to neurogenesis." *Science.* 251:1071–74.

7. Gartler, S. and Riggs, A. (1983). "Mammalian X-chromosome inactivation." *Ann. Rev.Genet.* 17:155–90.

8. Lyon, M.F. (1962). "Sex chromatin and gene action in the mammalian X chromosome." *Amer. J. Hum. Genet.* 14:135–48.

9. Lyon, M.F. (1990). "Evolution of the X chromosome." *Nature.* 348:585–86.

10. Morgan, T.H. (1910). "Sex limited inheritance in *Drosophila.*" *Science.* 32:120–22.

11. Roberts, L. (1988). "Zeroing in on the sex switch." *Science*. 239:21–23.

12. Shapiro, L., *et al.* (1979). "Noninactivation of an X-chromosome locus in man." *Science*. 204:1224–26.

13. Walker, C.L., *et al.* (1991). "The Barr body is a looped X chromosome formed by telomere association." *Proc. Nat. Acad. Sci.*, USA. 88:6191–95.

14. Wright, W.G. (1988). "Sex change in the Mollusca." *TREE*. 3:137–40.

15. Baker, B.S. (1989). "Sex in Flies: the splice of life." *Nature*. 340:521–524.

16. Benne, R. (1996) "RNA editing: how a message is changed." *Curr. Opin. Genet. Dev.* 6:221–231.

17. Cline, T.W. and Meyer, B.J. (1996) "Vive la difference: males vs females in flies vs worms." *Annu. Rev. Genet.* 30:637–702.

18. Goodfellow, P.N. and Lovell-Badge, R. (1993) "SRY and sex determination in mammals." *Annu. Rev. Genet.* 27:71–92.

19. McLaren, A. (1999). "Signaling for germ cells." *Genes Dev.* 13:373–376.

20. Swain, A. and Lovell-Badge, R. (1999). "Mammalian sex determination: a molecular drama." *Genes Dev.* 13:755–767.

21. Williamson, A. and Lehmann, R. (1996). "Germ cell development in *Drosophila*." *Annu. Rev. Cell Dev. Biol.* 12:365–391.

22. Wylie, C. (1999). "Germ cells." *Cell*. 96:165–174.

MODEL QUESTIONS

1. What is a sex switch?

2. How do genes regulate sex switches in *Drosophila*?

3. What are Barr bodies?

4. What is Lyonization?

5. Comment on genic balance. How is it helpful in sex determination in *Drosophila*?

6. What is dosage compensation?

7. With an example, explain various types of sex determination in animals.

8. How do genes determine sex of a human being?

9. Comment on compound chromosome type of sex determination with examples.

10. What is sex-limited trait? List various examples.

11. What are sex-influenced traits? Give examples.

EXTRANUCLEAR INHERITANCE

7

7.1 INTRODUCTION

Carl Correns was a plant geneticist. He did experiments to find the inheritance of leaf colour in four O'clock plant (*Mirabilis jalapa*). He was surprised to note that the progeny colour was always like that of female parent. Also the mendelian ratio 3 : 1 was not followed. But the answer to this question was found out later on by other scientists. This kind of unusual inheritance is due to DNA of chloroplast and not due to nuclear DNA. It is called non-nuclear inheritance, extranuclear inheritance, non-mendelian inheritance, cytoplasmic DNA inheritance or organellar DNA inheritance. Similar to extranuclear DNA (chloroplast DNA or cpDNA) in plants the animal cell also has extranuclear DNA (mitochondrial DNA or mtDNA). First, let us try to understand the architecture and functions of mitochondrial genome as well as chloroplast genome.

7.2   MITOCHONDRIAL INHERITANCE

All aerobic eukaryotic cells have mitochondria in their cytoplasm. The number varies according to cell function. These mitochondria function as powerhouses of the cell. They produce energy in the form of ATP molecules through oxidative phosphorylation. Mitochondrial genome is circular and has highly supercoiled double-stranded DNA molecules. The mitochondrial DNA (mtDNA) is not associated with histones or histone-like proteins. The size of the animal mitochondrial genome is less than 20,000 base pairs whereas the size of the mitochondrial genome in fungi and plants ranges above 80,000 base pairs. This difference is due to the presence of only essential genes in the animals, but in plants extra junk genome is present. The human mitochondrial genome has 16,569 base pairs. The genome contains 37 genes: 28 are encoded in one strand of the DNA and 9 on the other strand. These genes encode 22 transfer RNAs and 2 types of ribosomal RNAs and 13 different proteins that are involved in oxidative phosphorylation.

7.3   mtDNA MUTATIONS AND HUMAN DISEASE

Human cells contain hundreds of copies of mtDNA. Mitochondrial genome is maternally inherited. During fertilization, sperm contributes only nuclear material to the egg. Though male and female both inherit their mitochondria from their mother, males cannot transmit their mitochondria to the next generation.

Inherited and acquired mutations in the mitochondrial genome have been implicated in a wide range of human diseases, including neuromuscular disorders, deafness, cardiomyopathies, skin lesions, diabetes, aplastic anaemia, Parkinson's disease and also ageing in general.

7.4   CHLOROPLAST GENOME

Chloroplast is the site for photosynthesis in green plants and some of the protists. Chloroplast contains two membrane structures, the inner and the outer. The transport proteins are embedded in the inner membrane. The inner membrane surrounds the stroma, where metabolic enzymes are present. Thylakoids present in the chloroplast are responsible for electron transport, photosynthetic light capture and ATP synthesis. Chloroplast has its own genome. Chloroplast genome is circular, double-stranded, supercoiled and does not have any associated proteins. The chloroplast DNA (cpDNA) ranges between 80 kb and 600 kb. Number of copies of cpDNA also varies according to species. In the unicellular *Chlamydomonas*, one chloroplast contains 500–1500 cpDNA copies.

7.5 PARENTAL INHERITANCE IN *CHLAMYDOMONAS*

Chlamydomonas is a haploid unicellular alga and motile in nature. It has two flagella and a chloroplast. Chloroplast contains many copies of its own DNA. There are two mating types mt^+ and mt^-. Nuclear genes determine the mating type. *Chlamydomonas* has proved to be an excellent system for studying non-mendelian pattern of inheritance (for both chloroplast and mitochondrial mode of inheritance).

Let us see how characters are inherited through chloroplast in *Chlamydomonas*. Wild type *Chlamydomonas* is sensitive to erythromycin [ery^s] and the mutant type is resistant to erythromycin [ery^r]. As shown in Figure 7.1 when a cross of mt^+ [ery^r] × mt^- [ery^s] was made, about 95% of the

Str^s and str^r crosses

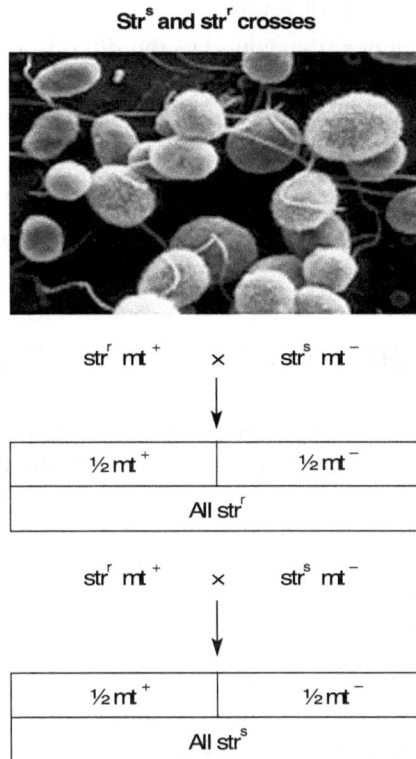

Figure 7.1   Uniparental inheritance of erythromycin-resistance in *Chlamydomonas*

progeny were erythromycin-resistant. Similarly when a reciprocal cross, mt⁻ [ery^r] × mt⁺ [ery^s] was made, 95% of the progeny were sensitive to erythromycin. These results are examples of uniparental inheritance. That is, the progeny always resembles the phenotype of the mt⁺ parent.

The other 5% of the zygotes show characters of both the parents. It is termed as **biparental inheritance**, because of the presence of chloroplast of both the parents. This type of zygote is also called **cytohet** (cytoplasmically heterozygous). Mostly the biparental zygotes segregate into one particular parental type during mitosis. That is, they are either resistant or sensitive to erythromycin.

7.6   RULES OF NON-MENDELIAN INHERITANCE

There are four scientific features that exist in non-mendelian inheritance. They are:

1.   Lack of meiosis-based gene or chromosome segregation, so the mendelian ratio is not found.

2.   Extranuclear genes transmitted to the offspring are from any one of the two parents (uniparental inheritance), that is, either from the mother (maternal inheritance) or from the father (paternal inheritance).

3.   These genes cannot be mapped to the nuclear genome.

4.   Even if the nucleus of a cell is substituted, the non-mendelian inheritance cannot be altered.

7.7   THE POKY MUTANT OF FUNGUS *NEUROSPORA*

The fungus *Neurospora* requires oxygen to grow and survive. The mitochondrial functions are essential for its day-to-day life. The three principal cytochromes which help in respiration are a+a3, b and c. The wild type has all three cytochrome compartments where as poky mutant lacks a+a3, b and a portion of cytochrome c. This affects the mutant in not generating sufficient ATP for its growth. The molecular defect in poky mutant is a 4 bp deletion in the promoter of 19S rRNA gene.

Mary and Herschel Mitchell crossed female poky mutant and male wild type *Neurospora* (Figure 7.2). This resulted into all poky mutant progeny. As in Figure 7.1 when reciprocal cross was made between wild type female and poky male, all the F1 progeny were wild type. This indicated that the

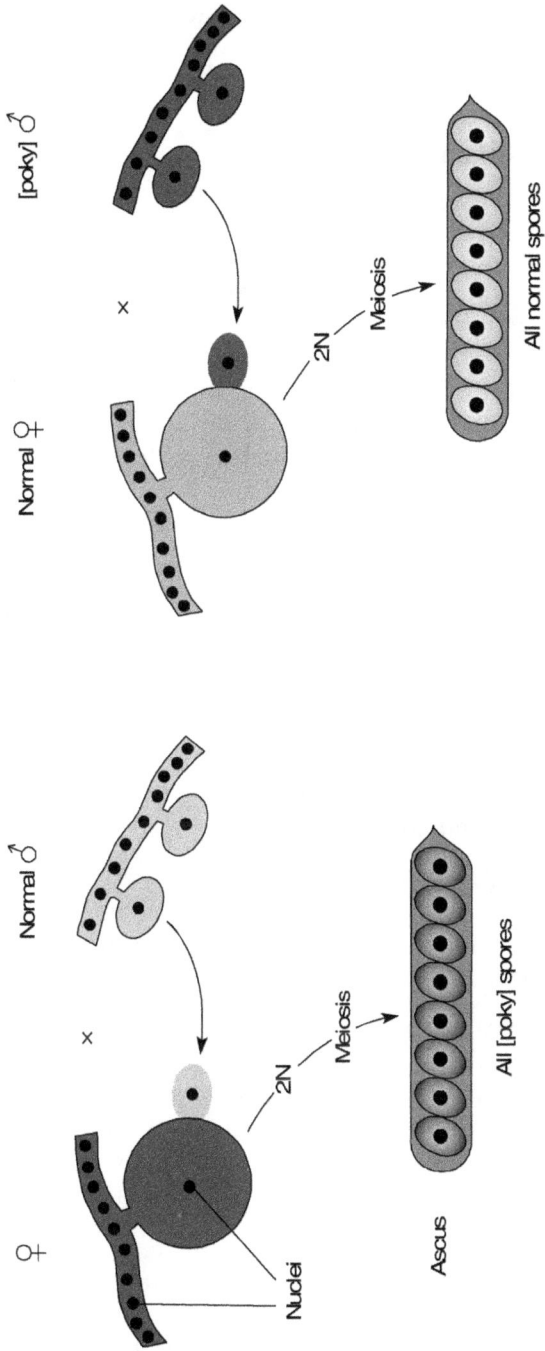

Figure 7.2   Poky mutant in *Neurospora*

phenotype of progeny was same as the mother. This also confirmed the maternal inheritance of phenotype. The slow-growing poky mutant was inherited in this fashion.

7.8   YEAST PETITE MUTANTS

Yeast as a single cell can grow with or without oxygen. On solid media, yeast cells cluster together as discrete colonies. In anaerobic condition, yeast cells obtain energy through fermentation (without involving mitochondria). In aerobic condition mitochondrial respiration takes place and this accelerates the growth rate. "Petites" are smaller colonies formed by few yeast cells on solid media. The wild type colonies are bigger and called "granules". The size of the petite colonies is small because of the slow growth rate. There are different classes of petites: the **nuclear petites, neutral petites** and **suppressive petites**.

Yeasts are haploid cells. A crossing occurs during fusion of mating type "a" cell and mating type "α" cell producing a diploid zygote. This zygote undergoes meiotic division (sporulation) and forms four haploid products (ascospores). Tetrad analysis can be done to find the segregation patterns.

When petites are crossed with wild type the outcome is four haploid spores. Among them, two are wild type and two are petite. These two petites are called nuclear petites since segregation pattern is similar to nuclear gene mutations (some mitochondrial proteins are encoded by nuclear genes). These nuclear petites are symbolized as pet⁻.

As shown in Figure 7.3, when a pet⁻ mutant is crossed with wild type (pet⁺), the diploid is pet⁺/pet⁻ (wild type colonies). When this pet⁺/pet⁻ undergoes meiotic division, the resulting tetrad produces 2 wild types and 2 nuclear petite phenotype.

The other two classes of petites such as neutral petites and the suppressive petites show deviation from mendelian ratio. As shown in Figure 7.3, when a neutral petite (rho N) is crossed with normal wild type yeast cell, the diploid produces wild type colonies. If these diploids go through meiotic process, all resulting tetrad products show a ratio of 0 : 4 petite : wild type respectively. Neutrals do not have an effect on the wild type. This result is a classical example of uniparental inheritance.

The third class of petites is suppressive petites (rho⁻ˢ). These petites have a suppressive effect on the wild type. When suppressive petites are crossed with wild type, the diploid [rho⁺]/[rho⁻ˢ] is formed. This has respiratory property in between normal and petites. This diploid mitotically

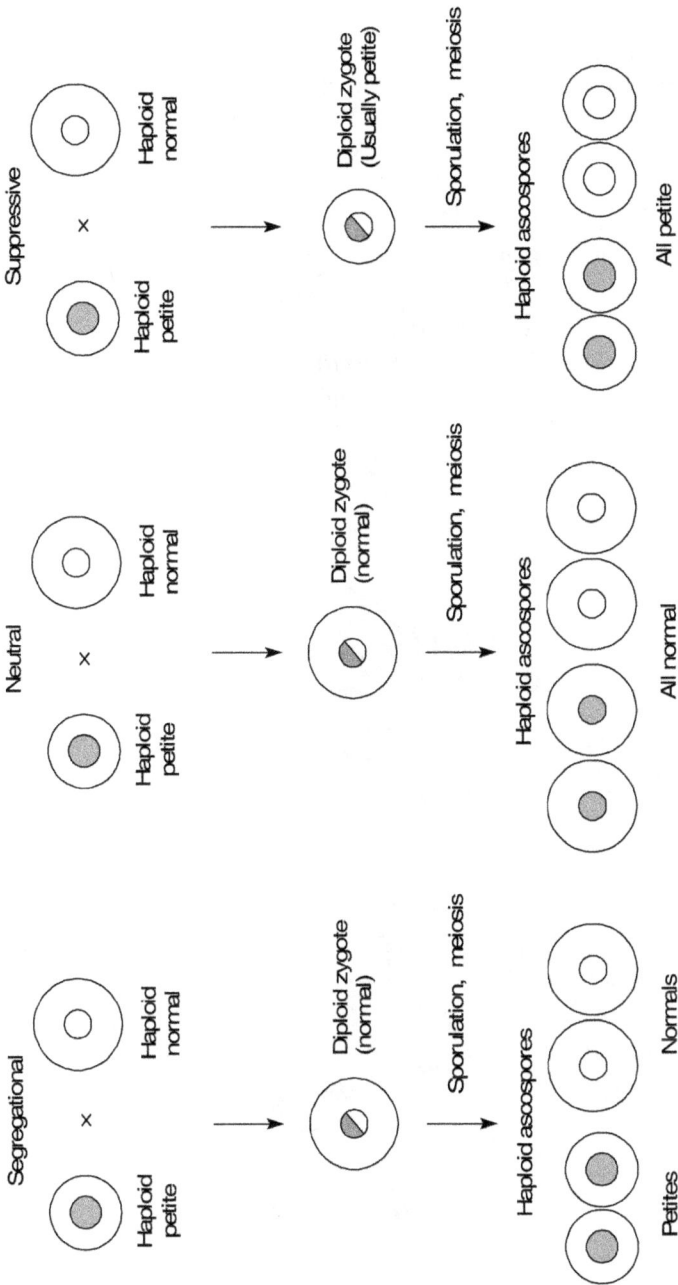

Figure 7.3   Petite mutant crosses

divides to produce 99 percent petites. If wild type diploid cell sporulates, it produces tetrads with a 0 : 4 ratio of petite: wild type. If suppressive diploid cell sporulates, it produces tetrads with a 4 : 0 ratio of petite: wild type. The suppressive petites act as a dominant condition. They produce the effect by irregular replication or by recombination effect, thereby altering gene organization.

7.9 SHOOT VARIEGATION IN THE FOUR O'CLOCK PLANT

A shoot of a plant consists of stem, leaves and flowers. Shoot variegation (Figure 7.4) in the four o'clock (*Mirabilis jalapa*) occurs through non-mendelian mode of inheritance. The wild type flowering plants have

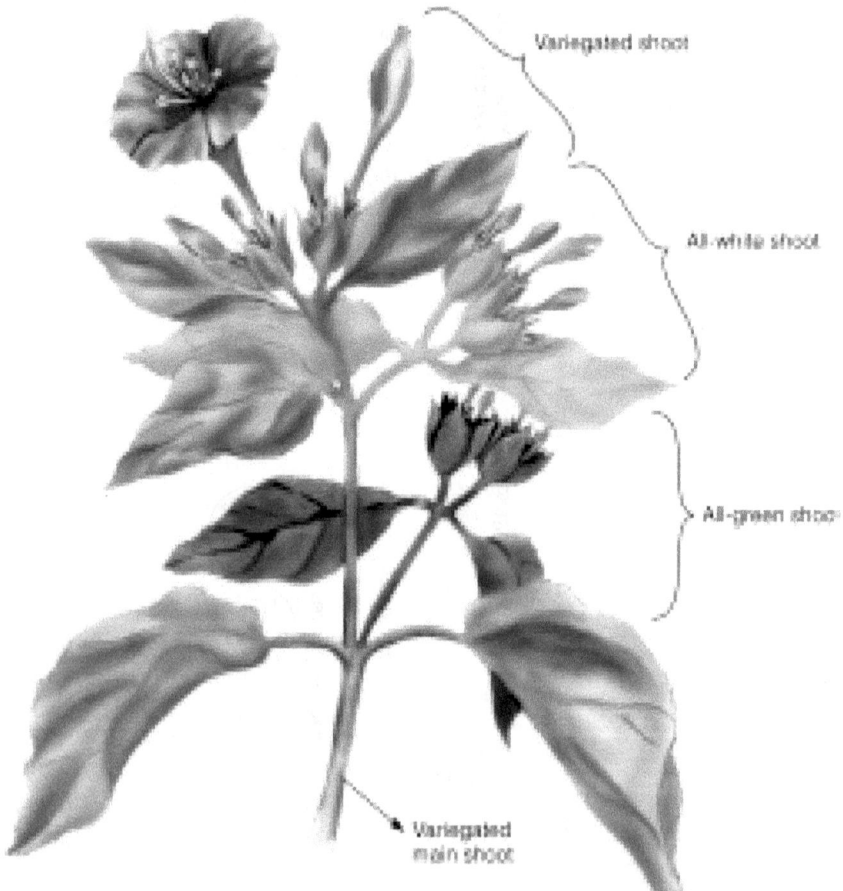

Figure 7.4 Shoot variegation in *Mirabilis jalapa*

green shoots because of the presence of the green pigment chlorophyll in chloroplast. White shoots have abnormal and colourless chloroplasts called leucoplast. These white shoots are not capable of carrying out photosynthesis. This is due to mutant gene of the chloroplast. White progeny always dies after seed germination due to lack of energy.

When a zygote receives both chloroplasts and leucoplasts from maternal cytoplasm it segregates into green tissue, white tissue or variegated tissue (mixture of green and white) in the progeny. This could be due to any of the following three reasons.

1. No contribution of cytoplasm of the pollen to the egg. So the extranuclear genetic determinant comes only from egg.

2. Due to segregation, though mixed plastids are found (chloroplast and leucoplasts) pure plastid line can be generated from an original mixed line.

3. The segregation of the plastid to daughter cell is random. Due to this some daughter cells receive chloroplast. Some receive leucoplast and some receive both the plastids. In some, the progeny phenotype is same as that of the maternal parent.

7.10 SHELL COILING PATTERN IN SNAIL

In snail (*Limnaea peregra*) the shell coiling phenotype is always determined by the genotype of the mother. This is called maternal effect. This maternal effect is different from maternal inheritance pattern of extranuclear genes. The maternal effect is the result of mRNA or proteins that are deposited in the oocyte before fertilization. In maternal effect the maternal nuclear genes specify the progeny.

In *Limnaea* snail, nuclear genes determine the shell coiling character. The dominant D allele is for coiling to the right or dextral coiling. The recessive d allele is for coiling to the left or sinistral coiling. When the true breeding dextral coiling female is crossed with true breeding sinistral male, the F1 progeny will have dextral shell coiling. Selfing the F1 produces a ratio of 1 : 2 : 1 (D/D, D/d and d/d genotypes). But all the F2 will be dextral (Figure 7.5).

Similarly when a sinistral (dd) female is crossed with a dextral male (DD), the F1 progeny (Dd) will have sinistral phenotype because the mother is having (d/d) genotype. When selfing is done in F1 snails, all F2 progeny will be dextral. But in both the cases selfing the F2 snails will give F3 progeny, which produces a ratio of 3 : 1, that is 3 dextral and one sinistral snails

(Figure 7.5). The molecular basis for this coiling is due to deposition of product of D allele in the cytoplasm of the egg which causes the next generation embryo to coil dextrally.

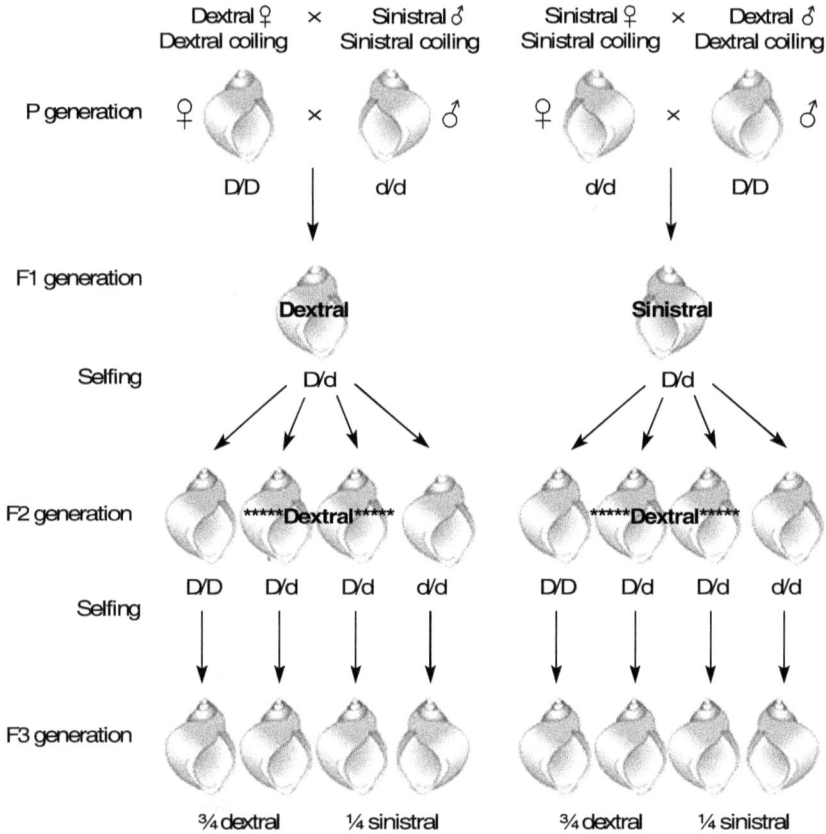

Figure 7.5   Different shell coiling in snail

7.11   KAPPA PARTICLES IN KILLER *PARAMAECIUM*

Sonneborn and others have found that certain stocks of *Paramecium* had the ability to kill the individuals of the other stock. The individuals causing death are termed "killers" and those that succumbed are termed "sensitives".

In *Paramecium*, there are two micronuclei and a macronucleus (Figure 7.6). The micronuclei are primarily reproductive nuclei. The macronuclei are for the vegetative function of the cell. During binary fission, that is cell division, the micronucleus undergoes mitotic division whereas the macronucleus constricts and is pulled in half. There are two types of nuclear rearrangements that take place in *Paramecium*, namely conjugation and autogamy.

In conjugation, both the paramecia come together and form a bridge between them. The macronucleus starts disappearing and both the micronuclei undergo meiotic division. Each *Paramecium* will have 8 nuclei; of them 7 will disappear and only one remains. This single nucleus undergoes mitotic division and forms two haploid nuclei. Now reciprocal exchange of nuclei occurs between the two paramecia so that each cell has one original and one migrant nucleus. These two nuclei fuse within a cell and form diploid nucleus. These diploid nuclei are genetically identical and undergo two mitotic processes. Four diploid nuclei per cell are formed. Among them two nuclei become macronuclei and two remain as micronuclei. During cell division the macronucleus separates so that each daughter cell gets one and the micronucleus undergoes mitosis. The two daughter cells are called exconjugates. Depending on bridging and amount of time, the exchange of cytoplasm may occur.

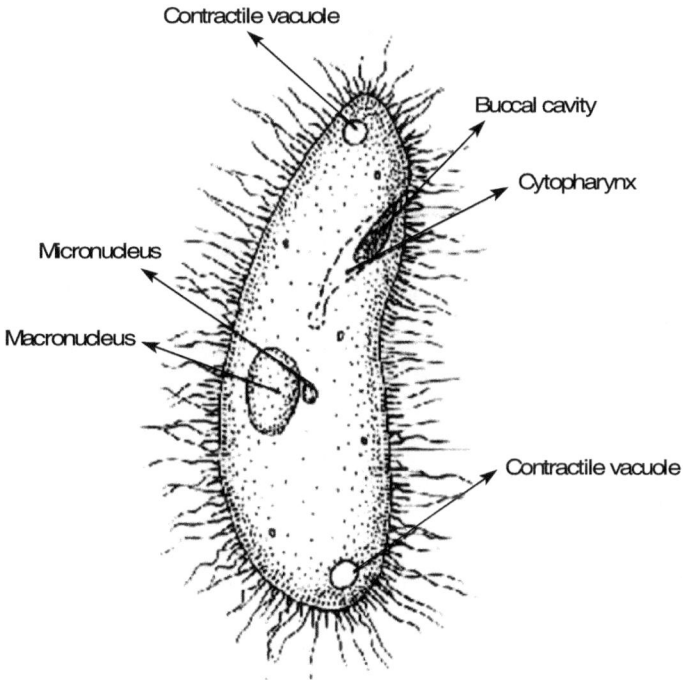

Figure 7.6 *Paramecium*

In autogamy, though only one *Paramecium* is involved, the nuclear processing is same as conjugation. All cells after autogamy are homozygous.

But the heterozygous exconjugate produces 50% killer and 50% sensitive cells in autogamy.

The kappa confers the killer specificity on the host. But the sensitive *Paramecium* is killed by the toxin called paramecin which is released by killer *Paramecium*.

In conjugation between killer cells and sensitive cells, exchange of cytoplasm and nucleus takes place through bridges. Both exconjugants are killers. This is due to the transfer of cytoplasmic particles such as "Kappa particles" from killer cells to other cells.

FOR ADDITIONAL READING

1. Costanzo, M.C. and Fox, T.D. (1990). "Control of mitochondrial gene expression in *Saccharomyces cerevisiae.*" *Annu. Rev. Genet.* 24:91–113.

2. Evans, R.J., Oakley, K.M. and Clark-Walker, G.D. (1985). "Elevated levels of petite formation in strains of *Saccharomyces cerevisiae* restored to respiratory competence. I. Association of both high and moderate frequencies of petite mutant formation with the presence of aberrant mitochondrial DNA." *Genetics.* 111:389–402.

3. Grun, P. (1976). *Cytoplasmic Genetics and Evolution.* Columbia University Press, New York.

4. Gyllensten, U., Wharton, D., Josefsson, A. and Wilson, A.C. (1991). "Paternal inheritance of mitochondrial DNA in mice." *Nature.* 352:255–57.

5. Hoffman, M. (1991). "How parents make their mark on genes." *Science.* 252:1250–51.

6. Palmer, J.D. (1985). "Comparative organization of chloroplast genomes." *Annu. Rev. Genet.* 19:325–54.

7. Palmer, J.D., Jorgensen, R.A. and Thompson, W.F. (1985). "Chloroplast DNA variation and evolution in *Pisum*: Patterns of change and phylogenetic analysis." *Genetics.* 109:195–213.

8. Preer, J.L., Preer, and Jurand, A. (1974). "Kappa and other endosymbionts in *Paramecium aurelia.*" *Bact. Rev.* 38:113–63.

9. Wickner, R.B., (1986). "Double-stranded RNA replication in yeast: the killer system." *Ann. Rev. Biochem.* 55:373–95.

MODEL QUESTIONS

1. Snail coiling is due to a maternal effect. Is it possible that it is caused by an allele at a sex-linked locus?

2. How would you determine that a segregative petite mutant in yeast is controlled by a chromosomal gene?

3. What results would be obtained by making all possible pair-wise crosses of the three types of yeast petites?

4. What are the similarities shared by mitochondria and plastids?

5. What evidence is there for the prokaryotic origin of mitochondria and chloroplasts?

6. What genetic tests could you conduct to show that the mate-killer phenotype in *Paramecium* requires a dominant allele at any one of two loci?

7. What type of asci do you expect if you cross a yeast strain carrying an antibiotic resistance gene in its mitochondria with a strain that has normal mitochondria?

8. A dextral snail (A) that resulted from a cross is self-fertilized and produces only sinistral progeny. What is the probable genotype of (A) and its parents?

MOLECULAR BIOLOGY OF THE GENE

8

8.1 NUCLEIC ACIDS

It has been known that genes are made up of DNA in eukaryotes but RNA in some microbes. The genetic materials such as deoxyribonucleic acid (DNA) and ribonucleic acid (RNA) are the molecular repositories for genetic information. In eukaryotes the DNA molecules are commonly packed with proteins into structures called **chromosomes**. A single chromosome typically contains thousands of individual **genes**. The sum of all the genes with associated materials is referred to as the cellular **genome**. Table 8.1 shows the nature of nucleic acid in different species.

The amount of genetic material varies according to species. **C-value** represents the total amount of DNA in a haploid set of genome. The C-value paradox is the amount of genome function not ascribed precisely. That is the discrepancy between size of the genome and genetic complexity of an organism. In most of the organisms DNA is the genetic material, however in a few bacteriophages and many viruses, RNA is the genetic material. As far as we know now, there are no prokaryotes or eukaryotes which have RNA as their genetic material.

Table 8.1   The genetic material in different species

Genetic material	Organism
Single-stranded RNA (linear)	Phage MS2, phage F2, phage QB, polio virus, retrovirus
Single-stranded RNA (circular)	Arena virus, bunyavirus, delta virus
Double-stranded RNA (linear)	Phage φ6, reovirus, rotavirus
Single-stranded DNA (linear)	Phage M13, Phage fd, phage f1, parvovirus (B19)
Single-stranded DNA (circular)	φX174
Double-stranded DNA (linear)	Phage T2, phage T4, phage T6, phage T7, herpesvirus, pox virus, adenovirus
Double-stranded DNA (circular)	Phage PM2, papova virus
Double-stranded DNA (incomplete circular)	Hepatitis B virus

8.2   EVIDENCE FOR DNA AS THE GENETIC MATERIAL

In 1928, **Frederick Griffith** reported that heat-killed S-bacteria could 'transform' living R-type bacteria into S-type bacteria. Griffith used two strains of *Streptococcus pneumoniae* bacterium, S-strain and R-strain. S-strain produces smooth, shiny colonies and is highly infectious (virulent). The R-strain produces rough colonies and is harmless. Griffith infected mice with different strains of the bacterium. When mice were infected with R-strain bacteria, bacteria did not affect the mice and mice were alive. When mice were infected with S-strain bacteria, bacteraemia was caused in mice. Hence all mice died. However, if the S-strain was heat killed then the mice survived.

Further to the above experiments, Griffith infected mice with a mixture of living R-strain bacteria and heat-killed S-strain bacteria. He was surprised to see that all mice died and the blood showed the presence of living S-strain bacterium. Griffith concluded that some R-strain bacteria had transformed into S-strain. He believed that an unknown agent (protein) is responsible for this change. He called the agent **transforming principle**.

In 1944, **Oswald T. Avery** along with his associates **Colin M.Macleod** and **Maclyn McCarty**, identified the nature of the transforming principle. In a test tube, they lysed cells and separated various macromolecular components such as lipids, polysaccharides, proteins and nucleic acids. They tested whether any one of those macromolecular components could transform living R-bacteria into S-bacteria. They identified that nucleic acids were the only components that could transform R-cells into S-cells.

Further to this experiment, Avery and his associates treated the nucleic acids with ribonuclease (RNase) and degraded the RNA (but not DNA). They still found transforming activity was present. However when they used deoxyribonuclease (DNase) and digested the DNA (not RNA), they found no transformation. They strongly suggested that DNA was the genetic material. But their work was criticized by saying that nucleic acids isolated from the bacteria were contaminated by proteins.

In 1953, using radioactive isotopes of phosphorus (^{32}P) and Sulphur (^{35}S), **A.D. Hershey and M. Chase** provided more evidence for DNA as the genetic material. They conducted experiments using bacteriophage T2 and the bacterium *Escherichia coli*. The bacteriophages were labelled with radioactive isotopes by growing them on bacteria grown in culture medium containing the radioisotopes ^{35}S or ^{32}P. The radioisotopes entered the bacterial cells and the next generation of the phages that burst out from the infected cells. Now they had two batches of bacteriophages: one had their proteins labelled with ^{35}S and the other had their DNA labelled with ^{32}P isotopes.

Now, Hershey and Chase mixed the ^{32}P-labelled T2 phages with unlabelled *E. coli* cells. They found that ^{32}P radioactivity was found within the bacteria after infection. When ^{32}S-labelled T2 phages were mixed with unlabelled *E. coli*, they found that isotopes stayed on the outside of the bacteria for most part. These experiments demonstrated that the DNA entered the cell and not proteins as evidenced by the presence of ^{32}P and the absence of ^{35}S. Hershey and Chase realized that the DNA is the genetic material that is responsible for the function and reproduction of phage T2 and not the protein. In 1969, Alfred Hershey shared the Nobel Prize in physiology or medicine for his discoveries.

8.3   CHEMICAL NATURE OF GENETIC MATERIAL

Genetic material is composed of three principal structures namely

- Bases (purine and pyrimidine)
- Sugars (ribose or deoxyribose)
- Phosphates

Complete hydrolysis of the nucleic acids yields nitrogenous bases (such as pyrimidine and purines), sugar components (ribose or deoxyribose) and phosphoric acid. Partial hydrolysis yields compounds known as nucleosides and nucleotides.

Having identified the genetic material as DNA or RNA, we shall now examine the chemical structure of these molecules. This will tell us how they function. The nitrogenous bases are heterocyclic rings of carbon and nitrogen atoms. They fall into two types. **Pyrimidines** which have a six member ring and **purines** which have fused five and six-member rings (imidazole pyrimidines rings respectively) (Figure 8.1).

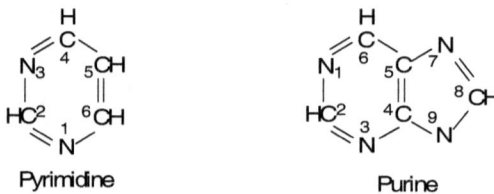

Figure 8.1   Structure of purines and pyrimidines

RNA and DNA contain the purine bases **adenine** (A) and **guanine** (G) as shown in Figure 8.2.

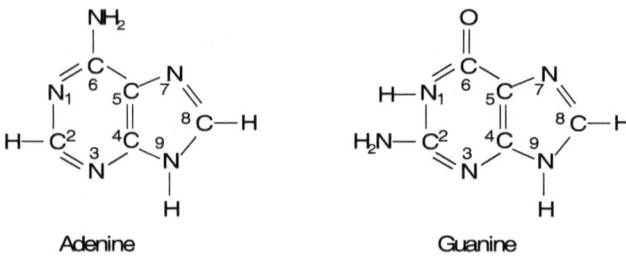

Figure 8.2   Structure of purines

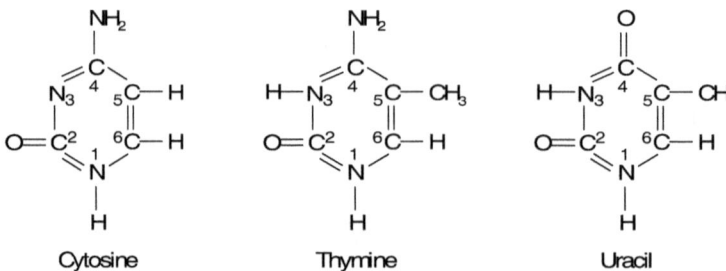

Figure 8.3   Structure of pyrimidines

The two pyrimidines in DNA are **cytosine** (C) and **thymine** (T) but in RNA thymine is replaced by **uracil** (U). The chemical structure of C, T and U are shown in Figure 8.3.

Some of the more important minor bases in tRNA are given in Table 8.2.

Table 8.2   Minor purine bases in tRNA

1-methyl adenine
2-methyl adenine
3-methyl adenine
1-methyl guanine
2-methyl guanine
7-methyl guanine
Hypoxanthine
2,2-Dimethyl guanine
Xanthine

The two pyrimidines in DNA are **cytosine** (C) and **thymine** (T) but in RNA thymine is replaced by **uracil** (U). The chemical structure of C, T and U are shown in Figure 8.3.

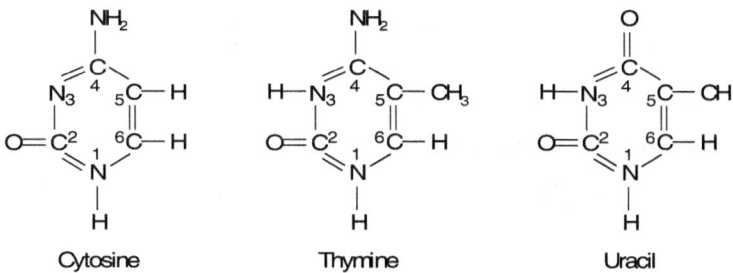

Figure 8.3   Structure of pyrimidines

Some of the minor pyrimidine bases found in tRNA are given in Table 8.3.

Table 8.3   Minor pyrimidine bases in tRNA

Dihydrouracil
5-hydroxy methyl uracil
2-thiouracil
2-thiocytosine
3-methyl cytosine
4-acetyl cytosine
5-carboxy methyl uracil

8.3.1   Sugar

There are two types of pentose sugars. In DNA the pentose is 2-deoxyribose, whereas in RNA it is ribose. The difference lies in the absence/presence of the OH group at position 2 of the sugar ring as shown in Figure 8.4.

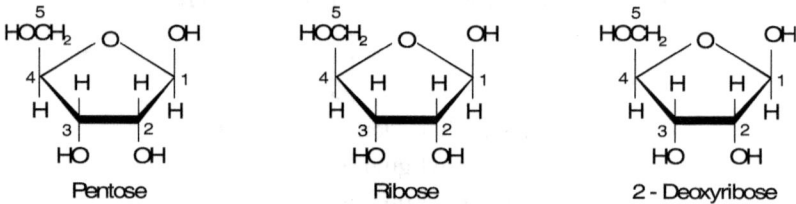

Figure 8.4   Structure of pentose sugars

8.3.2   Phosphates

The phosphates in DNA are either mono-, di- or triphosphates (Figure 8.5). The acidic nature of nucleic acid is due to the presence of phosphate esters. At neutral pH they exist in the forms shown in Figure 8.5.

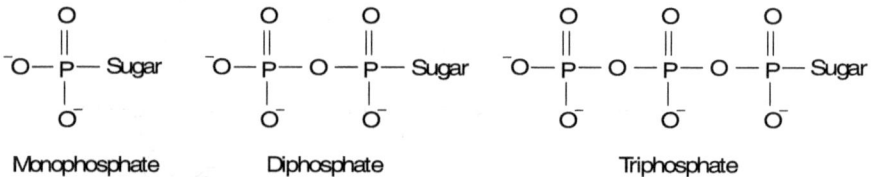

Figure 8.5   Structure of phosphates in DNA

8.3.3   Bonds

Genetic material is made up of three principal types of bonds. The strength of the bonds is important in the stability of the genetic material.

- Covalent bond exists between atoms that share electrons in their outermost shell.

- Hydrogen atom can form a weak electrostatic interaction (hydrogen bond) with an electronegative atom (usually nitrogen or oxygen).

- Ester bond involves covalent bonding, and is formed when an alcohol and an acid unite with elimination of water.

8.3.4 Nucleosides

- **Base** with **sugar** is termed **nucleoside.**
- The first carbon of the sugar ring is attached to the first nitrogen in the first position of pyrimidine or to the nitrogen in 9th position of purine through an N-glycosidic bond as shown in Figure 8.6.

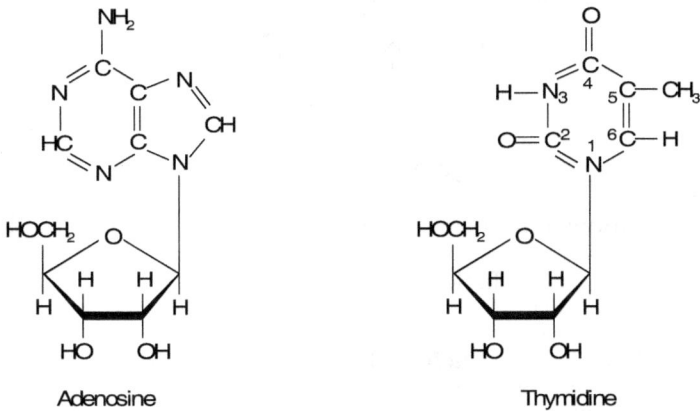

Figure 8.6   Nucleosides

8.3.5 Nucleotides

- **Base** with **sugar** and **phosphate** is termed **nucleotide** (Figure 8.7).
- Phosphate is attached to the 5′-carbon of the sugar.
- The 5′ hydroxyl group of one nucleotide unit is joined to the 3′ hydroxyl group of the next nucleotide by a phosphodiester linkage.
- A short nucleic acid is called an **oligonucleotide.**
- A longer nucleic acid is called a **polynucleotide.**
- Nucleotide is involved in numerous phosphate transfer reactions of ATP and other nucleoside triphosphates.
- UDP-Glucose and UDP-Galactose help in carbohydrate biosynthesis as high-energy intermediates for covalent bond synthesis.
- Nucleotide forms a portion of coenzymes such as FAD, NAD^+, $NADP^+$, and coenzyme A.
- ADP level regulates mitochondrial oxidative phosphorylation.
- Nucleotides act as allosteric regulators of enzyme activity.
- Cyclic AMP and cyclic GMP serve as second messengers.

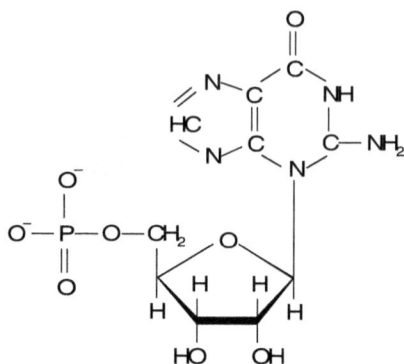

Figure 8.7   Guanylate

A list of various nucleosides and nucleotides are given in Table 8.4.

Table 8.4   Nomenclature

Base	Nucleosides	Nucleotides
Adenine	Adenosine	Adenylate
	Dioxyadenosine	Deoxyadenylate
Guanine	Guanosine	Guanylate
	Deoxyguanosine	Deoxyguanylate
Cytosine	Cytidine	Cytidylate
	Deoxycytidine	Deoxycytidylate
Thymine	Deoxythymidine	Deoxythymidylate
Uracil	Uridine	Uridylate

8.4   BIOLOGICALLY ACTIVE STRUCTURE

Though many experiments had proved that nucleic acid (DNA) was the genetic material, its actual structure was not known until 1953. Linus Pauling (1953) who received Nobel Prize for the discovery of α-helical structure of proteins was also investigating a three-stranded structure for the genetic material. But later on Watson and Crick decided that a two-stranded structure was more consistent based on three lines of evidence. This evidence was from the chemical nature of genetic components, X-ray crystallographic study and Chargaff's ratios.

We have already seen the chemical nature of nucleic acid. Now let us study the other features. The structure of DNA was analysed using X-ray crystallography by Maurice Wilkins, Rosalind Franklin and their associates. When a beam of parallel X-rays was aimed at DNA molecule, the atoms and special arrangements of the molecule diffracted the beam. The diffracted X-rays were recorded using a photographic plate; using such photographs they concluded that there were two distinct regular intervals of 0.34 nm and 3.4 nm along the axis of the molecule.

The most important third clue to the structure of DNA came from the work of **Erwin Chargaff** and his colleagues. Using quantitative chromatography, the DNA specimens of different species were analysed and the following conclusions were drawn.

- DNA extracted from different tissues of the same species have same base composition.
- Base composition varies between two different species.
- Base composition does not change with age, nutritional state or change in environment.
- The number of adenine residues is equal to the number of thymine (A = T) and the number of guanine is equal to the number of cytocine (G = C), i.e., A + G = T + C.

Watson and Crick (1953) postulated the most widely accepted three-dimensional model of DNA structure. According to them DNA has the following structure.

- It consists of two helical chains coiled around the same axis (right-handed double helix).
- It has hydrophilic backbones of deoxyribose and negatively charged phosphate groups found outside of the double ratio as shown in Figure 8.8.
- DNA has bases (purine and pyrimidine) stacked inside the double helix and are hydrophobic.
- It consists of two grooves, major groove and minor groove.
- It has two chains of the helix that are antiparallel and complementary.
- Bases inside the double helix would be 0.34 nm apart and the secondary repeat distance would be 3.4 nm.
- DNA contains 10.5 nucleotide residues in each complete turn of the double helix.

Figure 8.8   DNA double helical structure

8.5   ISOLATION OF DNA

- 20 ml of fresh blood must be collected in EDTA-coated vacutainer. The sample may be kept at 4°C for several days, in –20°C up to six months or in –70°C indefinitely.

- Add 30 ml of distilled water to fresh or thawed sample, mix and centrifuge at 3000 rpm for 10 min. Remove the supernatant.

- Then add 25 ml of 0.1% Nonidet P40 solution to the pellet, mix well and spin.

- Add 10 ml of 100 mM NaCl and 25 nM EDTA to the pellet and mix well.

- Now add 0.5 mg of proteinase K and 0.5 ml of 10 SDS and incubate at 37°C for 4–16 h.

- Carry out two phenol–chloroform extractions.

- Carry out two chloroform extractions.

- Precipitate the DNA at room temperature by adding 1 ml of 4M NaCl and 22 ml of absolute ethanol.

- Redissolve the DNA in 1 ml of either distilled water or 10 mM Tris-HCl (pH 7.5) and 1mM EDTA overnight at +4°C.

- Measure the DNA concentration at 260 nm using spectrophotometer.

8.6 COMPARISON OF DIFFERENT FORMS OF DNA

There are various forms of DNA and include A-DNA, B-DNA, Z-DNA, C-DNA and D-DNA. A comparison of these forms is given in Table 8.5.

Table 8.5   Comparison of different forms of DNA

A-DNA	B-DNA	Z-DNA	C-DNA	D-DNA
Right hand helix	Right hand helix	Left hand helix	Right hand helix	Right hand helix
11 residues/ turn	10.5 residues/ turn	12 residues/ turn	9.3 residues/ turn	8 residues/ turn
25 Å dia of helix	23 Å dia of helix	18 Å dia of helix	-	-
20° base tilt	–6° base tilt	7° base tilt	-	-
Axial rise 2.9 Å	3.4 Å	3.7 Å	-	-
75% RH	92% RH	-	66% RH	-
RNA–DNA hybrid	Normal form	Alternating purine and pyrimidine	A— T rich Lack G— C	A— T rich Lack G— C
Pitch complete turn 32 Å	34 Å	45 Å	-	-

8.7 PHYSICO-CHEMICAL PROPERTIES OF DNA

- DNA is highly acidic because secondary phosphate groups are negatively charged (ionized). It can bind to cations such as calcium and magnesium.

- A dilute solution of DNA is highly viscous. Increase or decrease in viscosity leads to denaturation.

- DNA can react with diphenylamine, glacial acetic acid and concentrated sulpuric acid. The blue colour can be observed under 660 nm.

- Heating or exposure to high salt concentration may disrupt the noncovalent forces that stabilize the DNA double helix. The two strands of a double helix separate entirely when all the hydrogen bonds between them are broken. This process of strand separation is called denaturation or melting.

- During denaturation state the viscosity of the solution decreases, optical rotation becomes more negative, the light absorption at 260 nm increases (hyperchromic effect) and buoyant density also increases.

- The ability of the two separated complementary strands to form into a double helix again is called renaturation.

8.8   TRIPLE HELICAL DNA STRUCTURE

Depending on ionic strength, superhelical stress and protein binding, the DNA can adopt a triple helical structure. Formation of an intramolecular DNA triple helix (H-DNA) occurs when polypyrimidine and polypurine sequences having mirror symmetry undergo conformational rearrangements due to torsional stress. In the human genome, near the promoter regions, oligopyrimidine/oligopurine mirror repeats are present.

FOR ADDITIONAL READING

1. Adams, R.L.P., Burdon, R.H., Campbell, A.M., Leader, D.P. and Smellie, R.M.S. (1981). *The Biochemistry of the Nucleic Acids.* Chapman and Hall Publication, London.

2. Berg, J.M., Tymoczko, J.L., Stryer, L. (2002). *Biochemistry,* 5th edn. W.H. Freeman and Company, New York.

3. Brown, T.A. (2002). *Genomes,* 2nd edn. John Wiley and Sons Inc., New York.

4. Lehninger, A.L., Nelson, D.L. and Cox, M.M. (1993). *Principles of Biochemistry,* CBS Publishers, India.

5. Lewin, B. (2000). *Genes VII,* Oxford University Press, Oxford.

6. Lodish, H., Baltimore, D., Berk, A., Zipursky, S.L., Matsudaira, P. and Darnell, J. (1995). *Molecular Cell Biology,* 3rd edn. Scientific American Books Inc., New York.

7. Murray, R.K., Granner, D.K., Mayes, P.A. and Rodwell, V.W. (2000). *Harper's Biochemistry,* 25th edn. McGraw-Hill, USA.

8. Sinden, R.R. (1996). *DNA Structure and Function.* Academic Press, New York.

9. Voet, D., Voet, J.G. and Pratt, C.W. (1999). *Fundamentals of Biochemistry.* John Wiley & Sons Inc., New York.

10. Watson, J.D., Hopkins, N., Roberts, J., Streitz, J.A. and Weiner, A. (1987). *Molecular Biology of the Gene,* 4th edn. Benjamin Cummings, California.

11. Zubay, G.L., Parson, W.W. and Vance, D.E. (2001). *Principles of Biochemistry*, Wm C. Brown Publishers, England.

12. Arscott, P.G. *et al.* (1989). "Scanning tunneling microscopy of Z-DNA." *Nature*. 339:484–86.

13. Avery, O.C. MacLeod, and McCarty, M. (1944). "Studies on the chemical nature of the substance inducing transformation of pneumococcal types." *J. Exp. Med.* 79:137–58.

14. Beebe, T.P *et al.* (1989). "Direct observation of native DNA structures with the scanning tunneling microscope." *Science*. 243:370–72.

15. Cairns, J. (1963). "The chromosomes of *E. coli.*" Cold Spr. Harb. Symp. Quant. Biol. 28:43–46.

16. Chargaff, E., and Davidson, J. (eds.). (1955). *The Nucleic acids*. Academic Press, New York.

17. Dickerson, R. (1983). "The DNA helix and how it is read." *Sci. Amer.*, Dec. 94–111.

18. Dickerson, R., *et al.* (1982). "The anatomy of A-, B- and Z-DNA." *Science*. 216:475–85.

19. Driscol, R.J., Youngquist, M.G. and Baldeschwieler, D.D. (1990). "Atomic scale imaging of DNA using scanning tunneling microscopy." *Nature*. 346:294–96.

20. Dulbecco, R. (1987). *The Design of Life*. Yale University Press, New Haven.

21. Fraenkel-Conrat, H. and Singer, B. (1957). "Virus reconstitution II. Combination of protein and nucleic acid from different strains." *Biochim. Biophys. Acta.* 24:540–48.

22. Lederberg, J. (1991). "The gene (H.J. Muller 1947)." *Genetics*. 129:313–16.

23. Moffat, A.S. (1991). "Triplex DNA finally comes of age." *Science*. 252:1374–75.

24. Rich, A., Nordheim, A. and Wang, A. (1984). "The chemistry of left-handed Z-DNA." *Ann. Rev. Biochem.* 53:791–846.

25. Trifonov, E.N. (1991). "DNA in profile." *Trends Biochem. Sci.* 16:467–70.

26. Watson, J.D. (1968). *The Double Helix*. Signet, New York.

27. Watson, J.D., and Crick, F.H.C. (1953). "Molecular structure of nucleic acids. A structure for deoxyribose nucleic acid." *Nature*. 171:737–38.

MODEL QUESTIONS

1. DNA
   a. is the hereditary material
   b. controls the biosynthetic properties of a cell
   c. both a and b
   d. neither a nor b

2. In DNA
   a. phosphorus is present
   b. sulphur is absent
   c. both a and b
   d. neither a nor b

3. The Watson and Crick DNA model shows
   a. that DNA is a double helix
   b. sugar phosphate backbones are outside
   c. paired bases are on the inside
   d. all of the above

4. Contrast the structures of A-DNA, B-DNA and Z-DNA.

5. Describe in detail the DNA structure of Watson and Crick model.

6. Write notes on unusual structures of the DNA.

7. Write a short description on triple helical nature of DNA.

8. Comment on various physiochemical properties of DNA.

9. Discuss in detail various experiments done to prove that DNA is the genetic material.

10. How is DNA extracted from the blood lymphocytes?

11. Is it correct to say that DNA is the genetic material always?

12. What is C-value and C-value paradox?

REPLICATION

9

9.1   INTRODUCTION

Replication is a process by which the genomic DNA gets duplicated and produces two daughter DNA. The entire genome multiplies before each cell division. In eukaryotes, replication occurs during S phase of the interphase whereas in prokaryotes replication occurs before each cell division.

Replication may be unidirectional or bidirectional. In **unidirectional** replication, one replication fork leaves the origin and proceeds along the DNA. In **bidirectional** replication, two replication forks are formed and they proceed away from the origin in opposite directions (Figure 9.1).

There are three models of DNA replication; they are **conservative, semiconservative** and **dispersive** models. In the semiconservative model, the double-helical DNA unwinds and the two strands separate. Each strand acts as a template for the synthesis of a new complementary strand of DNA. Here each progeny double-stranded DNA retains one parental strand. In conservative model, both the parental strands remain

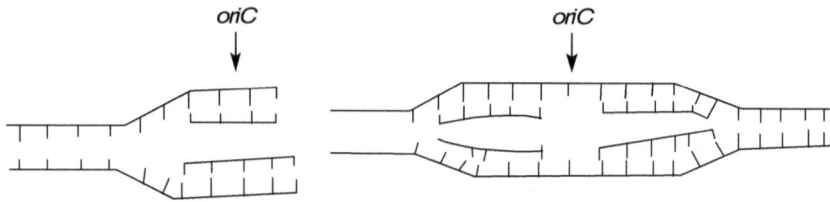

Figure 9.1   (a) Unidirectional replication and (b) Bidirectional replication

together but serve as a template. Here the progeny is of two kinds: One consists of the double-stranded DNA of the parent and the other consists of new material. In the dispersive model, the parental double-helical DNA is cleaved into small segments. New strands fill these gaps so that the parental and progeny segments are interspersed (Figure 9.2).

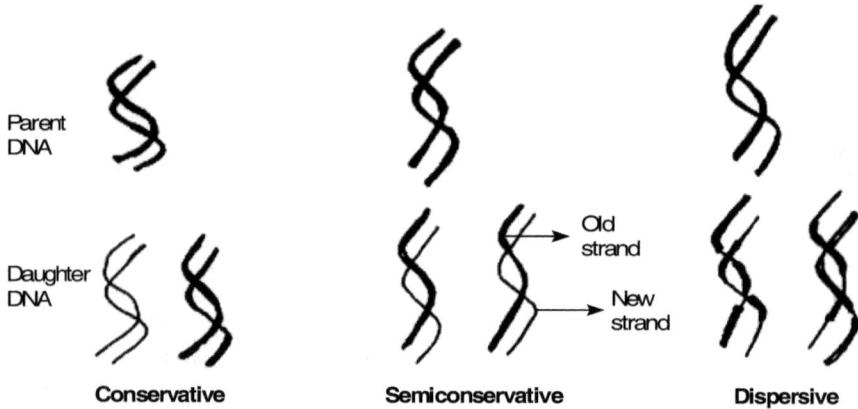

Figure 9.2   Models of DNA replication

9.2   EXPERIMENTAL PROOF

Mathew Meselson and **Frank Stahl** (1958) proved that the DNA undergoes semiconservative type of replication as shown in Figure 9.3. Cells were grown for many generations in a medium containing only heavy nitrogen ^{15}N, so that all the nitrogen in the DNA was ^{15}N. The cells were then transferred to a medium containing only light nitrogen, ^{14}N, and the density of the DNA was monitored closely for the next two cell generations. Cellular DNA was isolated after the first and second generations and centrifuged to equilibrium in a CsCl$_2$ density gradient. The ^{15}N DNA came to equilibrium at a lower position in the CsCl$_2$ gradient than ^{14}N DNA. Hybrid DNA equilibrated in an intermediate position. If DNA replication was conservative, each of the two heavy strands of parental DNA would be replicated to yield the original heavy duplex DNA and a DNA duplex containing two new light strands. Continuation of conservative replication would yield in the next generation one heavy DNA and three light DNAs but no hybrid DNAs.

The Meselson–Stahl experiment, however, showed that replication is semiconservative resulting in two daughter duplexes each containing one parental heavy strand and one new light strand. The next generation yielded two hybrid DNAs and two light DNAs.

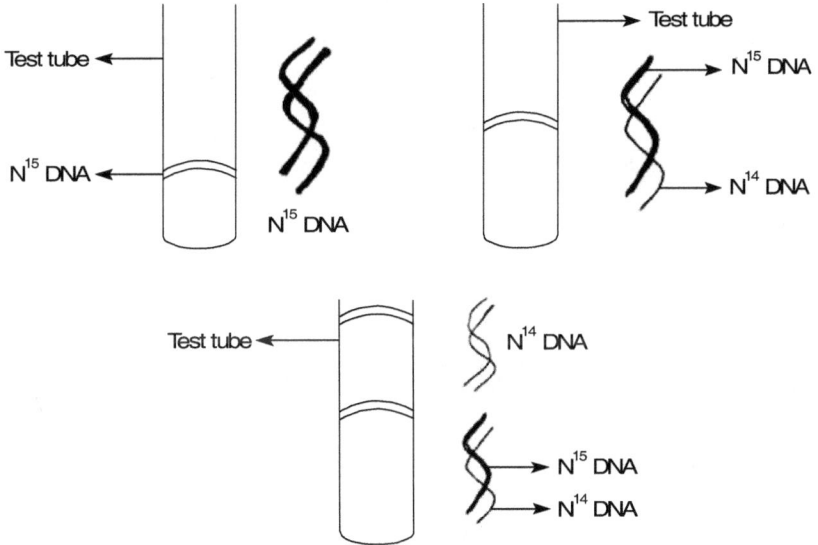

Figure 9.3   Meselson and Stahl experiment

There are three models of replication. They are

- Rolling circle model—rDNA in the *Xenopus* oocyte, fX174 and λ phage
- D-loop model—Mammalian mitochondiral DNA, chloroplast DNA
- Theta model—Circular DNA of polyoma virus

9.3   ROLLING CIRCLE MODEL

In this model of replication, a break (nick) in one of the phosphodiester bonds is made in one of the strands of the circular DNA. The free 3′ end is extended by polynucleotide addition and continued DNA synthesis "rolls off" a complete copy of the genome. The second strand synthesis converts the linear DNA to double-stranded DNA. This later circularizes when cleavage at the junction points between genomes occur. This model occurs in late stages of growth of λ phage and during rRNA production in *Xenopus* (Figure 9.4).

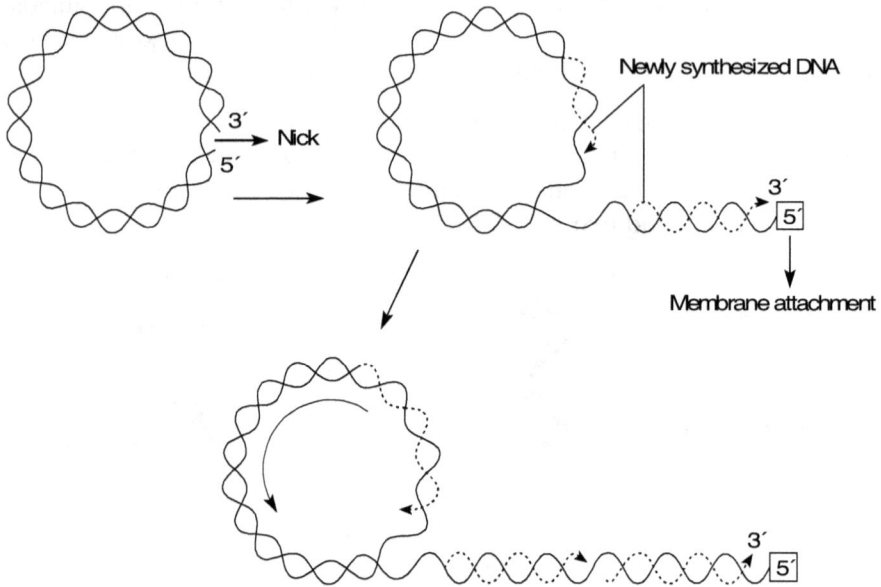

Figure 9.4   Rolling circle model

9.4   D-LOOP MODEL

Replication begins on one strand after displacing the other strand while forming displacement loop (D-loop). Replication continues until the process passes the origin of replication on the other strand. This initiates replication on the second strand in the opposite direction, which results in two circles (D-loop). Multiple D-loop formation is common among chloroplast or mitochondrial DNA replication (Figure 9.5).

9.5   COMPONENTS OF REPLICATION

The following components are required for replication.

1. DNA template
2. Enzymes
   a) DNA helicase
   b) DNA topoisomerase
   c) DNA gyrase
   d) DNA polymerase
   e) DNA ligase
   f) Telomerase

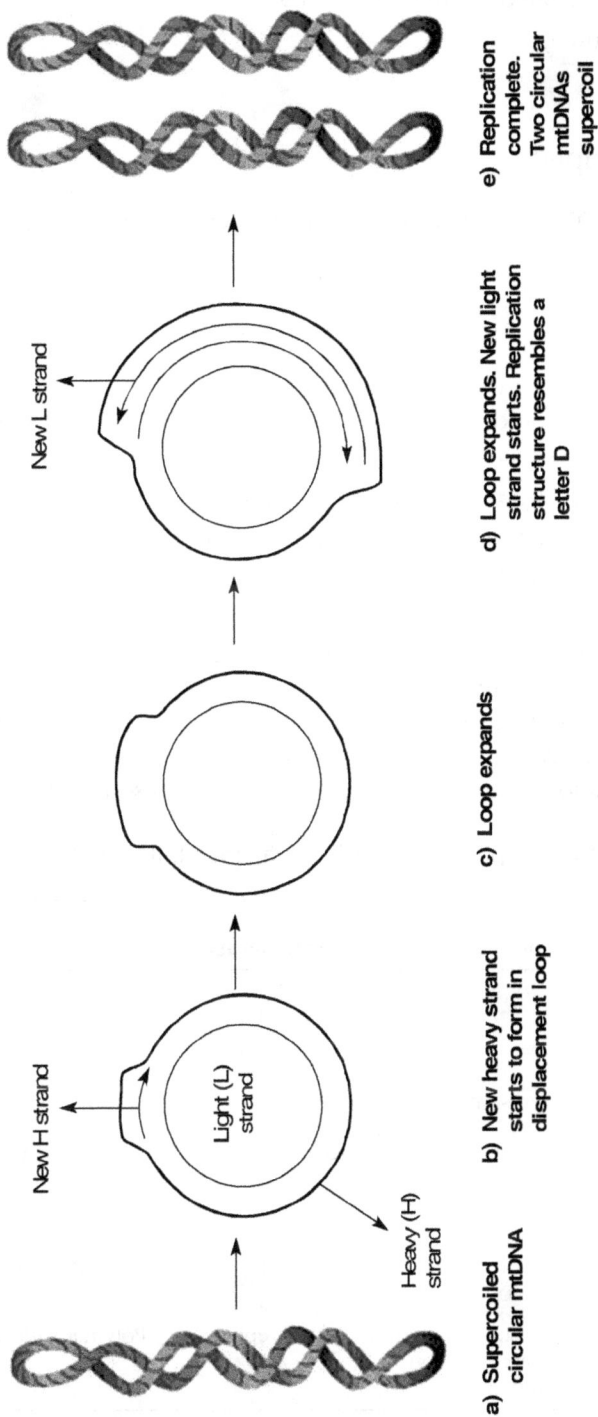

Figure 9.5 D-loop replication in chloroplast

3.   Various associated protein molecules

4.   Energy source

5.   Primer

9.5.1   DNA Template

DNA template is a DNA strand based on which a new complementary strand is synthesized in $5' \longrightarrow 3'$ direction.

9.5.2   Enzymes

DNA helicase   Helix-unwinding is accomplished by the enzyme called helicase. It breaks ATP and utilizes the free energy of hydrolysis to unwind the helix. Two ATP molecules are used per broken base pair.

DNA polymerase   In 1955 Arthur Kornberg and his colleagues found DNA polymerase from *E. coli* cells. In early 1970, DNA polymerase II and III were discovered. The properties of these three enzymes are given in Table 9.1.

DNA topoisomerase   Strand separation (unwinding) gives topological stress in the helical DNA structure, which is relieved by the action of topoisomerases. DNA gyrase (topoisomerase II) can introduce negative supercoils by using ATP. The superhelical density of bacterial DNA is balanced by regulation of the net activity of topoisomerase I and II.

Table 9.1   Properties of DNA polymerase I, II and III

	DNA polymerase I	DNA polymerase II	DNA polymerase III
Structural gene is	Pol A	Pol B	Pol C
Number of subunits	1	More or less 4	10
$3' \longrightarrow 5'$ exonuclease activity	Present	Present	Present
$5' \longrightarrow 3'$ exonuclease activity	Present	Absent	Absent
Polymerization rate	16–20 nucleotides/ second	7 nucleotides/ second	250–1000 nucleotides/second
Function	Proofreading and base-excision repair	DNA repair	Polymerization

There are four mammalian DNA polymerases and their properties are given in Table 9.2.

Table 9.2   Properties of various mammalian DNA polymerases

	DNA Polymerases			
	Alpha	**Beta**	**Gamma**	**Delta**
Found in	Nucleus	Nucleus	Mitochondria	Nucleus
Function	Replication and Priming	Repair	Mitochondrial replication	Replication
Subunits	4	1	1	1

DNA ligase   DNA ligase is a single-strand nick-closing enzyme; it uses ATP for energy.

9.6   REPLICATION OF *E. COLI* DNA

The process of synthesis of a DNA molecule can be divided into three stages; **initiation** (recognition of replication origin region), **elongation** (movement of replication fork and addition of ribonucleotides) **and termination** (completion, transcript formation and dissociation of RNA polymerase). These stages are distinguished by differences in their reactions and in the enzymes required, as shown in Figure 9.6.

9.6.1   Initiation

The replication origin region of *E. coli* is referred to as ***oriC*** and has about 245 base pairs, which are highly conserved, in most bacteria. This conserved segment contains two short repeat motifs. There are five copies of nine nucleotides and the other of a 13 base pair sequence, spread throughout the *oriC*.

A complex of **DNA A** proteins (about 30 numbers) binds to the replication origin region where the DNA is slightly negatively supercoiled and melts (opens) the DNA duplex by torsional stress. This process requires ATP hydrolysis and is probably facilitated by the AT-rich 13-nucleotide repeat regions. **HU** packaging proteins further promote melting. A complex of **DNA BC** proteins bind and form a **prepriming complex**. Once the DNA B (helicase) protein is bound and starts unwinding the DNA bidirectionally to two potential replication forks, the DNA C protein dissociates from the complex. The unwinding produces positive supercoils in the DNA and is relieved by **DNA gyrase** (topoisomerase II). Multiple molecules of **single strand binding protein** (SSB) bind to single-stranded DNA and stabilize the separated DNA strands by preventing reunion. DNA replication is regulated to occur only once in each cell cycle. Initiation at inappropriate time is prevented by the presence of the inactive DNA A–ADP complex and binding of a protein called loci A (inhibitor) to the 13 base-pair repeats.

a) **Replication origin region** *(oriC)*

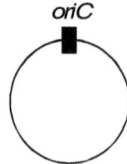

b) **Structure of *oriC***

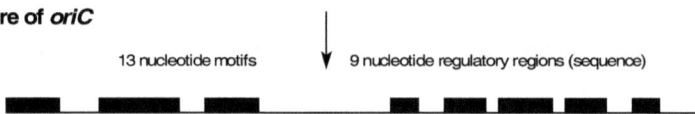

c) **Binding of DNA A proteins**

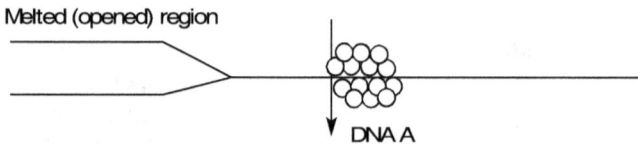

d) **Binding of HU proteins**

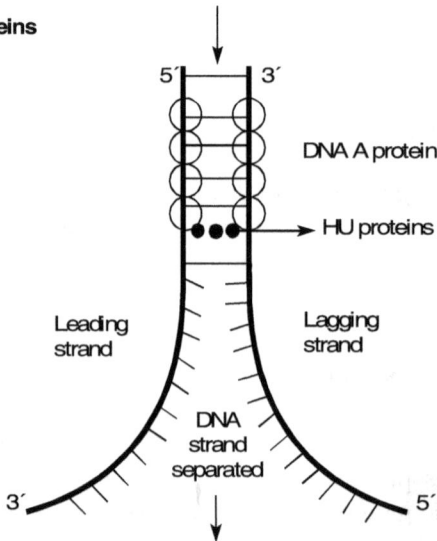

Figure 9.6   Schematic representation of DNA replication in *E. coli*

(Continues)

e) Binding of DNA BC proteins

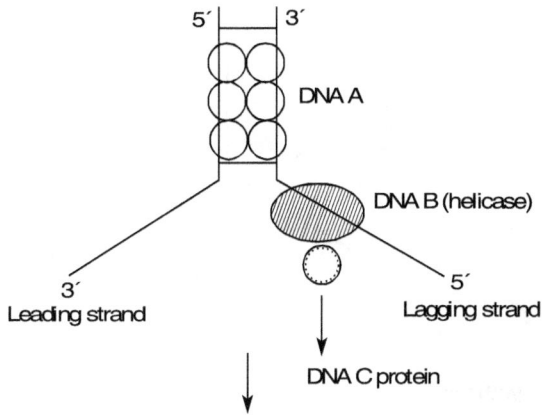

f) Binding of topoisomerase

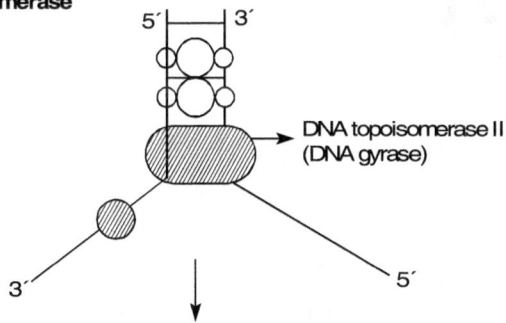

g) Binding of SSB

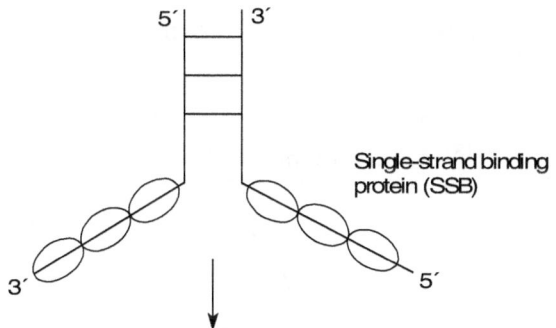

Figure 9.6   Schematic representation of DNA replication in *E. coli*

(Continues)

h) Primosome formation

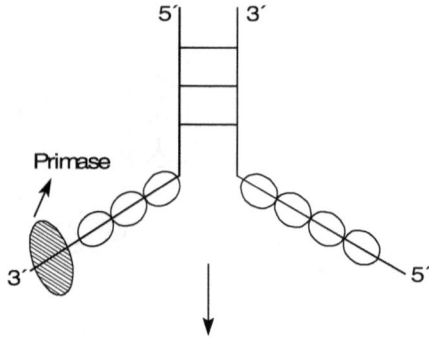

i) Primer synthesis

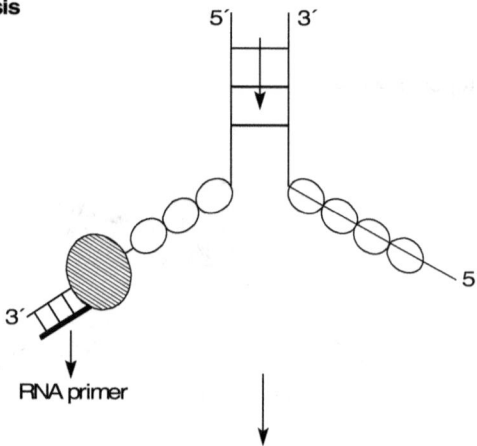

j) Elongation of new strand by DNA polymerase III

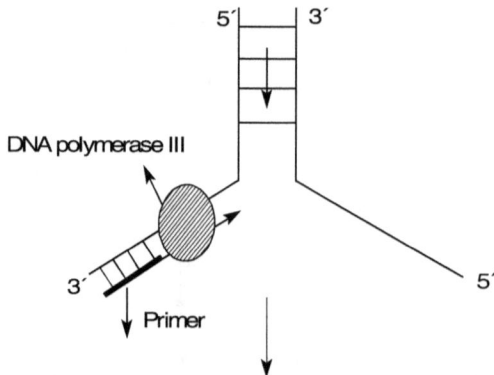

Figure 9.6   Schematic representation of DNA replication in *E. coli*

(Continues)

k)

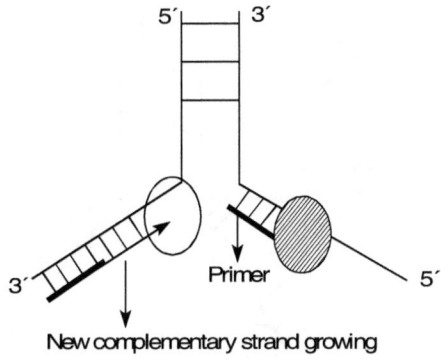

5′ 3′

Primer

3′

New complementary strand growing

5′

l)

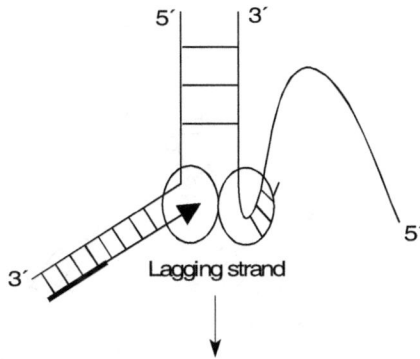

5′ 3′

3′

Lagging strand

5′

m)

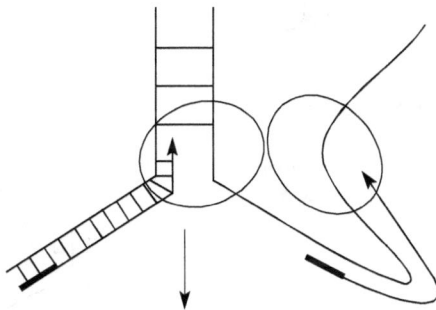

Figure 9.6 Schematic representation of DNA replication in *E. coli*

(Continues)

n) Binding of replication mediator proteins

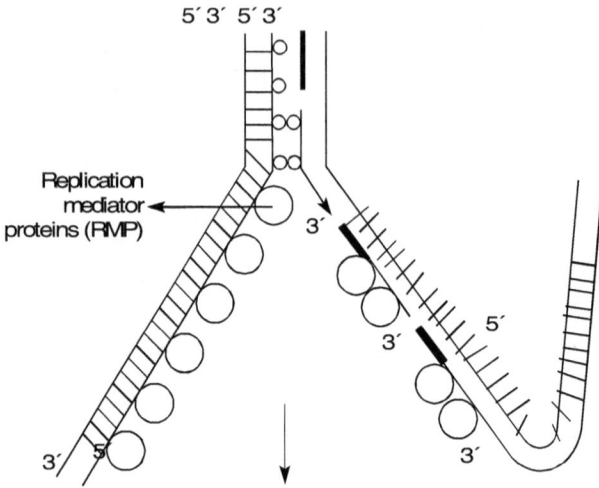

o) Removal of primers and nick sealing

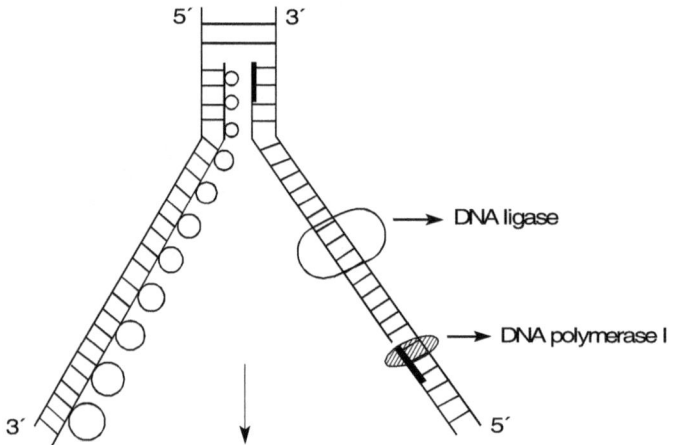

p) Termination process

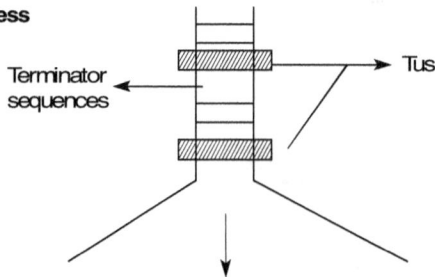

Figure 9.6   Schematic representation of DNA replication in *E. coli*

(Continues)

q) End copying and separating two duplex daughter strands

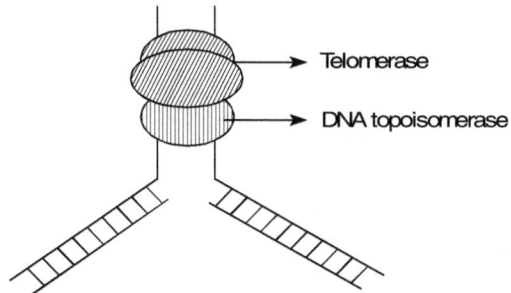

Figure 9.6   Schematic representation of DNA replication in *E. coli*

9.6.2   Elongation

Two important events occur in the elongation stage while the replication fork is progressing. DNA polymerase synthesizes complementary strands continuously in the leading strand and discontinuously in the lagging strand. This happens only when short primers are added, one to the leading strand and one for every segment of the lagging strand.

The prepriming complex attracts **primase** enzyme and becomes **primosome**. This primosome initiates replication of the leading strand by synthesizing short RNA primer (15–30 bases); to this, **DNA polymerase III** adds deoxyribonucleotides. As the replication complex approaches, the single-strand binding protein detaches from the template and **replication mediator proteins** (RMPs) bind to facilitate the copying. The replication processes proceed continuously in the leading strand.

Synthesis on the lagging strand begins when 1000–2000 nucelotides have been synthesized in the leading strand. In *E. coli*, synthesis of the leading and lagging strands may actually be coupled. The DNA polymerase III that is synthesizing leading strand is a dimeric unit (two copies) with a single gamma complex to interact with the beta subunit to control the attachment and removal of the enzyme from templates. The dimeric enzyme attaches to both the parent strands and moves in the same direction (3′ to 5′), for which the lagging strand forms a loop. The total complex containing dimer of DNA polymerase and the primosome, moving along the parent DNA during replication is called **replisome**. Once the replication is over, **DNA polymerase I** (having 5′ to 3′ exonuclease activity) proofreads and removes the RNA primer in both the leading and lagging strands. The primer is replaced with deoxynucleotides. The **DNA ligase** seals the single strand nick between two okazaki fragments (3′ hydroxyl end of one DNA fragment and 5′ phosphate end of another fragment) and forms a phosphodiester bond.

9.6.3   Termination

Normally there are two replication forks which begin from *oriC* region and meet at the other side of the circular *E. coli* genome during completion. In this end region seven **terminator sequences** have been identified in the *E. coli* genome. The **Tus protein** binds to this region and terminates the replication by blocking the passage of DNA B helicase, which is responsible for progression of the replication fork. Termination is followed by disassembly of the replisome. Further the extreme end of the lagging strand (3' region) is copied and sealed by the **telomerase** enzyme. Two interlinked daughter molecules are separated by **topoisomerase IV**.

FOR ADDITIONAL READING

1. Bielinsky, A.K. and Gerbi, S.A. (1998). "Discrete start sites for DNA synthesis in the yeast ARSI origin." *Science.* 279:95–98.

2. Bochkarev, A., Pfuetzner, R.A., Edwards, A.M. and Frappier, L. (1997). "Structure of the single-stranded DNA-binding domain of replication protein A bound to DNA." *Nature.* 385:176–181.

3. Champoux, J.J. (2001). "DNA topoisomerases: structure, function, and mechanism." *Ann. Rev. Biochem.* 70:369–413.

4. Cook, P. (1998). "Duplicating a tangled genome." *Science.* 281:1466–1467.

5. Cook, P.R. (1999). "The organization of replication and transcription." *Science.* 284:1790–1795.

6. Deshpande, A.M. and Newlon, C.S. (1996). "DNA replication fork pause sites dependent on transcription." *Science.* 272:1030–1033.

7. Diffley, J.F.X. and Cocker, J.H. (1992). "Protein-DNA interactions at a yeast replication origin." *Nature.* 357:169–172.

8. Falaschi, A. (2000). "Eukaryotic DNA replication: a model for a foxed double replisome." *Trends Genet.* 16:88–92.

9. Fangman, W.L. and Brewer, B.J. (1992). "A question of time—replication origins of eukaryotic chromosomes." *Cell.* 71:363–366.

10. Holmes, F.L. (1998). The DNA replication problem, 1953–1958. *Trends Biochem. Sci.* 23:117–120.

11. Kamada, K., Horiuchi, T., Ohsumi, K., Shimamoto, N. and Morikawa, K. (1996). "Structure of a replicator–terminator protein complexed with DNA." *Nature.* 383:598–603.

12. Kipling, D. and Faragher, R.G.A. (1999). "Ageing hard or hardly ageing?" *Nature.* 398:191–193.

13. Lima, C.D., Wang, J.C. and Mondragon, A. (1994). "Three-dimensional structure of the 67K N-terminal fragments of *E. coli* DNA topoisomerase I." *Nature.* 367:138–146.

14. Meselson, M. and Stahl, F. (1958). "The replication of DNA in *Escherichia coli.*" *Proc. Natl. Acad. Sci.*, USA. 44:671–682.

15. Myllykallio, H., Lopez-Garcia, P., *et al.* (2000). "Bacterial mode of replication with eukaryotic-like machinery in a hyperthermophilic archaeon." *Science.* 288:2212–2215.

16. Okazaki, T. and Okazaki, R. (1969). "Mechanisms of DNA chain growth." *Proc. Natl Acad. Sci.*, USA. 64:1242–1248.

17. Stewart, L., Redinbo, M.R., Qiu, X., Hol, W.G.J. and Champoux, J.J. (1998). "A model for the mechanism of human topoisomerase I." *Science.* 179: 1534–1541.

18. Takahashi, K. and Yanagida, M. (2000). "Replication meets cohesion." *Science.* 289:735–736.

19. Tye, B.K. (1999). "MCM proteins in DNA replication." *Ann. Rev. Biochem.* 68:649–686.

20. Waga, S. and Stillman, B. (1998). "The DNA replication fork in eukaryotic cells." *Ann. Rev. Biochem.* 67:721–751.

21. Wei, X., Samarabandu, J., Devdhar, R.S., Siegel, A.J., Acharya, R. and Berezney, R. (1998). "Segregation of transcription and replication sites into higher order domains." *Science.* 281:1502–1505.

22. Benkovic, S.J., Valentine, A.M. and Salinas, F. (2001). "Replisome mediated DNA replication." *Ann. Rev. Biochem.* 70:181–208.

23. DePamphilis, M.L. (ed.) (1996). *DNA Replication in Eukaryotic cells.* Cold Spring Harbor Laboratory Press, Cold Spring Harbor, NY.

24. Kelly, T.J. and Brown, G.W. (2000). "Regulation of chromosome replication." *Ann. Rev. Biochem.* 69:829–880.

MODEL QUESTIONS

1. Distinguish between conservative, semiconservative and dispersive types of DNA replication.

2. With illustration explain the conclusions that can be drawn from the results of Meselson–Stahl experiment.

3. Discuss in detail DNA topoisomerase activity in DNA replication.

4. With the aid of diagrams, explain displacement replication and rolling circle replication.

5. Discuss in detail the role of various replication origin regions.

6. What impact does the inability of DNA polymerases to synthesize DNA in the $3' \longrightarrow 5'$ direction have on DNA replication?

7. How is DNA replication primed in *Escherichia coli* and in eukaryotes?

8. Give a detailed description of the events in *Escherichia coli* replication.

9. In what ways does the eukaryotic replication fork differ from that of *Escherichia coli*.

10. Outline how replication terminates in *Escherichia coli*. What is currently known about the termination of replication in eukaryotes?

11. Describe how genome replication is regulated during S phase in eukaryotes.

10

10.1 INTRODUCTION

Transcription is a process by which the genetic information at DNA level is transferred to RNA level (synthesis of messenger RNA from the sense strand of double-stranded DNA) for their expression. This mechanism takes place inside the nucleus (except mitochondrial genome). Unlike replication, in transcription processes, only a particular gene or group of genes is transcribed. Following three major kinds of RNA are produced. **Messenger RNA** (mRNA) carries the sequences which code for amino acid sequence of a functional polypeptide chain; **transfer RNA** (tRNA) is an adapter that reads the information encoded in the mRNA and **ribosomal RNA** (rRNA) molecules associate with proteins to form the protein synthetic machinery.

The following components are required for trancription

- RNA polymerase
- Four different ribonucleoside 5′-triphosphates
- Mg^{2+}

10.2   RNA POLYMERASE

RNA polymerase is a large and complex enzyme system. Its molecular weight is 390,000 daltons. There are five core subunits (holoenzyme) and a sigma (σ) factor present in the complex (core enzyme). This sigma factor allows the RNA polymerase to recognize promoter region more effectively and quickly. The promoter region consists of two consensus sequences in the upstream regions such as 5´-TTGACA-3´ (-35 box) and 5´-TATAAT-3´ (-10 box). The spacing between the two boxes facilitates DNA binding status of RNA polymerase.

RNA polymerase requires DNA for activity and is most active with double-stranded DNA. Opening of the helix is due to DNA binding of the RNA polymerase (core enzyme) in the non-template strand (-10 box). Only one of the two strands is used as template (-ve strand or minus strand) and the other strand is a non-template strand (+ve strand or plus strand). They are copied in the 3´ →5´ direction. According to the base-pairing rule the uridylate residue is inserted in the RNA transcript against adenylate residue of the DNA template. The adenylate is inserted against thymidylate and guanylate for cytidylate (vice versa) in the growing RNA strand. There are three kinds of RNA polymerase in eukaryotes, viz. RNA polymerase I, RNA polymerase II and RNA polymerase III.

- RNA polymerase I   synthesizes ribosomal RNA (rRNA) and is insensitive to alpha-amanitin.
- RNA polymerase II   synthesizes messenger RNA (mRNA) and is extremely sensitive to alpha-amanitin.
- RNA polymerase III   synthesizes tRNA and 5S rRNA and is sensitive to alpha-amanitin, but only at high levels.

Four different ribonucleotides are added to the new growing RNA strand as per the base-pairing rules. During each ribonucleotide addition, the beta and gamma phosphates are removed from the incoming nucleotide and the hydroxyl group is removed from the 3´-carbon of the end nucleotide. The RNA polymerase covers about 30 base pairs of the template strand including transcription bubble. It also holds the template as well as growing RNA tightly till termination. Transcription termination occurs in bacteria either by **intrinsic terminators** (inverted palindrome, hairpin loop) or by **rho-dependent factor.** The interaction between RNA–DNA hybrid is weakened more by the A–U base-pairing at the end so that termination is favoured. Rho is a helicase, which actively breaks the base pair between the template and transcript. This results in termination of transcription.

10.3 REVERSE TRANSCRIPTASE (RNA-DEPENDENT DNA POLYMERASE)

Retroviruses contain an enzyme called **reverse transcriptase**. It catalyses the synthesis of complementary DNA strand using a single strand of viral RNA (as the template). It also degrades the RNA strand in the RNA–DNA hybrid and replaces it with DNA. The duplex DNA formed in this way often becomes incorporated into the eukaryotic genome. Reverse transcriptase has become the most important enzyme in constructing complementary DNA (cDNA) library and RT-PCR reactions.

10.4 RNA REPLICASE (RNA-DIRECTED RNA POLYMERASE)

Certain bacteriophages such as f2, Ms2, Qb and R17 have RNA as their genome. RNA replicase enzyme catalyses the synthesis of mRNA using their single-stranded RNA (template). RNA replicase has four subunits and its molecular weight is about 210,000 dalton.

10.5 POST-TRANSCRIPTIONAL MODIFICATIONS

Almost all of the bacterial mRNA do not undergo any significant modification after synthesis. In most of the cases transcription is coupled with translation processes. However the mRNAs of eukaryotes undergo various post-transcriptional modifications such as **capping, tailing** and **intron splicing**.

10.5.1 Capping

The **guanylyl transferase** adds an extra guanosine to the extreme 5´ end of the RNA transcript. Further to this, **guanine methyltransferase** converts new terminal guanosine into 7´-methylguanosine by attachment of a methyl group to the 7th position.

10.5.2 Tailing

All eukaryotic mRNAs have a series of adenosines, about 250 numbers at the 3´ends. This reaction is catalysed by polyadenylate polymerase enzyme.

10.5.3   Splicing

Introns are the intervening sequences present in between (or sometimes within) exons (coding regions). The introns are removed from primary transcript (pre-mRNA or hnRNA) by a mechanism called **splicing**. Table 10.1 shows different genes and their introns.

Table 10.1   Genes and their introns

Genes	Number of introns
Insulin	2
Serum albumin	13
Factor VIII	25
Dystrophin	78
Type IV collagen	117

There are two schools of thought on the origin of introns. They are

- *Introns late* which are introns accumulated in the eukaryotic genome and evolved recently.
- *Introns early* which are introns which are ancient and which have been lost in most prokaryotic genome.

The phylogenetic relationships among major groups of introns are presented here. **Self-splicing** tRNA introns arose early (around 3500 million years ago), whereas **spliceosomal** introns arose much more recently (about 1000 million years ago). The **protein-spliced** introns that are restricted to nuclear tRNA and to the tRNA and rRNA genes of archaebacteria may be of intermediate antiquity (about 1700 million years ago).

Some of the prokaryotes that have introns in their genome are

1. Sulphur-dependent anaerobic bacteria
   Desulfurococcus mobilis) - 23S rRNA gene
   Sulfolobus solfatariccus - tRNA^{Leu} and tRNA^{Ser} genes
2. *Halobacterium volcanii* - tRNA^{Trp} gene
3. T4phage thymidylate synthase gene

The location of introns between protein structural domains facilitates recombination within those regions and assorts domain functions independently. This reduces the problem of folding of the new protein made easier. This kind of exon shuffling is an important feature in protein evolution.

There are eight classes of introns. They are

1. Nuclear pre-mRNA introns of the GU-AG class
2. Nuclear pre-mRNA introns of the AU-AC class
3. Self-splicing group 1 introns
4. Pre-tRNA gene introns of eukaryotes
5. Group II introns of fungi, plants and prokaryotes
6. Group III introns
7. Twintrons
8. Archaeal introns

Introns begin with GU and end with AG. In between these two doublet codons, a pyrimidine-rich consensus sequence is present, which gives signal for splicing.

The introns fall under three major categories. The Group 1 introns interrupt between the codons. Most of the genes of this type undergo autosplicing reaction which requires formation of a specific secondary structure involving four short consensus sequences. The guanosine residue acts as a cofactor for transesterification reaction without any input of energy. The guanosine breaks the bond at left 5´ end of the intron. The hydroxyl at the free end of the exon then attacks the right 3´ intron–exon junction. The intron loses the guanosine and the terminal 15 bases by circularizing. Although autosplicing occurs *in vitro*, the *in vivo* splicing reaction requires participation of some proteins. The virusoid RNA can undergo self-cleavage at a "hammer head" structure, where RNase P ribonucleoprotein does the catalytic activity.

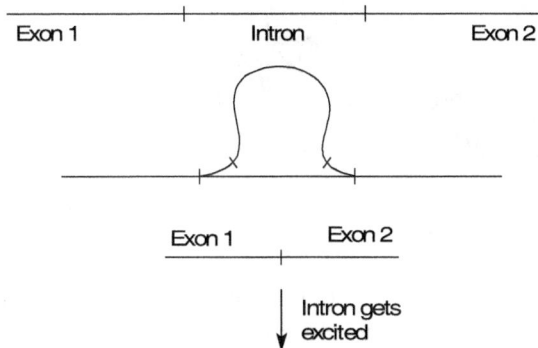

Figure 10.1   Intron looping and excision

In eukaryotes, some group 1 and group 2 (intron that follows GU-AG rule) mitochondrial introns code for a regulatory protein called maturase, which is involved in splicing. The proteins coded by some group 1 introns are endonucleases that create double-stranded cleavage in target sites in DNA; the cleavage initiates gene conversion process in which the sequence of the intron itself is copied into the target site.

Some proteins have both maturase and endonuclease activity. But in yeast tRNA, splicing occurs in two steps. In the first step the endonuclease recognizes the secondary structure of the precursor and cleaves both the ends of the intron. The next step involves ligation in the presence of ATP (Figure 10.1).

Introns are removed from the nuclear RNAs of higher eukaryotes by recognizing short consensus sequences conserved at exon–intron boundaries and within the left 5´ splicing junction. This involves formation of a lariat loop by joining the GT end of the intron, through a 5´ - 2´ linkage to the adenine base at position 6 of the branch. Then the right 3´ splicing junction is cut and the two exons are ligated. The yeast consensus sequence TACTAAC has catalytic action similar to that of mammalian nuclear RNAs. There are two types of systems by which consensus sequences are recognized;

1. enzyme recognizing the consensus sequence or
2. enzyme recognizing the secondary structure formed by the pairing of RNA bases.

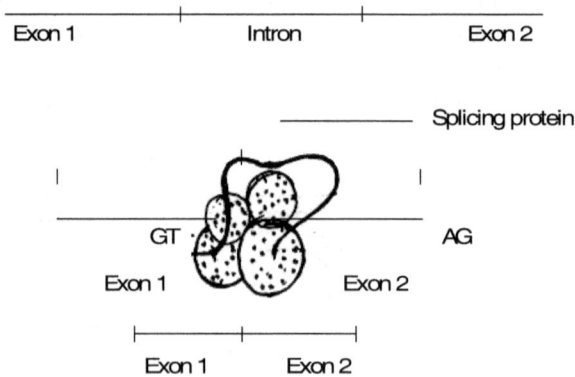

Figure 10.2   Spliceosome

Splicing in nuclear RNA requires small ribonucleoproteins (snRNPs) and other components that generate the spliceosome. This spliceosome recognizes the consensus sequence and makes them into a reactive conformation. Splicing is usually intramolecular (*cis*-splicing), but some cases of intermolecular splicing (*trans*-splicing) have also been reported. For example, in the yeast, *Saccharomyces cerevisiae*, the intron sequences are recognized, aligned and catalysed by pre-mRNA transacting factors such as four small ribonucleoprotein particles (snRNPs) U1, U2, U4, U6 and U5, and proteins (PRP 5, PRP 2, PRP 16, etc.). ATP is required at least for three steps in this pathway (Figure 10.2).

FOR ADDITIONAL READING

1. Geiduschek, E.P. and Kassavetis, G.A. (2001). "The RNA polymerase III transcription apparatus." *J. Mol. Biol.* 310:1–26.

2. Kornberg, R.D. (1999). "Eukaryotic transcriptional control." *Trends. Genet.* 15:M46–M49.

3. Lilley, D.M.J. (eds.). (1995). *DNA–protein. Structural Interactions.* IRL Press, Oxford.

4. Alberts, B., Bray, D., Lewis, J., Raff, M., Roberts, K. and Watson, J.D. (1994). *Molecular Biology of the Cell*, 3rd edn. Garland Publishing, New York.

5. Ambros, V. (2001). "Dicing up RNAs." *Science.* 293:811–813.

6. Bonen, L. and Vogel, J. (2001). "The ins and outs of group II introns." *Trends Genet.* 17, 322–331.

7. Conaway, J.W. and Conaway, R.C. (1999). "Transcription elongation and human disease." *Ann. Rev. Biochem.* 68:301–309.

8. Conaway, J.W., Shilatifard, A., Dvir, A. and Conaway, R.C. (2000). "Control of elongation by RNA polymerase II." *Trends Biochem. Sci.* 25:375–380.

9. Copertino, D.W. and Hallick, R.B. (1993). "Group II and Group III introns of twintrons: potential relationships with nuclear pre-mRNA introns." *Trends Biochem. Sc.* 18:467–471.

10. Doherty, E.A. and Doudna, J.A. (2000). "Ribozyme structures and mechanisms." *Ann. Rev. Biochem.* 69:597–615.

11. Klug, A. (2001). "A marvelous machine for making messages." *Science.* 292:1844–1846.

12. Lee, T.I. and Young, R.A. (2000). "Transcription of eukaryotic protein coding genes." *Ann. Rev. Genet.* 34:77–137.

13. Newman, A. (2001). "RNA enzymes for RNA splicing." *Nature.* 413, 695–696.

14. Nilsen, T.W. (1996). "A parallel spliceosome." *Science.* 273, 1813.

15. Scott, J. (1997). "RNA editing: message change for a fat controller." *Nature.* 387:242–243.

16. Strachan, T. and Read, A.P. (1999). *Human Molecular Genetics*, 2nd edn. BIOS Scientific Publishers, Oxford.

17. Tollervey, D. (1996). "Small nucleolar RNAs guide ribosomal RNA methylation." *Science.* 273:1056–1057.

18. Turner, P.C., McLennan, A.G., Bates, A.D. and White, M.R.H. (1997). *Instant notes in Molecular Biology.* BIOS Scientific Publishers, Oxford.

19. Wente, S.R. (2000). "Gatekeepers of the nucleus." *Science.* 288:1374–1377.

20. Cech, T.R. (1990). "Self-splicing of group I introns." *Ann. Rev. Biochem.* 59:543–568.

21. Geiduschek, E.P. and Kassavetis, G.A. (2001). "The RNA polymerase III Transcription apparatus." *J. Mol. Biol.* 310:1–26.

22. Gott, J.M. and Emeson, R.B. (2000). "Functions and mechanisms of RNA editing. "*Ann. Rev. Genet.* 34:499–531.

23. Losick, R.L. and Sonenshein, A.L. (2001). "Turning gene regulation on its head." *Science.* 293:2018–2019.

24. Maxwell, E.S. and Fournier, M.J. (1995). "The small nucleolar RNAs." *Ann. Rev. Biochem.* 64:897–934.

25. Venema, J. and Tollervey, D. (1999). "Ribosome synthesis in *Saccharomyces cerevisiae.*" *Ann. Rev. Genet.* 33:261–311.

MODEL QUESTIONS

1. Describe how transcription is initiated and terminated in bacteria.

2. Outline the significant features of the elongation phase.

3. Discuss various steps involved in capping of mRNA.

4. Discuss in detail the splicing mechanism.

5. Classify introns based on their existence in the gene.

6. Comment on spliceosome.

7. Contrast different types of RNA polymerases.

8. What is reverse transcriptase? Comment on its functions.

9. What is RNA replicase? Discuss with examples.

10. How does transcription occur in eukaryotic genome?

TRANSLATION AND GENETIC CODE

11

11.1 INTRODUCTION

Translation is the process in which the mRNA is decoded (translated) into amino acids of the functional polypeptide chain. This process occurs in the cell cytoplasm (eukaryotes). The main source of energy is ATP which provides energy for activation of amino acids and GTP which provides the energy for protein synthesis.

11.2 TRANSFER RIBONUCLEIC ACID (tRNA)

Transfer RNAs are a group of small RNAs synthesized in the nucleus, transferred to the cytoplasm and get charged in the cytoplasm. Each tRNA contains 75 to 85 nucleotides and serves as a molecular adapter during protein synthesis. Its molecular weight is found to be between 2400 and 3100 Dalton. There is at least one kind of tRNA for each amino acid and some amino acids like serine and proline have two or more specific tRNAs. In most prokaryotes, tRNA has a cloverleaf structure. Each tRNA has five different arms. They are the acceptor arm (seven base pairs between the 5´ and 3´ end, having CCA terminal sequence where amino acids are attached), D-arm, anticodon arm (this base-pairs with mRNA) and V loop and pseudouridine arm (T Ψ C) (folds the tRNA into an L-shaped structure). The aminoacyl-tRNA-synthetases charge tRNA with their specific amino acids. There are two different groups of aminoacyl-tRNA-synthetases, viz. class-1 (charges amino acids to the 2´-OH of the terminal nucleotide) and class-2 (charges amino acid to the 3´-OH of the terminal nucleotide).

11.3   RIBOSOMES

The ribosome is the main site for protein synthesis. It is a spheroidal particle consisting of a large and a small subunit. In prokaryotes (*E. coli*), the intact particle has a sedimentation coefficient of 70S (S = Svedberg unit) where the small subunit is 30S and the large subunit is 50S. Each 30S subunit contains one 16S ribosomal RNA (rRNA) molecule and 21 different proteins. Each 50S subunit contains one 5S rRNA molecule, one 23S rRNA molecule and 32 different proteins. In eukaryotes, the intact particle is 80S and dissociates to give 40S small subunit and 60S large subunit. Each 40S subunit contains one 18S rRNA and 30 proteins. Each 60S subunit contains 28S rRNA, 5.8S rRNA, 5S rRNA and 40 proteins. Each subunit contains two transfer-RNA-binding sites—A-site (aminoacyl site) and P-site (peptidyl site). During protein synthesis more ribosomes attach to a single messenger RNA molecule and form a polysome. It also protects the mRNA during protein synthesis.

11.4   STEPS INVOLVED IN PROTEIN SYNTHESIS

In prokaryotes *(E. coli)* there are 5 major steps involved in protein synthesis.

    Step 1   Activation of amino acids and tRNA charging
    Step 2   Initiation
    Step 3   Elongation
    Step 4   Termination
    Step 5   Post-translational modification

Step 1   Activation of Amino acids and Charging of tRNA

The components required in this step are the following.

* 20 types of amino acids

- 20 different types of aminoacyl-tRNA synthetase
- 20 or more tRNAs
- ATP as energy source
- Mg^{2+}

Mechanism

- Activation of amino acid takes place in the cytosol.
- Each of 20 different amino acids is esterified by aminoacyl-tRNA synthetase to one or more corresponding tRNA at the amino acid accepting site. This occurs in two steps.

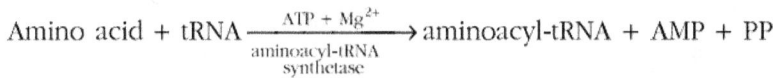

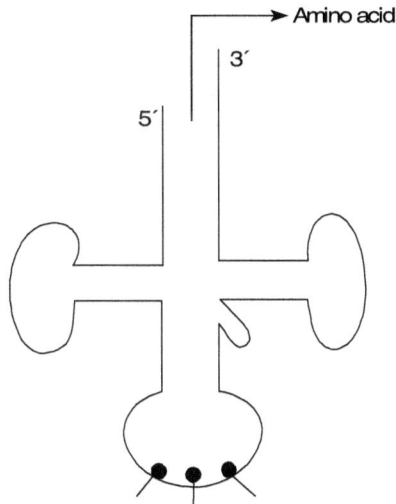

$$\text{Amino acid} + \text{tRNA} \xrightarrow[\substack{\text{aminoacyl-tRNA} \\ \text{synthetase}}]{\text{ATP} + Mg^{2+}} \text{aminoacyl-tRNA} + \text{AMP} + \text{PP}$$

Figure 11.1   Charging of the tRNA

First, AMP is added to the amino acid (adenylation) by cleavage of one ATP as shown in Figure 11.1. Then the AMP is used to catalyse the addition of the amino acid onto one or more corresponding tRNA as shown in Figure 11.1. Both steps are carried out by aminoacyl-tRNA synthetase with high specificity to the process to prevent mispairing.

Step 2   Initiation

The components required for initiation are

- 50S and 30S ribosomal subunits

- mRNA
- Initiation factors (IF1, IF2 and IF3)
- N-formylmethionine tRNA
- GTP
- Mg^{2+}

Mechanism

- A pyrimidine-rich sequence near the 3′ end of the 16S RNA of the 30S subunit is complementing to the **Shine–Dalgarno sequence** (also known as ribosome-binding site) of the mRNA in the 5′ end.

- This interaction fixes the initiator codon AUG to the precise location (P-site) on the 30S subunit.

- The 30S subunit binds to the initiation factor 3 (IF3) to prevent 30S and 50S subunits from combining.

- The complex consisting of 30S subunit, IF3 and mRNA binds to IF2–GTP-fmet-tRNA^{met} complex. The codon–anticodon interaction involving the initiating AUG is fixed in the P-site as shown in Figure 11.2.

- The above large complex binds with the 50S ribosomal subunit with the help of IF1. Simultaneously GTP is hydrolysed to GDP and Pi. Now IF3 and IF2 are released from the ribosomal subunits. This results in the functional initiation complex and completion of the initiation phase.

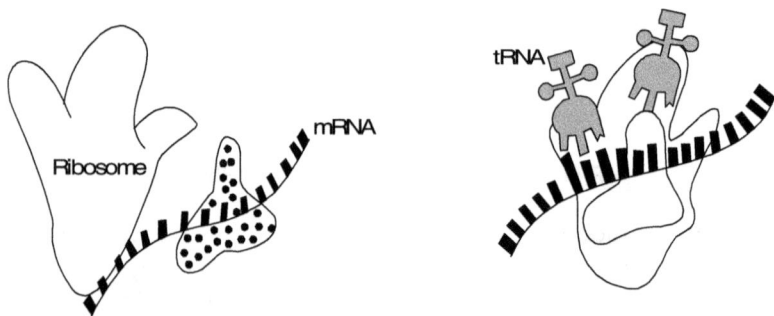

Figure 11.2   Initiation

Step 3   Elongation

The components required for the process of elongation are

- Functional ribosomal subunits

- Different types of charged tRNAs
- Elongation factors (EF-Tu, EF-Ts, EF-G)
- Peptidyl transferase
- GTP
- Mg^{2+}

Mechanism

- The charged tRNA is bound to EF-Tu–GTP complex to form a large complex (tRNA–EF-Tu–GTP). Then tRNA (bearing anticodon against mRNA) is attached in the A-site of the functional ribosomal complex using energy received from GTP hydrolysis. The Ef-Tu–GDP complex is released and further recycled by EF-Ts.

- A **peptide bond** is formed between the amino acid bound in the tRNA of the A-site and amino acid attached to the tRNA of P-site and results in dipeptidyl-tRNA found in the A-site and uncharged tRNA in the P-site (Figure 11.3).

- Now the ribosome moves one codon towards the 3´ end of the mRNA so that the second codon of mRNA is in the P-site and third codon in the A-site. Also, the dipeptide-tRNA in the A-site moves to the P-site. The deacylated tRNA in the P-site moves to the exit site or gets ejected from the ribosome. The movement requires EF-G (translocase) and energy (from the hydrolysis of GTP). The elongation cycle continues till the A-site receives terminator codon.

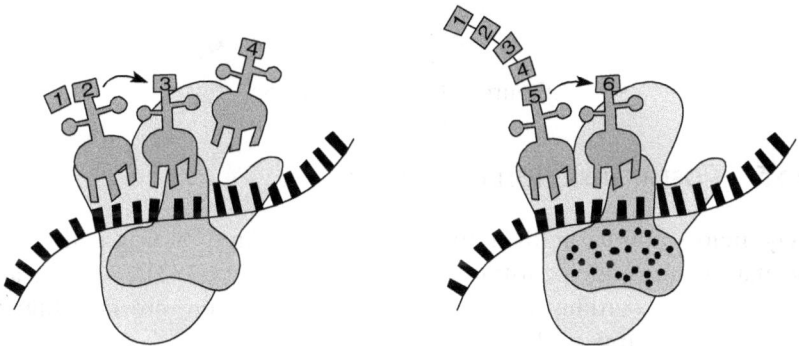

Figure 11.3   Elongation

Step 4   Termination

The components required for the process of termination are

- Terminator codon in mRNA (UAA, UAG, UGA)

- Release factors (RF1, RF2, RF3)
- ATP, GTP

Mechanism

- Termination of polypeptide synthesis is signalled by one of the following terminator codons, viz. UAA, UAG, UGA. There are no charged tRNAs having anticodon that is capable of recognizing such termination signal. In prokaryotes, releasing factors RF1 and RF2 are capable of recognizing a termination signal residue in the A-site (Figure 11.4).

- RF1 recognizes termination codon 5´-UAG-3´ and 5´-UAA-3´, whereas RF2 recognizes 5´-UGA-3´ and 5´-UAA-3´. The RF3 stimulates the release of RF-1 and RF-2 from the ribosome after termination. The ribosome recycling factor (RRF) may enter the P- or A-site and disassemble the ribosome using GTP as the source of energy.

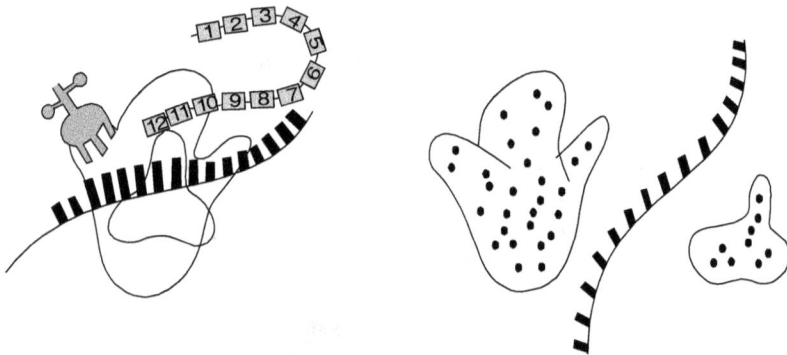

Figure 11.4 Termination

Step 5   Post-translational Modification

Newly made amino acid chains attain their final biologically active confirmation through post-translational modifications. The modification pattern varies according to the kind of proteins. There are four kinds of processing that happen, they are:

- Protein folding—tertiary structure formation
- Proteolytic cleavage—chain is cut short by proteases
- Chemical modification—individual amino acids are modified
- Intron splicing—intervening amino acid (extra amino acids) sequences are removed

Some of the important chemical processes which take place during post-translation are

- acetylation of amino terminal residue
- loss of signal sequence
- removal or modification of carboxy terminal residue
- modification of individual amino acids
- attachment of carbohydrate side chains
- formation of disulphide cross links
- addition of isoprenyl groups in cystein residues

11.5   GENETIC CODE

It is interesting to understand how the four letters of DNA (A,T,G and C) group to code for an amino acid.

Codon is a triplet of nucleotides (three-letter symbol) coding for a specific amino acid. The DNA alphabets (A, T, G and C) in groups of three yield 64 different combinations (codons). Genetic experiments conclusively proved the following important characteristics of the genetic codes for amino acids.

Triplet   A single amino acid is represented in the form of three nucleotides in mRNA. Hence the codon is a triplet.

Universal   Genetic code is universal. A code for an amino acid remains the same for all plants and animals.

Commaless   There is no comma or punctuation between adjacent codons. Each codon is followed by the next codon leaving no space between them.

Linear arrangement   The codons are linearly arranged in the mRNA.

Degeneracy   Some amino acids are represented by more than one codon. For example, leucine and serine have six codons, glycine and alanine have four and glutamate, tyrosine, and histidine have two. Only methionine and tryptophan have single codons. Degeneracy is due to the lack of specificity in the third base of codon. This third base is called a wobble base. The first base of the anticodon present in tRNA is formed of abnormal bases (inosinate I). These abnormal bases are able to base-pair with any base (U, C and A) present in the third place of codon. This phenomenon is called wobble hypothesis.

Initiation codon   AUG codon signals the beginning of all polypeptide chains in both prokaryotes and eukaryotes. It also codes for methionine residue in the internal position of the polypeptide.

Termination codon   UAA, UAG and UGA are termination codons. They do not represent any known amino acids but signal the end of the polypeptide chain synthesis. They are also called nonsense codons. These nonsense codons are named amber, ochre and opal respectively.

11.6   INHIBITORS

Table 11.1   Inhibitors of replication, transcription and protein synthesis

Antibiotics/ inhibitors	Action
Streptomycin	Inhibits initiation and causes misreading of mRNA
Tetracycline	Binds to the 30S subunit and prevents binding of aminoacyl-tRNA to the A-site
Chloramphenicol	Inhibits the peptidyl transferase activity of the 50S ribosomal subunit during elongation
Cycloheximide	Inhibits the peptidyl transferase activity of the 60S ribosomal subunit
Erythromycin	Binds to the 50S subunit and inhibits translocation
Puromycin	Causes premature chain termination by acting as an analog of aminoacyl-tRNA (prokaryotes and eukaryotes)
Rifampicin	Binds to β subunit of RNA polymerase and inhibits first phosphodiester bond in the RNA chain
Actinomycin D	Inhibits RNA polymerase
Nalidixic Acid	Inhibitor for bacterial DNA gyrase
Macrolides	Bind to 50S ribosome and prevents translocation
Aminoglycosides	Bind to 30S ribosome and cause depletion of ribosomal pool, and misreading of mRNA
Thiostrepton	Prevents binding of EF-G–GTP complex to ribosome and inhibits elongation process
Kirromycin	Prevents dissociation of EF-Tu–GDP complex from ribosome and inhibits elongation process
Trimethoprim	Inhibits synthesis of N^{10}-Formyl-H_4 folate, thereby preventing formation of fMet-tRNA

FOR ADDITIONAL READING

1. Adams, R.L.P., Burdon, R.H., Campbell, A.M., Leader, D.P. and Smellie, R.M.S. (1981). *The Biochemistry of the Nucleic Acids.* Champman and Hall Publication, London.

2. Berg, J.M., Tymoczko, J.L., Stryer, L. (2002). Biochemistry, 5th edn., W.H. Freeman and Company, New York.

3. Brown, T.A. (2002). *Genomes,* 2nd edn., John Wiley and Sons Inc., New York.

4. Buratowski, S. (1994). "The basics of basal transcription by RNA polymerase II." *Cell.* 77:1–3.

5. Kornberg, A. and Baker, T. (1992). *DNA Replication,* 2nd edn., W.H. Freeman and Company, New York.

6. Lehninger, A.L., Nelson, D.L. and Cox, M.M. (1993). *Principles of Biochemistry.* BS Publishers, India.

7. Lewin, B. (2000). *Genes VII.* Oxford University Press, Oxford.

8. Lodish, H., Baltimore, D., Berk, A., Zipursky, S.L., Matsudaira, P. and Darnell, J. (1995). *Molecular Cell Biology,* 3rd edn. Scientific American Books Inc., New York.

9. Murray, R.K., Granner, D.K., Mayes, P.A. and Rodwell, V.W. (2000). *Harper's Biochemistry.* 25th edn., McGraw-Hill, USA.

10. Sinden, R.R. (1996). *DNA Structure and Function.* Academic Press, New York.

11. Voet, D., Voet, J.G., Pratt, C.W. (1999). *Fundamentals of Biochemistry.* John Wiley & Sons Inc., New York.

12. Watson, J.D., Hopkins, N., Roberts, J., Streitz, J.A. and Weiner, A. (1987). *Molecular Biology of the Gene,* 4th edn. Benjamin Cummings, California.

13. Zubay, G.L., Parson, W.W., Vance, D.E. (2001). *Principles of Biochemistry,* Wm C. Brown Publishers, England.

14. Crick, F.H.C. (1966). "Codon–anticodon pairing: The wobble hypothesis." *J. Mol. Biol.* 19:548–555.

15. Crick, F.H.C., Brenner, S., and Watts-Tobin, R.J. (1961). "General Nature of the genetic code for proteins." *Nature.* 192:1227–1232.

16. Dingwall, C., and Laskey, R.A. (1986). "Protein import into the cell nucleus." *Annu. Rev. Cell Biol.* 2:367–390.

17. Garen, A. (1968). "Sense and nonsense in the genetic code." *Science.* 160:149–159.

18. Khorana, H.G. (1966–67). "Polynucleotide synthesis and the genetic code." *Harvey Lect.* 62:79–105.

19. Brenner, S., Jacob, F., and Meselson, M. (1961). "An unstable intermediate carrying information from genes to ribosomes for protein synthesis." *Nature.* 190:576–581.

20. Jackson, R.J., and Standart, N. (1990). "Do the poly(A) tail and 3´ untranslated region control mRNA translation?" *Cell.* 62:15–24.

21. Kozak, M. (1983). "Comparison of initiation of protein synthesis in prokaryotes, eucaryotes, and organelles." *Microbiol. Rev.* 47:145.

22. Kozak, M. (1989). "Context effects and inefficient initiation at non-AUG codons in eukaryotic cell-free translation systems." *Mil. Cell. Biol.* 9:5073–5080.

23. McCarthy, J.E.G., and Brimacombe, R. (1994). "Prokaryotic translation: The interactive pathway leading to initiation." *Trends Genet.* 10:402–407.

24. Meyer, D.I. (1982). "The signal hypothesis: A working model." *Trends. Biochem. Sci.* 7:320–321.

25. Nirenberg, M., and Leder, P. (1964). "RNA code words and protein synthesis." *Science.* 145:1399–1407.

26. Silver, P.A. (1991). "How proteins enter the nucleus." *Cell.* 64:489–497.

27. Verner, K., and Schatz, G. (1988). "Protein translocation across membranes." *Science.* 241:1307–1313.

28. Watson, J.D. (1963). "The involvement of RNA in the synthesis of proteins." *Science.* 140:17–26.

29. Wimberly, B.T., Brodersen, D.E., Clemons, W.M., Morgan-Warren, R.J., Carter, A.P., Vonhein, C., Hartsch, T. and Ramakrishnan, V. (2000). "Structure of the 30S ribosomal subunit." *Nature.* 407:327–339.

MODEL QUESTIONS

1. Decode the following information and identify the aminoacids
   i.   AGAUUUAUACGCAGC   ii.   UAUUGUUCUUTUAUA

2. Write an elaborate account on tRNA.

3. What are the functions of ribosomes?

4. Explain the mechanism of initiation and elongation processes in translation.

5. Define translocation.

6. Why do proteins undergo post-translational modification?

7. What are the major chemical components found in DNA?

8. Is it correct to say that DNA is always the genetic material?

9. Define the following terms:
   i.   Replication   ii.   Transcription   iii.   Translation
10. What are the three principal structural differences between DNA and RNA?
11. Comment on the role of inhibitors.
12. List the various inhibitors of transcription and translation?

REGULATION OF GENE EXPRESSION IN MICROBES

12

12.1 INTRODUCTION

It is with the help of microbes such as bacteria and viruses that we have entered the modern era of molecular genetics. **Viruses** are non-living, obligate intracellular parasites that cannot make energy or proteins independently. They have either DNA or RNA as genetic material and their genome is packed into a protein coat (capsid) or a membrane (an envelope). The **viroids** are infectious small plant RNA molecules and function independently without being encapsidated by any protein coat. The **virusoids** (also called satellite RNAs) are infectious RNA but are encapsidated by plant viruses. They are packed together with a viral genome and replicate with the assistance of host virus. **Phages** are viruses that infect bacterium. **Bacteria** are free-living unicellular organisms that have a single copy of chromosome. If a bacterium can grow in a minimal medium then it is said to be a prototroph. If a bacterium requires organic substances for growth to occur, then it is termed an auxotroph. In culture, the bacterial growth can be represented in three phases as shown in Figure 12.1.

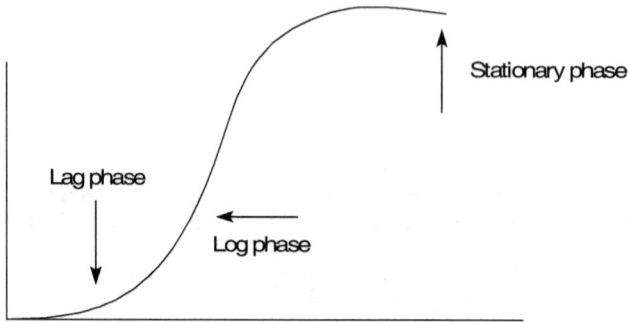

Figure 12.1   Bacterial growth curve

Lag phase   When bacteria are inoculated in a liquid medium, during initial hours slow growth occurs.

Log phase   A rapid growth phase in which cell density doubles repeatedly at an optimum temperature.

Stationary phase   When there is no further bacterial growth, the cell number becomes constant.

12.2   DNA TRANSFER BETWEEN BACTERIAL CELLS

The transfer of genetic information from one bacterium to the other can occur in any one of the following three methods—conjugation, transformation and transduction. The most important consequence of DNA transfer is that these processes spread antibiotic resistance genes from one bacterium to another.

Conjugation is the mating of two bacterial cells during which DNA is transferred from the donor to the recipient cell. The mating process is controlled by an **F plasmid** (fertility factor), which carries the genes for conjugation. The **sex pilus** (conjugation tube) is formed by pilin protein. Mating begins when the pilus of the donor male bacterium with F factor (F^+) attaches to the recipient female bacterium, which lacks F factor (F^-). Now both the cells have direct contact to each other through the pilus. One strand of the F factor DNA is transferred across the conjugal bridge into the recipient cell. Once the complementary strand synthesis to form a double-stranded F factor plasmid in the donor and recipient cells is accomplished, the conjugation process is complete. The recipient is now an F^+ male cell and is capable of transmitting the plasmid further. It is important to note that only the F factor has been transferred and not the bacterial chromosome.

Certain F⁺ bacterial cells have their F plasmid integrated into the bacterial DNA (chromosome). These cells acquire the capability of transferring the entire modified chromosome into another bacterium. These cells are called **Hfr (high-frequency recombination)** cells. During this transfer, a single long DNA strand having F factor at the leading end followed by the bacterial chromosome and then the remaining of the F factor enters the recipient F⁻ cell. The number of donor cell genes that are transferred also vary, since the F plasmid can integrate at several different sites in the bacterial DNA. This results in the transfer of only a portion of the donor chromosome, because the attachment between the two cells can break any time.

Transformation is the transfer of DNA itself from one cell to another. It occurs by any of the following two mechanims:

1. In nature, dying bacteria may release their DNA and some recipient cells may take this up.

2. In the laboratory, one can extract the DNA from one type of bacteria and introduce it into genetically different bacteria.

In 1944, it was shown that DNA extracted from smooth pneumococci bacteria could transform rough pneumococci into encapsulated smooth organisms (virulent). This was the first evidence that DNA was the genetic material.

Transduction is a method in which phage (as vector) carries and transfers host DNA into the bacterium. A piece of bacterial DNA is incorporated into the viral genome and this is carried into the recipient cell (host) at the time of infection. There are two types of transduction. The **generalized** type occurs when the virus carries a segment of the bacterial chromosome and incorporates it into the recipient. In the **specialized** type, the phage DNA integrates into the bacterial cell DNA, then excises it and carries it to an adjacent part of the cell DNA that is usually specific to that phage.

12.3   REPRODUCTION IN λ PHAGE

Bacteriophage λ is a temperate phage having medium-sized DNA of 48,502 base pairs. The DNA of the invading phage has two possible fates: either phage particles multiply within the host and lyse them for their release (lytic cycle) or phage integrates into the host chromosome and replicates passively with the host DNA for many generations (lysogeny). The option of entering into lytic and lysogenic cycles is governed by the interaction of five regulatory proteins namely CI, CII, Cro, N and Q. These proteins regulate a number of promoter regions of the phage DNA. The CI and Cro proteins are repressors and the CII protein is an activator. The N and Q proteins interact directly with the *E. coli* RNA polymerase and transcribes certain transcription termination sequences of the phage DNA for **antitermination activity.**

Selection of lysogenic cycle or lytic cycle by *E. coli* cells is partly decided based on the nutritional status. When the *E. coli* cells are growing in a rich medium, their proteases become more abundant. The CII proteins are subject to rapid degradation by their proteases. Absence of CII protein restricts CI protein production and *E. coli* cells are in favour of the lytic cycle. When *E. coli* cells are starved, the level of CII protein is elevated and this results in the production of CI protein that favours the lysogenic cycle.

12.3.1   Lytic Cycle

The regulatory mechanism begins when the phage infects *E. coli* and its DNA enters *E. coli*. RNA polymerase initiates transcription at two promoters, namely P_R and P_L. The mRNA synthesis is limited in both cases by transcription terminators. These terminators are designated as t_{R1} and t_{R2} for transcripts originating at P_R and t_L for transcripts originating at P_L. The mRNA from P_R includes the *cro* gene and the *cII* gene. The transcript from P_L includes the *N* gene. Once N protein is sufficiently synthesized, it triggers other genes in the cascade. The N protein interacts with RNA polymerase and modifies it to override the three termination signals such as t_L, t_{R1}, and t_{R2}. This leads to the production of long mRNA of the Q gene and genes for proteins required in viral replication. Once the Q proteins are produced, they interact with RNA polymerase and antagonize transcription termination at other sites. They also trigger transcription of genes for the structural proteins needed to assemble a virus particle. Once the new phages are assembled, the *E. coli* is lysed and the virus particles are freed.

Though the functions of N and Q proteins are similar, their mechanisms of action sometimes vary. The N protein binds to specific DNA sequences named *nutL* and *nutR*, which are located in the upstream of transcription terminators. When RNA polymerase reaches *nutL* and *nutR* sequences they are modified in a reaction requiring N and three host cell proteins. This makes the RNA polymerase to transcribe through many kinds of terminators. The Q protein interacts with RNA polymerase at a sequence where the transcription pauses shortly after initiation. It modifies the RNA polymerase in a reaction requiring Q and one other protein. The mechanism of N and Q action remains still obscure and paves the way for research in this new arena.

12.3.2   Lysogenic Cycle

The CII protein regulates the lysogenic cycle. The CII protein is an activator that stimulates transcription from P_{RE} and P_{int} promoters. The mRNA of P_{RE} contains the *cI* gene, which encodes for the CI protein. This CI protein is a repressor and is called the λ repressor. If this repressor protein is produced earlier, it will suppress virtually all other bacteriophage transcriptions except that originating at the P_{RM} promoter. The transcript from P_{int} includes INT and XIS protein genes required for the integration of viral DNA into the host chromosome through site-specific recombination. Once integration is over, the expression of the bacteriophage genes is repressed by CI protein. The CI protein is autoregulated (produced in higher concentration when needed to establish lysogeny and produced in lower concentration when required to maintain it). The bacteriophage genome is replicated only within the host chromosome.

The lysogenic cycle can continue for many cell generations unless interrupted by a DNA-damaging agent. The sudden reduction in the CI protein concentration triggers the phage particle to come off from the lysogenic state. Removal of CI protein in turn produces Cro protein. The Cro protein is also a repressor, but it antagonizes the activity of CI protein. The Cro protein binds to the same operators as CI protein and blocks further CI protein synthesis. This leads to the exit of the viral DNA from the bacterial chromosome and a lytic cycle.

12.4   REGULATION OF GENE EXPRESSION

Though there are many genes in the cell genome, only a few of them are expressed at any given time, based on the cellular need. The regulation of gene expression is highly essential for the optimal use of available energy and also to maintain structural/functional differences among cells. This can be achieved under 4 different levels.

- Synthesis of primary transcript
- Post-transcriptional modification
- mRNA degradation
- Translation

Gene regulation may be either for a short term or long term. Housekeeping genes are the genes of central metabolic pathways (citric acid cycle) expressed almost all the time and constitutive genes are organ-specific and are expressed when need arises.

12.4.1   Regulatory Components and Mechanism

- **Induction**   Gene products that increase in concentration under prescribed molecular circumstances are referred to as inducible, and the process is known as induction.

- **Repression**   Gene products that decrease in concentration in response to a molecular signal are referred to as repressible, and the action is referred to as repression.

- **Upstream**   It is the region which contains all regulatory sequences, generally present before structural genes.

- **Down stream**   This is a positive region, which contains structural genes for expression.

- **Promoter**   Region found in the DNA near the downstream and it is the site for binding of RNA polymerase.

- **Operator**   In prokaryotes, the binding sites of repressors are called operators.

- **Activator**   Found near a promoter region, enhances the RNA–promoter interaction.

- **Negative regulation**   Regulation by means of a repressor protein, which binds to the DNA and blocks transcription.

- **Positive regulation**   Activator binds to sites adjacent to promoters and enhances the binding of RNA polymerase for transcription.

- **Repressor**   A protein which prevents transcription of genes.

12.4.2   Lactose Operon

In 1960, Francois Jacob and Jacques Monad described lactose gene regulation in *E. coli*. There are three structural genes *lacA, lacY* and *lacZ*. All these three genes are transcribed into a single mRNA and translate into products such as β-galactosidase (*Z*), galactoside permease (*Y*) and thiogalactoside transacetylase (*A*). The products of '*Z*' and '*Y*' genes are involved in lactose hydrolysis and transport respectively. They are coordinately regulated by the genetic element located adjacent to them as shown in Figure 12.2.

The *lac* repressor (encoded by the *lacI*) is a homotetrameric protein which has four functional subunits—an N-terminal helix-turn-helix motif, a linker for DNA binding, a site for inducer binding and a C-terminal alpha helix for quaternary structure maintenance. In the absence of lactose (inducer), the *lac* repressor binds to the operator (O) region with very high affinity. This prevents RNA polymerase binding to the operator for transcription processes. All structural genes of the *lac* operon are repressed. But a few copies of β-galactosidase normally remain.

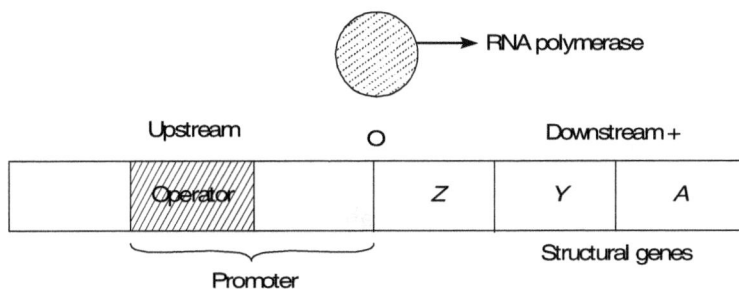

Figure 12.2   Transcription initiation

When the cells are given lactose, it triggers the *lac* operon. The inducer allolactose (an isomer of lactose) binds to a specific site on the repressor and causes conformational changes in the repressor. This leads to dissociation of repressor from the operator. Now the RNA polymerase binds to the operator and transcribes structural genes to catabolize lactose. The lactose is isomerized to allolactose by a few copies of β-galactosidase. The isopropyl thiogalactoside (IPTG) is an effective but nonmetabolizable inducer of the *lac* operon.

When glucose and lactose are present together, the other mechanism called catabolic repression functions to keep the genes for catabolism of lactose. The repressive effect of glucose is mediated by cAMP and catabolic gene activator protein (CAP). When glucose is absent, CAP–cAMP complex (positive regulatory element) binds near the *lac* promoter and stimulates *lac* operon structural gene transcription. When glucose is present, the concentration of cAMP declines and this prevents CAP binding, which in turn decreases the expression of *lac* operon. For strong induction, both the presence of lactose (to inactivate the repressor) and the absence or presence of low glucose level (for CAP binding) is required.

12.4.3   Arabinose Operon

In *E. coli*, arabinose is converted into xylulose 5-phosphate using three different enzymes such as arabinose isomerase, ribulose kinase and ribulose 5-phosphate epimerase (encoded by *araA*, *araB* and *araD* respectively). The *ara* operon consists of three structural genes, an operator site (*araO*$_1$ and *araO*$_2$), a site for regulatory protein binding (*araI*) and a promoter site (Figures 12.3 and 12.4).

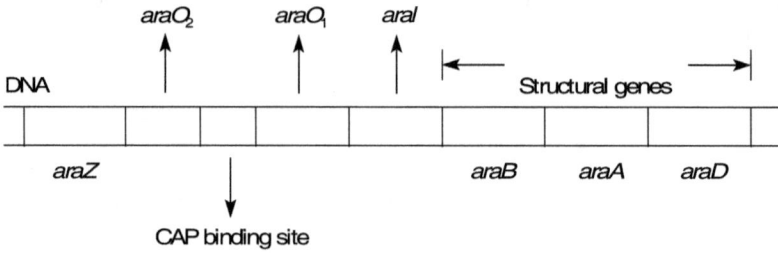

Figure 12.3   Arabinose operon

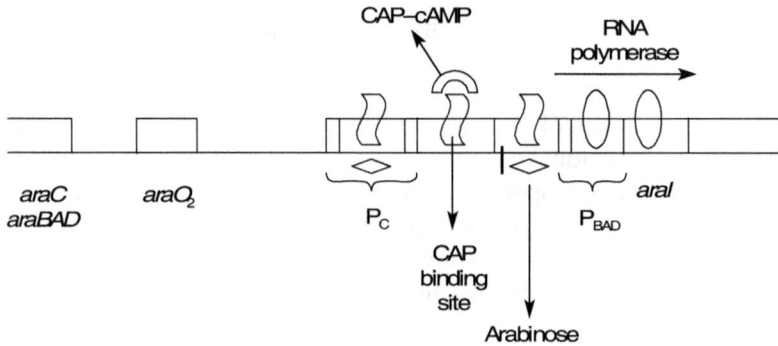

Figure 12.4   Arabinose operon and structural genes

When the concentration of *araC* regulatory protein exceeds about 40 copies per cell it binds to *araO$_1$* and represses transcription of the *araC* gene. AraC protein acts as both a positive and a negative regulator of the *araBAD* genes by binding to *araO$_2$* and *araI*. When glucose is present and arabinose is less or nil, one *araC* protein is bound to *araO$_2$* and second to the *araI*. These two *araC* proteins in turn bind to each other and form a DNA loop of about 210 base pairs. This represses the transcription from the promoter for the *araBAD* genes.

When the presence of arabinose and glucose level is nil or low, CAP–cAMP becomes abundant and binds to the site adjacent to *araI*. Simultaneously arabinose binds to the *araC* protein and alters its conformation. This opens the DNA loop. Now the *araC* protein binds at araI region and becomes an activator. This induces transcription of the *araBAD* genes for arabinose metabolism.

In the conditions where arabinose and glucose are both abundant or absent the operon remains repressed.

12.4.4 Tryptophan Operon

In *E. coli*, tryptophan is synthesized by five enzymes, viz. anthranilate synthetase (Component I), anthranilate synthetase (Component II), anthranilate isomerase, tryptophan synthetase (β subunit) and tryptophan

Schematic representation of tryptophan operon

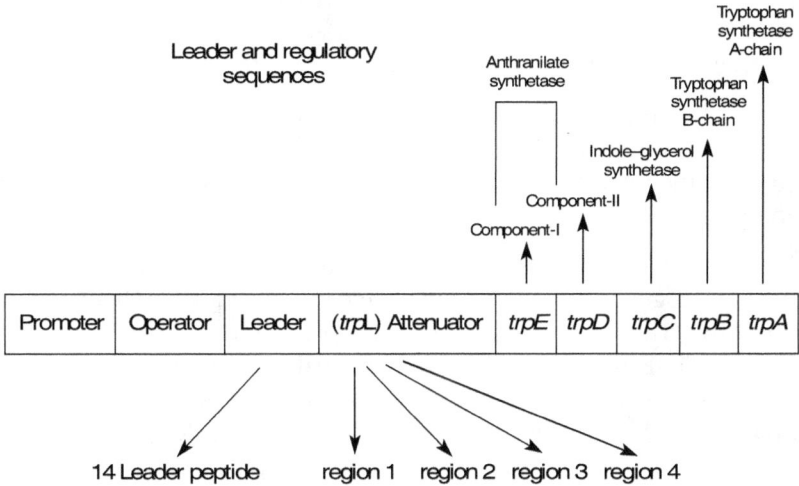

When tryptophan is abundant

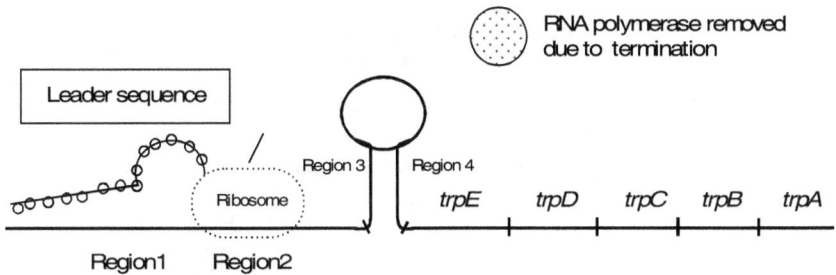

When tryptophan is scarce

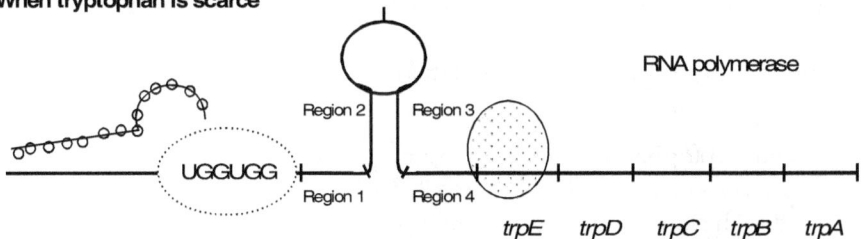

Figure 12.5 Tryptophan operon and structural genes

synthetase (α subunit). These enzymes are encoded by *trpE*, *trpD*, *trpC*, *trpB* and *trpA* genes (Figure 12.5). These five *trp* operon genes are coordinately expressed under the control of repressor protein. Tryptophan operon is controlled by either repressor or attenuator. The *trp* repressor is a homodimer, with each subunit containing 107 amino acid residues. When tryptophan is abundant, it binds to the *trp* repressor, causing a conformational change that permits the repressor to bind its operator. The *trp* operator site overlaps the promoter, and binding of the repressor blocks binding of RNA polymerase. In this operon, the tryptophan acts as a corepressor.

When the tryptophan level is low, repression is removed and the transcription is triggered. When the level of tryptophan increases as it binds to repressor, the rate of transcription decreases and this results in premature termination.

There is an additional transcriptional control segment located upstream of the structural gene *trpE* in a 162-nucleotide leader sequence (**transcription attenuation**). The *trp* operon attenuation process uses signal encoded in four regions (1–4) within a 162-nucleotide leader sequence at the 5′ end of the mRNA. At the end of the leader sequence there is a region called the **attenuator** which functions as an intrinsic terminator. Regions 3 and 4 of the attenuator base-pair to form a $G \equiv C$-rich stem and loop structure, which is followed by several sequential uridylate residues. This stem and loop structure resembles a transcription terminator and it stops the transcription of five contiguous structural genes.

Formation of the attenuator stem and loop depends on translation of a short leader sequence of the mRNA that encodes for a leader peptide of 14 amino acids. This short leader sequence gene is found before the four regulatory regions (attenuator). Translation of the leader peptide begins immediately after transcription. The leader peptide is the key element, which senses the level of tryptophan in a bacterium. Termination of attenuator depends on the level of tryptophan.

When tryptophan level is high, the supply of charged tryptophan tRNA (Trp–tRNA^{Trp}) is also high. In this case the ribosomes synthesize the leader peptide and continue progressing through region 1 and 2 of the regulatory sequence which prevents region 2 pairing with region 3. This leads to the pairing of regions 3 and 4, and generates terminator hairpin. The transcription is stopped.

When tryptophan level is low, the ribosome stalls (ribosome stalling) at the two Trp codons of region 1 due to lack of charged tRNA^{Trp}. Since region 1 is found within the ribosome, region 2 and region 3 base-pair and the RNA polymerase continues transcription of structural genes past the attenuator.

FOR ADDITIONAL READING

1. David Friefelder, (2002). *Microbial Genetics.* Narosa Publishing House.

2. Gaston, K.A. *et al.* (1990). "Stringent spacing requirements for transcription activation by CRP." *Cell.* 62:733–743.

3. Gralla, J.D. (1996). "Activation and repression of *E.coli* promoters." *Current Opinion in Genetics and Development.* 6:526.

4. Hartl, D.J. and Jones, E.W. (1998). *Genetics: Principles and Analysis,* 4th edn. Jones and Bartlett Publishers, London.

5. Lee, F. and Yanofsky, C. (1977). "Transcription termination at the *trp* operon attenuators of *E. coli* and *S.typhimurium*: RNA secondary structure and regulation of termination." *Proc. Nat. Acad. Sci.,* USA. 74:4365–4368.

6. Lehninger, A.L., Nelson, D.L. and Cox, M.M. (1993). *Principles of Biochemistry.* CBS Publishers, India.

7. Lewin , (2000). *Genes VII.* Oxford University Press, Oxford.

8. Lewis, M. *et al.* (1996). "Crystal Structure of the Lactose Operon Repressor and "Its Complexes with DNA and Inducer." *Science.* 271:1247–1254.

9. Maniatis, T. and Ptashne, M. (1976). "A DNA Operator-Repressor System." *Scientific American* (January).

10. Miller, J.H. and Reznikoff, W.S. (eds.). (1980). *The Operon,* 2nd edn. Cold Spring Harbor Laboratory Press, New York.

11. Miller, J. and Reznikoff, W., (eds.). (1978). *The Operon.* Cold Spring Harbor Laboratory, New York.

12. Oehler, S. (1990). "The three operators of the *lac* operon cooperate in repression." *EMBO J.* 9:973–979.

13. Ptashne, M. and Gilbert, W. (1970). "Genetic Repressors." *Scientific American* (June).

14. Weber, I.T., McKay, D.B. and Steitz, T.A. (1982). "Two Helix DNA binding Motifs of CAP found in *lac* Repressor and *gal* Repressor." *Nucleic Acids Research.* 10:5085–5102.

15. Yanofsky, C. (1981). "Attenuation in the control of expression of bacterial operons." *Nature.* 289:751.

16. Zurawski, G. *et al.* (1978). "Translational control of transcription termination at the attenuator of the *E. coli* tryptophan operon." *Proc. Nat. Acad. Sci.,* USA. 75:5988–5991.

MODEL QUESTIONS

1. Define the following terms

   i.   inducer   ii.   promoter   iii.   operator

   iv. activator   v.   repressor   vi.   TATA box

2. What are housekeeping genes?

3. Discuss in detail short-term regulation and long-term regulation.

4. What is downstream in a gene?

5. How does positive regulation function in *lac* operon?

6. What are attenuators?

7. Write notes on Homeobox.

8. Discuss in detail

   i.   *lac* operon   ii.   arabinose operon   iii.   tryptophan operon

9. What are leader sequences?

10. List various regulatory components and their roles.

11. Discuss the methods of genetic material transfer between microbes.

12. What is transduction? Explain with example.

13. What are the modes of reproduction in λ phage?

14. Write a short description of the way in which λ phages select reproductive cycles.

REGULATION OF GENE EXPRESSION IN EUKARYOTES

13

13.1 INTRODUCTION

Multicellular organisms (eukaryotes) are made up of different cell types; each cell type differs in both structure and function. Though mammalian hepatocytes, cardiomyocytes and lymphocytes have same genome, they differ in all aspects. This is because of cell differentiation and uniqueness in their gene expression. The patterns of gene expression vary in response to external environment. Accordingly expression of individual gene is switched on and off in cells by regulatory proteins. In prokaryotes, these regulatory proteins bind to the adjacent sites of RNA polymerase and regulate transcription (expression), whereas in eukaryotes, regulation is more complex and regulatory sites are at a distance.

There are two different classes of regulatory phenomena in gene expression: they are short-term (or reversible) regulation and long-term (or irreversible) regulation. In the short-term regulation, in response to external stimuli or the environment, the activities or concentration of enzyme or hormone levels rise and fall. Long-term regulation is associated with determination, differentiation and development of the cells or organ systems. In this chapter let us understand various types of gene regulation in various organisms, such as;

- the *qa* gene cluster in fungi (*Neurospora*)
- the *gal* gene cluster in yeast
- heat-shock response in *Drosophila*
- *eve* gene regulation in *Drosophila*
- *β*-globin gene regulation in human

13.2   *qa* GENE CLUSTER IN *NEUROSPORA*

A *Neurospora* cell contains all the enzymes required for the synthesis of aromatic molecules. These aromatic molecules are precursors for tyrosine and phenylalanine production. When the level of aromatic molecules is in excess in a cell, several scavenging enzymes are induced to catabolize them. The inductions of scavenging enzymes are regulated by quinic acid. The three scavenging enzymes are dehydroquinase, dehydrogenase and *dehydrase*. They are encoded by the linked genes *qa-2*, *qa-3* and *qa-4* (Figure 13.1). The *qa* gene appears to be under positive control. That is, the inducer, quinic acid, combines with a regulator product of *qa-1* gene to turn on the *qa* gene expression. If there are any deletions in *qa-1* gene then the constitutive enzyme synthesis is blocked. The active *qa-1* gene product has two functional domains: an initiator region and an inducer-binding region. Mutation in initiator region would prevent activation of transcription, whereas mutation in inducer-binding region will simply prevent inducer binding but switch on the transcription all the time.

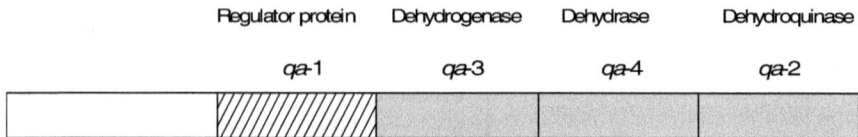

Regulator protein	Dehydrogenase	Dehydrase	Dehydroquinase
qa-1	*qa*-3	*qa*-4	*qa*-2

Figure 13.1   *qa* gene cluster in *Neurospora crassa*

13.3   *gal* GENE CLUSTER IN YEAST

Fermentation of galactose in yeast is carried out by three enzymes, kinase, transferase and epimerase. These enzymes are encoded by the linked loci known as *gal-1*, *gal-7* and *gal-10*. The yeast *gal* cluster differs from *qa*, where the *gal* cluster is controlled by the product of an unlinked gene.

Figure 13.2 shows *gal* gene cluster in yeast. The *i* repressor of yeast acts on another unlinked gene, *gal-4*. The *gal-4* locus appears to encode a positive regulator protein (similar to qa–1 protein), which stimulates (includes) transcription of the *gal* gene cluster. This is termed as positive regulation. The *gal-4* locus is negatively controlled by the *i* gene product. There is a region in *gal-4* locus, called *c*, which behaves like an operator locus. When *i* repressor is unable to bind to the mutated *c* locus, unregulated gene expression occurs.

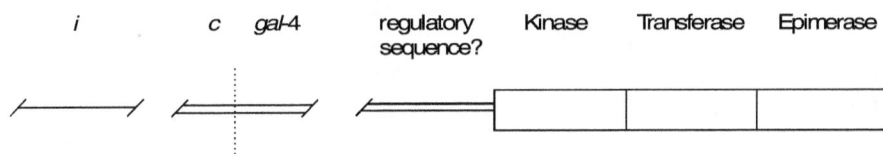

Figure 13.2   *gal* gene cluster in yeast

13.4   TRANSCRIPTION CONTROL IN DROSOPHILA

When *Drosophila* or cultured cells of *Drosophila* are insulated by raising the temperature or by exposing to metabolic inhibitors, a small number of heat-shock genes are transcribed rapidly. The transcription of active genes is stopped and translation of active gene mRNA is stalled. The newly synthesized heat-shock mRNAs are translated into heat-shock protein to protect the cells.

In *Drosophila* larva, when the temperature is raised form 25 to 37°C, the salivary gland chromosome undergoes heat-induced change in the puffing patterns. In normal conditions, puffing of a band in a polytene chromosome indicates high transcriptional activity of active genes. But within one minute of the temperature raise, most pre-existing puffs begin to regress and nine new puffs are detected. Thus, primary response to heat shock causes the co-ordinated induction of nine scattered gene loci.

Another important example is the multi-component genetic switch that controls the transcription of the *Drosophila* even-skipped (*eve*) gene. The *eve* gene plays a significant role in the development of the *Drosophila* embryo. If any mutation inactivates *eve* gene, many parts of the *Drosophila* embryo fail to form, which in turn leads to death of an embryo in early development. The regulatory region of the *eve* gene is very large and stripe 2 is a regulatory module found in the regulatory region of *eve* gene. This module region contains recognition sequence for two regulatory protein genes such as Bicoid and Hanchback, which activate eve gene transcription. There are two other regulatory proteins such as Kruppel and Giant, which repress the *eve* gene transcription. The relative concentration of these proteins determines whether stripe 2 module turns on the transcription of the *eve* gene.

13.5 HUMAN β-GLOBIN GENE REGULATION

In humans, the complex regulatory control region is best understood by learning β-globin gene expression in red blood cells (erythrocytes). Here the gene regulation is controlled by a group of activator and repressor proteins. Moreover the concentration of these regulatory proteins changes during erythrocyte development and triggers gene expression. The human β-globin gene is one among the cluster of globin genes. There are five genes found in this cluster (Figure 13.3) and expressed in different stages of red blood cell lineage and in different organs. They are:

ε-globin gene is expressed in the embryonic yolk sac

γ-globin gene is expressed in the yolk sac and foetal liver and

δ and β-globin genes are expressed in the adult bone marrow

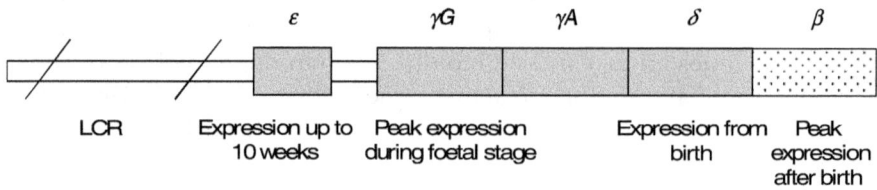

Figure 13.3   Human globin gene cluster

Specific regulatory proteins turn on these genes at the appropriate time in specific tissue. Besides this, Locus Control Region (LCR) regulates the entire gene cluster. In upstream the LCR appears far from the gene cluster.

In brain and skin cells, the globin genes are not expressed. The whole gene cluster is tightly packed into chromatin. But in erythroid, though gene cluster is folded into nucleosomes, the chromatin of the globin gene locus becomes decondensed and regulatory proteins bind to the DNA. This assemblage helps in expression of individual genes. The LCR appears to act by controlling the chromatin condensation. Deletion of LCR in part or complete can cause a certain type of thalassemia. The exact mechanism behind thalassemia, is lack of gene regulatory proteins that bind to the LCR and cause DNA looping to interact with proteins that regulate globin gene expression.

13.6 DNA BINDING MOTIFS AND GENE REGULATION

It is interesting to study how a cell decides which of its genes should be expressed. As described earlier, the gene expression is controlled by a regulatory region (sequence) before transcription begins. Thousands of such regulatory sequences have been identified so far. A different regulatory protein group recognizes each of these sequences by involving ionic bond, hydrogen bond or hydrophobic interactions. These kinds of protein–DNA interactions are weak. Let us study some of the motifs which are involved in controlling gene expression in both prokaryotes and eukaryotes.

Helix-rurn-helix motif is one of the most common DNA binding motifs. This motif is constructed from two α-helices connected by a short chain of amino acids (Figure 13.4). These amino acid chains constitute the "turn" and both helices are held at a fixed angle. The C-terminal helix is also called as recognition helix. This helix fits into the major groove of the DNA. The amino acid's side chains differ from protein to protein and play an important role in recognizing specific DNA sequence to which the protein binds. Some of the helix-turn-helix proteins are tryptophan repressor, lambda Cro, lambda repressor fragment and catabolic activator protein (CAP fragment). Principles of gene regulation established in bacteria are relevant to higher organisms (eukaryotes) also. In *Drosophila*, a special class of helix-turn-helix protein such as homeodomain proteins was identified. Structural studies have shown that yeast and *Drosophila* homeodomain proteins have similar conformation and DNA recognition sites.

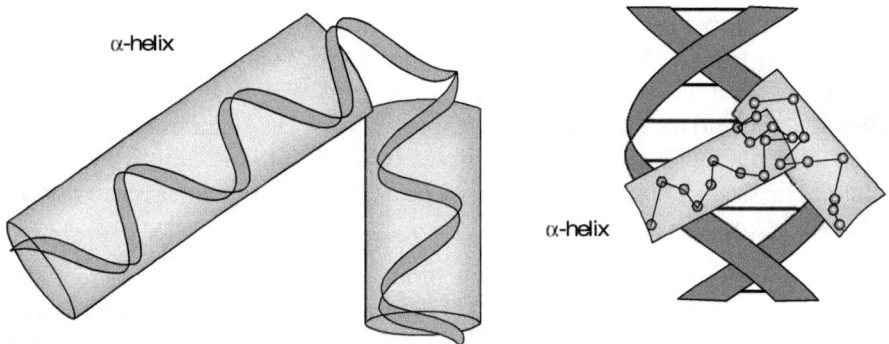

Figure 13.4   Helix-turn-helix

A second type of DNA-binding motif is referred to as Zinc finger motif. This was first discovered in 1985 by Klug and his colleagues in the transcription factor TFIIIA in *Xenopus*. The motif consists of an α-helix and a β sheet held together by the zinc ion. The four amino acids that bind to the zinc (Cys 3, Cys 6, His 19 and His 23) also hold α-helix and β sheet together (Figure 13.5). This protein recognizes the DNA using three zinc fingers that are arranged as direct repeats.

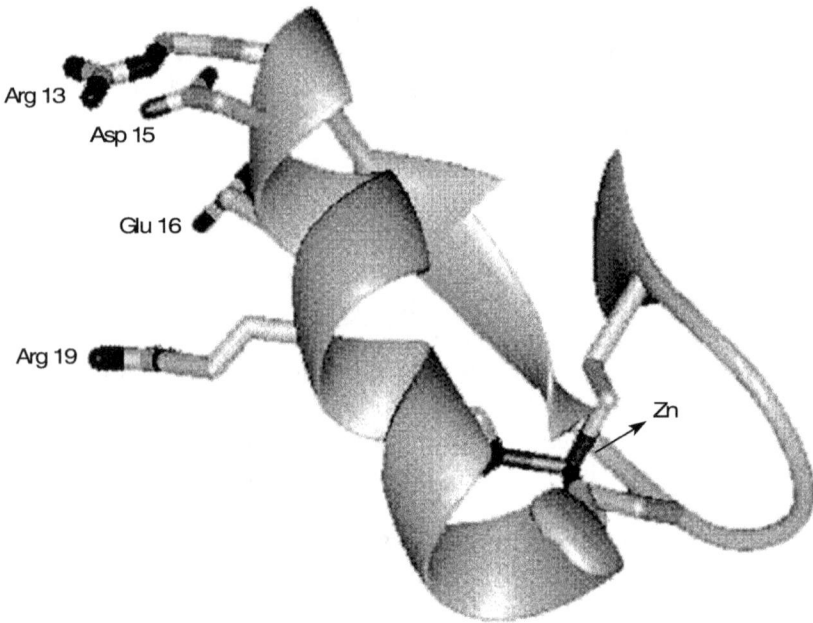

Figure 13.5   Zinc finger

Leucin zipper motif has two α-helices; one from each monomer is joined together to form a short-coiled coil (Figure 13.6). This motif combines two functions, that is, one portion of the protein which is responsible for dimerization and the other which is responsible for DNA binding. The helices are held together by interactions between hydrophobic amino acid side chain. The best example is Yeast Gcn 4 protein, which regulates transcription in response to the environment.

Another important DNA binding motif, similar to leucin zipper, is the helix-loop-helix (HLH) motif. An HLH consists of a short α-helix connected by a longer α-helix (loop like structure). Both helices fold back and pack against each other (Figure 13.7). This folding helps in DNA binding as well as for dimerization (binding to another HLH motif).

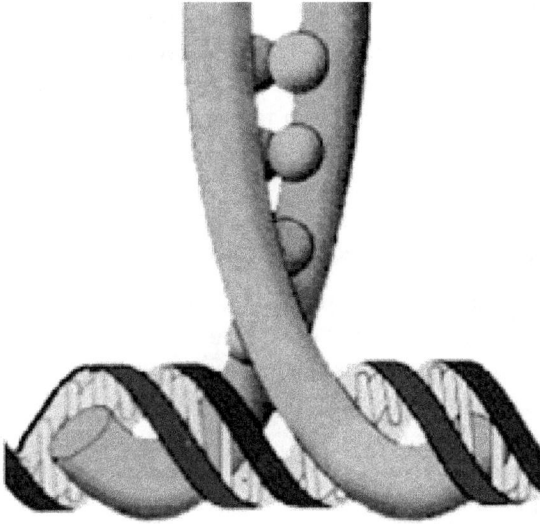

Figure 13.6   Leucin–Zipper motif

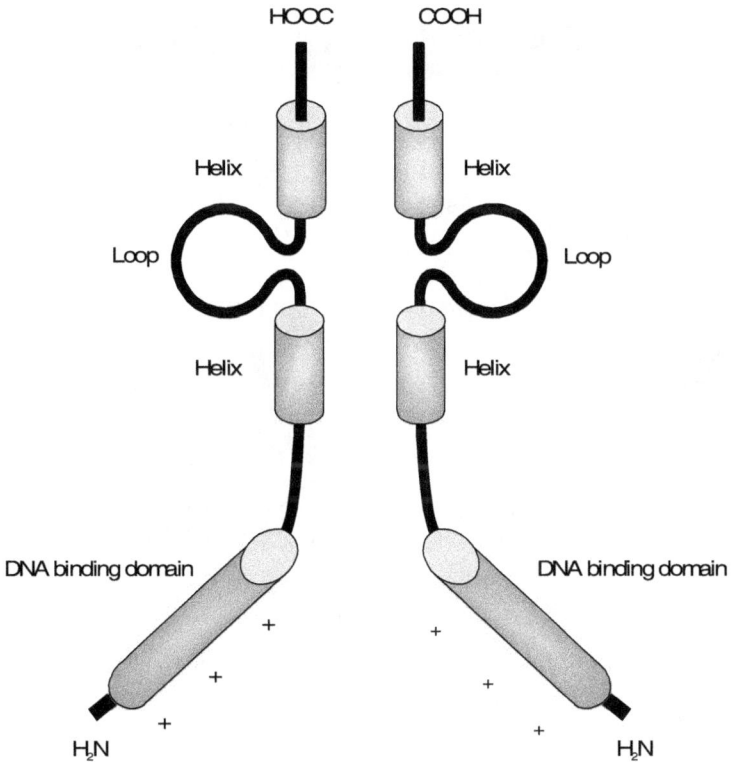

Figure 13.7   Helix-loop-helix

Copper fist is another type of DNA-binding motif. Yeast ACE 1 regulatory protein is the example. Copper, a potential heavy metal, enters the cell and poisons it when it is in excess quantity. Cystein residues in the ACE 1 protein actually bind to the copper ion like a fist (many projection like). The fist is positively charged and interacts with DNA. The ACE 1 protein binds at the promoter region and acts to enhance the transcription of the yeast metallothionein gene, *Cup* 1 protein. The product of the *Cup* 1 gene binds to excess copper ions and protects the cell.

Methylation and Z-DNA formation also play an important role in controlling transcription in eukaryotes. The degree of methylation of DNA is related to its transcription status. The DNA is undermethylated when genes are active and fully methylated when genes are inactive. For example the X chromosome genes are deactivated by methylation. But using 5-azacylidine, we can reactivate it. Structure of Z-DNA has been detailed in earlier part (Chapter 9) of this book. If the gene is to be repressed (turned off), the CG sequences are methylated to form a Z-DNA. Methylation also controls the formation of nucleosome.

FOR ADDITIONAL READING

1. Harrison, S.C. (1991). "A structural taxonomy of DNA-binding domains." *Nature.* 353:715–719.

2. Laity, J.H., Lee, B.M. and Wright, P.E. (2001). "Zinc finger proteins: new insights into structural and functional diversity." *Curr. Opin. Struc. Biol.* 11:39–46.

3. Carey, M. and Smales, S.T. (2000). *Transcriptional regulation in Eukaryotes: concepts, strategies and technique.* Cold Spring Harbor Lab Press, NY.

4. McKnight, S.L. (1991). "Molecular zippers in gene regulation." *Sci. Am.* 264:54–64.

5. Adhya, S. and Gottesman, M. (1978). "Control of transcription termination." *Ann. Rev. Biochem.* 47:967–996.

6. Craig, R.K., Bathhurst, I.C. and Herries, D.G. (1980). "Post-transcriptional regulation of gene expression in guinea pig tissues." *Nature.* 288:618–619.

7. De Crombrugghe, B., Busby, S. and Buc, H. (1984). "Cyclic AMP receptor protein: Role in transcription activation." *Science.* 224:831–838.

8. Weiringa, B., Meyer, F., Reiser, J. and Weissmann, C. (1983). "Unusual splice sites revealed by mutagenic inactivation of an authentic splice site of the rabbit β-globin gene." *Nature.* 301:38–43.

MODEL QUESTIONS

1. How does *qa* gene cluster in *Neurospora* get controlled?

2. What is a homeobox?

3. What is a homeodomain?

4. What are transcriptional regulators?

5. What is the significance of homeobox and domain?

6. What is the helix-turn-helix motif of DNA binding?

7. What is the helix-loop-helix motif of DNA binding?

8. Explain other motifs with appropriate example.

9. How is globin gene cluster in humans regulated?

10. What are the various transcriptional control systems in *Drosophila*?

11. Comment on the significance of DNA methylation.

12. How does *gal* gene transcription control functions in yeast?

DNA MUTATION, REPAIR AND DETECTION

14

14.1   INTRODUCTION

A mutation is any heritable change in the genetic material of an organism or cell. Mutations can be classified into two major types, viz. **germ cell mutation and somatic cell mutation**. The mutation occurring in the germ cell or reproductive cell is termed as germ cell mutation. On the other hand mutation occurring in body cell or somatic cell is termed as somatic cell mutation. The germ cell mutation mostly leads to various kinds of birth defects and syndromic conditions in humans. The somatic cell mutation generally leads to cancer in man.

Mutation can be caused by a **mutagen** (mutation causing agent) and termed as **induced mutation** or may occur in the absence of any known mutagenic agent and it is known as **spontaneous mutation**. There are different types of physical (UV radiation, ionizing radiation, heat) and chemical mutagens (ethidium bromide, ethyl methane sulphonate, nitrous acid). Their action differs in altering the genetic material. Some act as base analogs and are used up during DNA synthesis, some react directly with the DNA and block various actions, but some mutagens indirectly affect the DNA by producing secondary products that affect the DNA.

There are various categories of environmental agents that cause damage to living cells; they are **carcinogens** (causing cancer), **clastogens** (causing chromosome fragmentation), **mutagens** (causing mutations), **oncogens** (inducing tumour formation) and **teratogens** (causing birth defects).

There are seven important types of DNA damages that can occur in a cell. They are **base loss, base substitution, photo damage by dimer, replication errors, interstrand cross links, DNA protein cross links** and **strand breaks**. The simplest type of mutation is base substitution, in which nucleotide pair in a DNA duplex is replaced with a different nucleotide pair. The base substitutions which replace one pyrimidine base with the other or one purine base with the other, are called **transition mutation**.

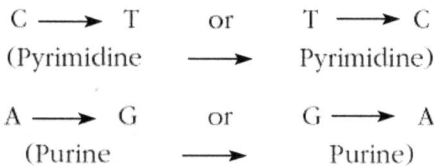

$$C \longrightarrow T \qquad or \qquad T \longrightarrow C$$
$$(Pyrimidine \longrightarrow Pyrimidine)$$

$$A \longrightarrow G \qquad or \qquad G \longrightarrow A$$
$$(Purine \longrightarrow Purine)$$

The base substitution, which replaces a pyrimidine with a purine or purine with a pyrimidine, is called **transversion mutation**.

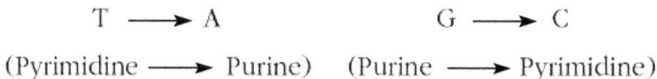

$$T \longrightarrow A \qquad\qquad G \longrightarrow C$$
$$(Pyrimidine \longrightarrow Purine) \quad (Purine \longrightarrow Pyrimidine)$$

Most transition mutations in the third codon position do not change the amino acid sequence, and these kind of mutations are known as **silent mutations**. Most base substitutions in coding regions do result in changed amino acid pattern and this is known as **missense mutation** (For example sickle cell anaemia results from the replacement of glutamic acid by valine at the 6th position in the β-globulin chain). Sometimes the base substitution creates a new stop codon, called a **nonsense mutation**. Nonsense mutation causes premature chain termination; the translated polypeptide fragment is almost always non-functional.

The genomes of higher eukaryotes contain large tandem repeats of short nucleotide sequences called **simple tandem repeat polymorphism** (STRP). The STRP insertions or deletions which occur in regulatory or coding region have a significant effect. Unless the number of nucleotides added or deleted is an exact multiple of three, the insertion or deletion shifts the phase in which the ribosome reads the triplet codons and alters all of the amino acids. Such mutations are called frameshift mutations.

14.2   MUTATION REPAIR

Spontaneous damage to DNA in human cells takes place at a rate of approximately 1 event per billion nucleotide pairs per minute. Under normal circumstances, damage done to the DNA by spontaneous chemical reactions in the nucleus, chemical mutagens and radiation are repaired through specific mechanisms. There are five ways by which the DNA gets repaired. They are: **mismatch repair, photoreactivation, excision repair, post-replication repair** and **SOS repair**.

Mismatch repair consists of the excision of a segment of a DNA strand that contains a base mismatch, followed by repair synthesis. In microbes the cleavage sites are 5 nucleotides to the 3´ end and 8 nucleotides to the 5´ end of the damage, yielding an excised fragment of 13 nucleotides. In eukaryotes, the cleavage sites are 5 nucleotides to the 3´ end and 24 nucleotides to the 5´ end, yielding an excised fragment of 29 nucleotides. After excision, DNA polymerase uses the remaining strand as template and fills the gap thereby eliminating the mismatch. This mismatch repair system also corrects most single base insertions or deletions.

The UV-induced pyrimidine dimers are reversed by **photoreactivation** system. The enzyme photolyase breaks the bonds that join the pyrimidines in the dimer and restores original bases. The enzyme binds to the dimer in the dark, and then utilizes the energy of blue light to cleave the bond.

In **excision repair**, a stretch of a damaged DNA strand is removed from a duplex molecule and replaced by a new strand using the template of normal strand. The endonuclease recognizes the distortion produced by the DNA damage and makes one or two cuts in the sugar phosphate backbone, several nucleotides away from the damage on either side. A 3´-OH group is produced at the 5´ cut, in which DNA polymerase uses a primer to synthesize a new strand. The joining of newly synthesized strand is done by DNA ligase.

Sometimes DNA damage persists rather than being reversed or removed. When DNA polymerase reaches a damaged site it stops the synthesis of strands. After a brief period, synthesis is reinitiated beyond the damage and chain growth continues, producing a gap in the damaged spot. The gap can be filled by strand exchange with the parental strand that has the same polarity, and the secondary gap produced in the strand can be filled by repair synthesis. This process is called **post-replication repair**.

SOS repair, found in *E coli*, is a complex process, which includes a bypass system that allows DNA replication to take place across the pyrimidine dimer or DNA distortions. The RecA protein has several functions in SOS repair. It directly inhibits the editing function of polymerase III by binding tightly in the region of the distortion resulting from a pyrimidine dimer. The presence of RecA at the dimer site inhibits editing and causes the mispaired base to remain in the daughter strand as a mutation.

14.3   MUTATION DETECTION

Understanding the relationship between a phenotype and genotype depends on the ability to isolate and characterize an individual gene or genes. In general, there are four approaches used to find a human gene: functional cloning, candidate cloning, positional cloning and positional-candidate cloning. With a cloned gene, experiments that test the actions and interactions of the gene product and determine how various mutant gene products undermine normal processes can be devised. In addition, diagnostic tests for specific gene mutations can be developed from information based on the DNA sequence of normal and mutated forms of a gene. This also helps to find how different mutations in different exons are responsible for the disease.

Three kinds of alterations can be detected in the DNA, they are

- Deletions or insertions
- Single base substitutions
- Trinucleotide repeat expansions

The mutation detection technique is applied either for

- Scanning (which searches unknown mutations) or for
- Diagnosis (test known mutations).

A number of simple and inexpensive assays are used in mutation detection. They are listed below. Some of them are explained in detail.

- Single-strand conformational polymorphism (SSCP)
- Denaturing gradient gel electrophoresis (DGGE)
- Heteroduplex analysis (HA)
- Cleavage using RNase
- Chemical mismatch cleavage (CMC)
- Mutation detection using T4 endonuclease VII
- DNA detection and sequence distinction through oligonucleotide probes
- Detection of sequence variation using primer extension
- Detection of mutations by hybridization with sequence-specific oligo-nucleotide probes
- Protein truncation test (PTT)

14.4   SINGLE-STRAND CONFORMATION POLYMORPHISM

Single-strand conformation polymorphism is a commonly used procedure for detecting gene mutations. By using a specific primer (forward and reverse), exon or region of a gene is amplified by PCR from the DNA of patient or normal individuals. Each pair of primers is determined from sequences that flank each exon or from the terminal ends of each exon. This helps to search for mutations in the 5´ upstream region or in the 3´ downstream region of a gene. This can also be extended to find splice junctions of exons and introns. Each set of primer is tested with human DNA to ensure that they can amplify only single copy regions. That is, they produce a single band when electrophoresis is done. For good results, each primer pair should amplify ~200 bp of DNA.

After PCR amplification of DNA of patient and normal individuals, the products are denatured, then cooled and electrophoresed. Each denatured single-stranded DNA molecule assumes a three-dimensional conformation based on primary nucleotide sequence. The conformation is the consequence of intrastrand base pairing and other bonds. Because of these, the two single strands of a double-stranded DNA molecule have different nucleotide sequences. Each strand has a specific three-dimensional conformation. The different conformations have different rates of migration in a gel during electrophoresis. That is why two bands can be seen using DNA stain.

If two DNA molecules from two different sources representing the same segment of a gene differ by a single nucleotide pair, then their conformations differ. Each of the four strands will migrate during electrophoresis at its own distinctive rate. By comparing the DNA of patient and normal individuals, the differences in the migration of single-stranded DNA molecules are easily detected. The SSCA localizes a nucleotide alteration to a specific region or exon of a gene. But the nature of the mutational difference is only obtained by DNA sequencing. The SSCA can detect about 90% of the single base pair mutations in PCR products that are less than 200 bp.

14.5   DENATURING GRADIENT GEL ELECTROPHORESIS

The intact PCR products of the exons or regions of a gene of a patient and normal individuals are electrophoresed individually in a gel through an increasing concentration gradient of urea or formamide (DNA denaturants). As a double-stranded DNA molecule migrates through the concentration gradient, denaturation is initiated for breaking hydrogen bonds between complementary bases. The regions with a large number of AT base pairs separate at a lower concentration than the regions of GC base pairs. This differential denaturation separates segments of strands of a DNA molecule by altering its conformation and also by its mobility through the gel.

A single base-pair alteration can change the point in the gradient during denaturation and modify its rate of migration in the gel. To ensure that differentially denatured DNA molecules can be resolved readily by DGGE, one of each set of primers has attached to its 5′ end a string of 40 GC units. After the amplification, this provides the product with a stretch of 40 G : C base pairs that will not denature at the highest concentration of denaturants in the gradient. The DGGE method can detect more than 95% of the single-base differences in PCR products that are ~600 bp. The bands in the gel are visualized by staining the DNA.

14.6   HETERODUPLEX ANALYSIS

A double-stranded DNA molecule with one or a few mismatched nucleotide pairs is called heteroduplex DNA. A DNA molecule without nucleotide mismatches is homoduplex DNA. In principle, a single nucleotide mismatch in a DNA molecule is sufficient to alter the mobility in electrophoresis in comparison to homoduplex DNA molecules. A nucleotide mismatch causes a bulge that distorts the conformation of the DNA molecule and retards its migration when electrophoresed.

For heteroduplex analysis testing, DNA samples from a patient and normal individual are combined and amplified by PCR, using primers for the exons and flanking regions of a cloned putative disease gene. If an amplified DNA segment from the different DNA samples has a nucleotide difference, heteroduplex DNA molecules will form during the renaturation step after PCR amplification.

Following PCR, the amplified product is heated to 95°C and slowly cooled to maximize heteroduplex DNA formation. Then, each DNA sample is electrophoresed in a lane of a special gel that enhances the difference in mobility between homoduplex and heteroduplex DNA molecules. Now, both the homoduplex molecules will migrate to the same extent, because they have the same conformation and length. The heteroduplex DNA molecules will migrate more slowly and form a single band. The bands with heteroduplex and homoduplex DNA molecules can be visualized by DNA staining. Heteroduplex analysis can detect more than 95% of the single nucleotide mismatches in DNA fragments of 300 bp or less in length.

14.7 CHEMICAL MISMATCH CLEAVAGE

The chemical mismatch cleavage (CMC) mutation detection assay is a different kind of heteroduplex analysis. DNA samples from patient and normal individuals are amplified by PCR with primers for the exons of a putative disease gene. The PCR product in one of the DNA samples and not the other is labelled during amplification. For the CMC assay, the PCR products from each reaction are mixed, denatured, and slowly cooled. Heteroduplex DNA molecules form when there is a nucleotide difference.

A renatured sample is divided and kept in two different aliquots. The aliquot is treated with hydroxylamine and osmium tetroxide respectively. The hydroxylamine treatment modifies mismatched cytosine residues but not those that are matched. The osmium tetroxide treatment selectively modifies mismatched thymidine units, while matched thymidine residues survive unscathed. Next, both samples are treated with piperidine. This treatment creates a nick in the DNA strand by cleaving a modified cytosine or thymidine residue. Denaturing gel electrophoresis separates the DNA of each sample. If cleavage has occurred in a strand, then two fragments are produced in a gel electrophoresis. Strands that do not contain a modified cytosine or thymidine residue remain intact. Thus, cleavage products denote a mismatch, which signifies a nucleotide difference between the DNA samples. The pattern of DNA bands can be visualized by chemiluminescent detection assay or by an autoradiography. The CMC assay detects more than 95% of single mismatched nucleotides in pieces of DNA up to 1,700 bp in length.

14.8 PROTEIN TRUNCATION TEST

The protein truncation test (PTT) is used to detect nonsense mutation, out-of-frame deletion and insertion, in-frame deletion (>25 bp), and splice-site alterations that skip an exon (>25 bp). The PTT is based on reverse transcriptase-PCR. Total RNA is extracted from tissue that expresses the disease gene, and the first strand cDNAs are synthesized from the mRNAs by reverse transcriptase with oligo(dT) as the primer. Then, primers for the first (5´) and last (3´) exons are used to produce a full-length cDNA corresponding to the putative disease gene. For the PTT, the upstream (5´) primer contains a sequence for an RNA polymerase promoter and a eukaryotic initiation of translation recognition site. This recognition site helps in the transcription and translation of the amplification product *in vitro* by a cell-free extract containing both an RNA polymerase that binds to the promoter sequence and all the components that support translation.

The newly synthesized protein is labelled with either a biotinylated amino acid or a radioactive substance. The proteins of the cell-free extract are separated by sodium dodecyl sulphate-polyacrylamide gel electrophoresis (SDS-PAGE). The synthesized protein is detected by either streptavidin-based chemiluminescence or autoradiography protocol, depending on the labelling of the protein. A full-length template free of mutation codes for a full-length labelled protein. If an internal deletion greater than 25 bp, a nonsense mutation, or a splice-junction (exon-skipping) mutation is present in the gene, then a shortened (truncated) protein is produced. Sequencing is required to establish the nature of the nucleotide change that causes protein truncation.

FOR ADDITIONAL READING

1. Edkins, E. and Forrest, S. (1998). *Mutation Detection – A Practical Approach*. Cotton, R.G.H., (ed.). IRL Press at Oxford University Press, New York.

2. Chicurel, M. (2001). "Can organisms speed their own mutation?" *Science*. 292:1824–1827.

3. Cotton, R.G.H. (1997). "Slowly but surely towards better scanning for mutations." *Trends Genet.* 13:43–46.

4. Foster, P.L. (1999). "Mechanism of stationary phase mutation: a decade of adaptive mutation." *Ann. Rev. Genet.* 33:57–88.

5. Goodman, M.F. (2000). "Coping with replication 'train wrecks' in *Escherichia coli* using Pol V, Pol II and RecA proteins." *Trends Biochem. Sci.* 25:189–195.

6. Hanaoka, F. (2001). "SOS polymerases." *Nature*. 409:33–34.

7. Johnson, R.E., Prakash, S. and Prakash, L. (1999). "Efficient bypass of a thymine-thymine dimer by yeast DNA polymerase, Pol." *Science*. 283:1001–1004.

8. Kolodner, R.D. (1995). "Mismatch repair: mechanisms and relationship to cancer susceptibility." *Trends Biochem. Sci.* 20:397–401.

9. Kolodner, R.D.E. (2000). "Guarding against mutation." *Nature*. 407:687–689.

10. Lehmann, A.R. (1995). "Nucleotide excision repair and the link with transcription." *Trends Biochem. Sci.* 20:402–405.

11. Lindahl, T. and Wood, R.D. (1999). "Quality control by DNA repair." *Science*. 29/86:1897–1905.

12. Mellon, I., Rajpal, D.K., Koi, M., Boland, C.R. and Champe, G.N. (1996). "Transcription-coupled repair deficiency and mutations in human mismatch repair genes." *Science*. 272:557–560.

13. Richardson, C. and Jasin, M. (2000). "Frequent chromosomal translocations induced by DNA double-strand breaks." *Nature*. 405:697–700.

14. Seeberg, E., Eide, L. and Bjroas, M. (1995). "The base excision repair pathway." *Trends Biochem. Sci.* 20:391–397.

15. Shannon, M. and Weigert, M. (1998). "Fixing mismatches." *Science.* 279:1159–1160.

16. Walker, J.R., Corpina, R.A. and Goldberg, J. (2001). "Structure of the Ku heterodimer bound to DNA and its implications for double strand break repair." *Nature.* 412:607–614.

17. Wood, R.D., Mitchell, M., Sgouros, J. and Lindahl, T. (2001). "Human DNA repair genes." *Science.* 291:1284–1289.

18. Buermeyer, A.B., Deschenes, S.M., Baker, S.M. and Liskay, R.M. (1999). "Mammalian DNA mismatch repair." *Ann. Rev. Genet.* 33:533–564.

19. Harfe, B.D. and Jinks-Robertson, S. (2000). "DNA mismatch repair and genetic instability." *Ann. Rev. Genet.* 34:359–39.

20. Kunkel, T.A. and Bebenek, K. (2000). "DNA replication fidelity." *Ann. Rev. Biochem.* 69:497–529.

MODEL QUESTIONS

1. What is mutation?

2. What is a spontaneous mutation?

3. What is an induced mutation?

4. What is mismatch repair and how does it happen?

5. Discuss various types of mutation repair systems.

6. List various mutation detection systems.

7. How is protein truncation test applied for the detection of nonsense mutation?

8. Comment on SSCP and heteroduplex analysis.

9. Explain in detail the method and application of denaturing gradient gel electrophoresis.

10. Give an account of the primer extension method to detect sequence variation?

11. What is the application of RT–PCR in protein truncation test?

12. How is T4 endonuclease VII used in detecting mutation?

13. How can errors in DNA replication lead to mutation?

14. Describe various types of DNA repair systems that are known.

15. Is there any association between DNA mutation, repair and human disease?

TRANSPOSABLE ELEMENTS

15

15.1 INTRODUCTION

In the 1940s, the American geneticist Barbara McClintock made a remarkable discovery. She found that certain DNA segments in the chromosomes of prokaryotes and eukaryotes can move (jump) from one location to another in the genome. These mobile genetic elements are called transposable elements (transposons) or insertion sequences. This finding gave a new dimension while developing evolutionary theories of genomes. This chapter will help you to learn more about transposons and the changes they cause.

15.2 FEATURES OF TRANSPOSABLE ELEMENTS

Transposable elements are normal components of the genomes of prokaryotes and eukaryotes. Transposable elements fall into two categories based on how they move. In the first category, the transposable elements encode for proteins that move them to a new position. In the second category, it replicates the DNA to produce new elements that interact with the genome and get integrated. The above two categories are generally found in both prokaryotes and eukaryotes. Besides these, there are other classes of transposable elements related to retroviruses. They encode for a reverse transcriptase for making DNA copies of their RNA transcript. These DNA copies subsequently integrate at a new location in the genome. Transposable elements of this category are present only in eukaryotes.

In prokaryotes, transposable elements can jump to new locations on the same chromosome (only one chromosome is found) whereas in eukaryotes they can insert either on the same chromosome or on different chromosomes. Insertion of transposable elements into the genome can cause

- a nonhomologous recombination during cell proliferation
- gene mutations
- increase or decrease in gene expression
- important contributions to the evolution of the genomes

15.3   TRANSPOSABLE ELEMENTS IN PROKARYOTES

As mentioned earlier there are two types of transposable elements in prokaryotes; they are insertion sequence (IS) elements or transposons (Tn). An insertion sequence element or IS element is the simplest mobile genetic element found in prokaryotes. An IS element has a complete set of genes required to insert the element into a chromosome at new locations. The IS element was first identified in *E. coli*.

E. coli contains a number of IS elements such as IS1, IS2 and IS10R. IS1 is 768 bp long and is present in 0 to 10 copies on the *E. coli* and in one copy on the F plasmid. The IS10R is found in a class of R plasmids that can replicate in *E. coli*. All IS elements have terminal inverted repeats (IRs) of 9 to 41 bp (Figure 15.1).

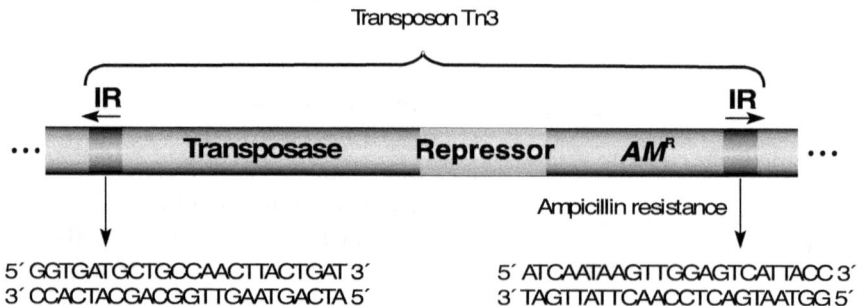

Figure 15.1   Insertional sequence of *E. coli*

When an IS element transposes, first it replicates using the host replication enzyme. A copy of the IS element inserts into a chromosome at the new location. The original IS element remains in the same location. During actual transposition, an enzyme transposase as well as the inverted repeat sequences are essential (Figure 15.2).

First a staggered cut is made at the target site and the IS element is then inserted. DNA polymerase and DNA ligase fill the gaps and produce an integrated IS element.

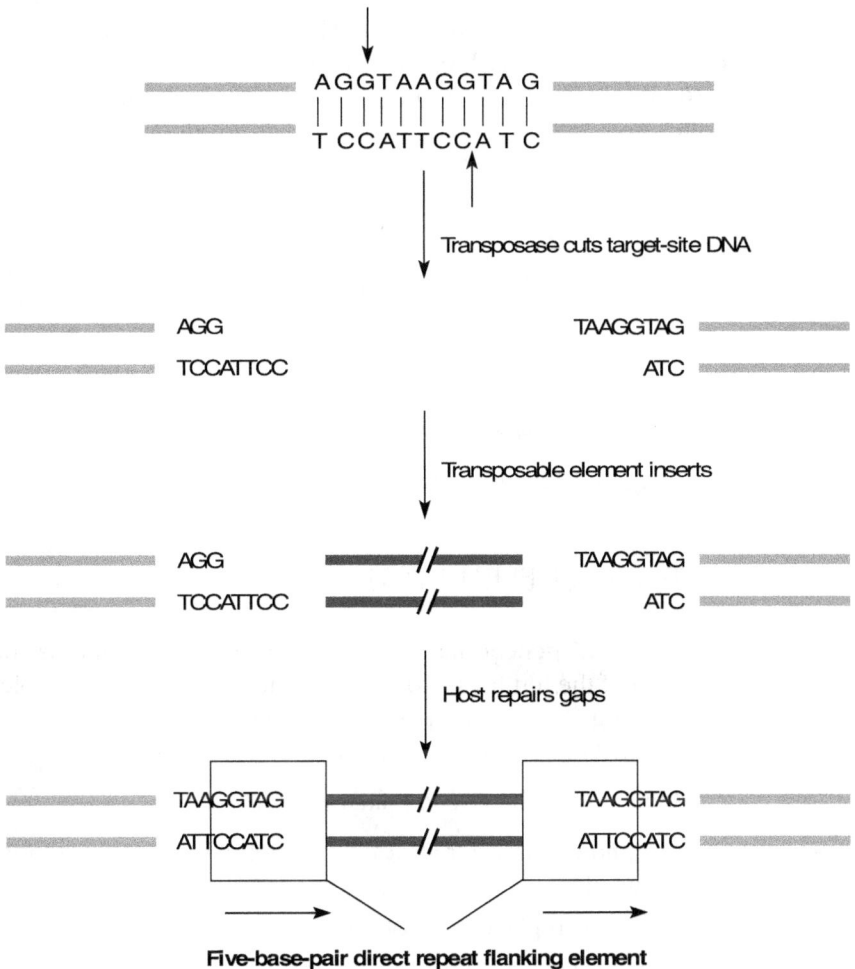

Figure 15.2   Inverted repeats

The other type of transposable element in the prokaryote is the transposon (Tn). Transposons contain genes for the insertion of the DNA segment into the chromosome and mobilization of the element to other locations on the chromosome. Prokaryotic transposons are further subdivided into composite transposons and noncomposite transposons. Composite transposons (Tn10) are complex and thousands of base pairs long. Contract region of composite transposon consists of genes for antibiotic resistance (Tetracycline). Both the ends of transposons are flanked by IS element modules such as inverted repeats.

Like composite transposons, noncomposite transposons also contain genes for drug resistance, but the IS element module is lacking at both the ends. Tn3 is the best example of a noncomposite transposon. Tn3 has 4,957 base pairs, 38-bp terminal inverted repeats and genes for transposase (*tnpA*), resolvase (*tnpB*) and β-lactamase (*βla*). β-lactamase destroys the antibiotics such as penicillin and ampicillin (Figure 15.3). Tn3 produces 5 base pair target site duplication during insertion into the genome.

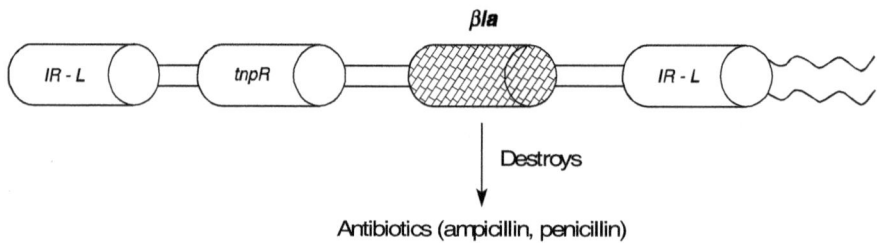

Figure 15.3   Noncomposite transposons

15.4   TRANSPOSABLE ELEMENTS IN PLASMIDS

In *E. coli*, the transfer of genetic material between two conjugants is the result of the function of the fertility factor F. The F factor is a circular double-stranded self-replicating DNA, called plasmid. Certain plasmids are capable of integrating into bacterial chromosomes called episomes.

The genetic elements of the *E. coli* F factor are as follows:

- *tra* genes required to transfer DNA from donor to recipient bacterium during conjugation
- Genes coding for replication proteins
- Four IS elements such as two copies of IS3, one of IS2 and one of an insertion sequence (γδ)

The *E. coli* genome (chromosome) also has copies of these four IS elements at various locations. This helps in the integration of F factor into the *E. coli* chromosomes at specific sites and in different orientations through conventional genetic recombination. Similarly R plasmid in bacteria (*Shigella*) has medical significance. The administration of various antibiotics (tetracycline, penicillin, sulphanilamide, chloramophenicol, streptomycin) in patients with dysentery in Japanese hospitals were found to be ineffective. This was due to the multiple resistance phenotypes. This occurred due to the accumulation of resistance gene in R plasmids through transposons.

15.5   TRANSPOSABLE ELEMENTS IN BACTERIOPHAGE

Mu is a bacteriophage that infects *E. coli*. It is also able to move by transposition and cause mutation in the host genome. That is why this temperate phage is named Mu (mutator).

Mu genome is a 37-kb linker piece of DNA with a small piece of unequal lengths of host DNA at the two ends. When *Mu* infects *E. coli* and enters the lysogenic state, the *Mu* genome integrates into the *E. coli* chromosome by non-replicative (conservative) transposition to produce the integrated prophage DNA, flanked by a 5-bp direct repeat of the target site sequence. During this integration, phage-encoded repressor prevents the *Mu* gene expression. The *Mu* prophage replicates when the *E. coli* chromosome replicates. Transposition by *Mu* can cause various mutations such as deletions, inversions and translocations.

15.6   TRANSPOSABLE ELEMENTS IN EUKARYOTES

Transposable elements have been identified and studied well in many eukaryotes such as yeast, *Drosophila*, corn and humans. They resemble bacterial transposons in anatomical features and transposition properties. Barbara McClintock was awarded Nobel Prize (1983) in physiology or medicine for the discovery of these mobile genetic elements (transposons).

The eukaryotic mobile elements (transposons) have genes which encode enzymes required for transposition and also for integrating them into the chromosomes at various sites.

15.7   TRANSPOSONS IN PLANTS

Many genes are required for the synthesis of red anthocyanin pigment, which gives the corn kernel a purple colour. There are two important forms of transposons identified in corn family; they are autonomous elements (which can transpose by themselves) and nonautonomous (which require presence of autonomous element to make up missing functions).

Barbara McClintock's classic experiment showed that the kernel rather than being purple or white, exhibited spots of purple pigment on white kernel due to the result of controlling elements (now we call transposons). Figure 15.4 shows purple kernels as a result of active C gene. When Ac transposon (activator) activates D2 transposition, the D2 transposon gets inserted into the C gene and causes mutation. This makes the kernel colourless due to gene activity (mutant C gene). Ac transposons can again activate transposition of Ds transposons out of C gene in a few cells during kernel development. This causes spotted kernel. The remarkable fact of McClintock's conclusion was the control of phenotype by transposable elements, when everyone thought that the genome is static with respect to gene location.

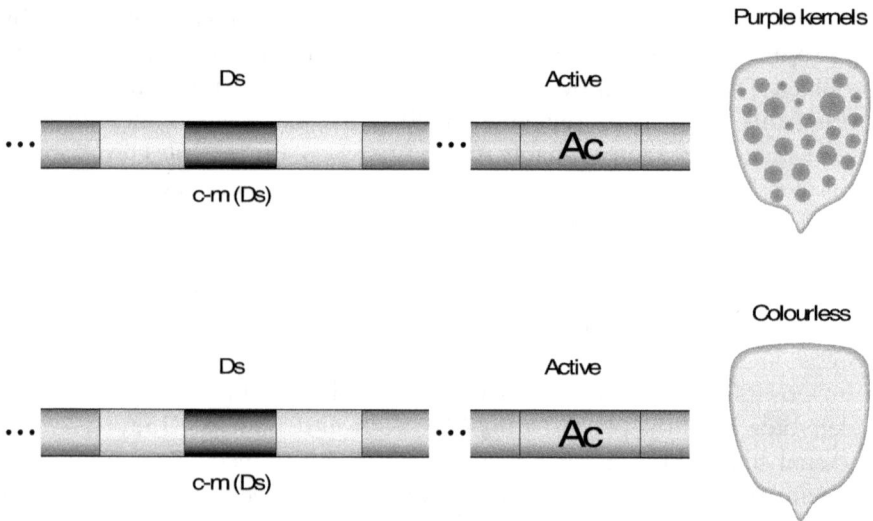

Figure 15.4   Colouration of kernel by transposons

15.8   Ty ELEMENT IN YEAST

Ty transposable elements of yeast have terminal repeat sequence, integration in nonhomology site and target site duplication (5 bp) upon insertion. These structural properties are also common in bacterial transposons. A Ty element is about 5.9 kb long and includes two long terminal repeats of 334 bp (LTR) or otherwise called as delta (δ). Each Ty element encodes a single mRNA (5,700 nucleotides). The mRNA contains two ORF (open reading frames) such as TyA and TyB. Transposition of Ty elements is found to be similar to retroviruses. They move around the genome via RNA intermediates; so they are also called retrotransposons.

15.9 TRANSPOSONS IN *DROSOPHILA*

In *Drosophila* about 15 percent of the genome is mobile. Figure 15.5 shows the structure of a copia transposable genetic element of *Drosophila*, similar to Ty elements of yeast; the copia element consists of long terminal repeats of 276 bp at both ends of a 5000 bp DNA segment. Copia transposes via RNA intermediate by using a reverse-transcriptase-catalysed process.

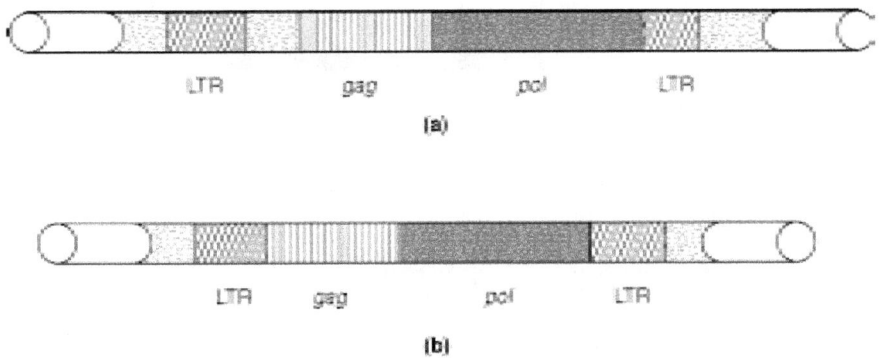

Figure 15.5   a) Ty1 in yeast and b) Copia in *Drosophila*

In *Drosophila melanogaster*, the haploid genome has about 40 copies of a family of transposons called P elements. The P elements vary in length from 500 to 2,900 bp. Each has 31 bp terminal inverted repeats. Smaller elements are formed by internal deletion of the longest P element. The longest P element is similar to Ac in corn. They encode for a transposable enzyme which can help transposition of longer as well as shorter segments. The P elements are also used as vectors for transferring genes into the germ line of *Drosophila* embryo.

15.10 HUMAN RETROTRANSPOSONS

Retrotransposons are also present in mammalian genomes. In mammalian genome there are two kinds of repetitive class of sequences found. They are SINEs (Short Interspersed Sequence) and LINEs (Long Interspersed Sequences). SINEs are 100–300 bp repeated sequences whereas LINEs are

5000 to 35,000 bp long. In humans a very abundant SINE family is the Alu family. A 300 bp long DNA segment is repeated 3,00,000 to 5,00,000 in total human genome (3%). During evolution, the insertion/deletion of the sequence is random so that the individuals in the family are related but not identical. Similarly one mammalian LINEs family (L1 elements) also transposes via RNA intermediate (retrotransposons). The maximum length of the L1 element is 6500 bp. The full length of L1 elements contains a large open reading frame that has homology with reverse transcriptase. One good example is insertional mutagenesis of L1 element into the factor VIII gene causing haemophilia (OMIM 306700). Molecular analysis of parent DNA sample showed that the insertion was not present in either of them.

FOR ADDITIONAL READING

1. Peter, J. Russell. (2001). *Genetics.* Benjamin Cummins Publishing Company, New York. pp. 573–595.

2. Berg, D.E. and Howe, M.M. (1989). *Mobile DNA.* American Society of Microbiology. No. 2 Washington, DC.

3. Boeke, J.D., Garfinkel, D.J., Styles, C.A. and Fink, G.R. (1985). "Ty elements transpose through an RNA intermediate." *Cell.* 40:491–500.

4. Cohen, S.N., and Shapiro, J.A. (1980). "Transposable genetic elements." *Sci. Am.* 242:40–49.

5. Engles, W.R. (1983). "The P family of transposable elements in *Drosophila.*" *Annu. Rev. Genet.* 17:315–344.

6. Federoff, N.V. (1989). "About maize transposable elements and development." *Cell.* 56:181–191.

7. Garfinkel, D.J., Boeke, J.D. and Fink, G.R. (1985). "Ty element transposition: Reverse transcriptase and virus-like particles." *Cell.* 42:507–517.

8. Iida, S., Meyer, J. and Arber, W. (1983). *Prokaryotic IS elements in Mobile Genetic elements.* J.A. Shapiro (ed.). Academic Press, New York. pp. 159–221.

9. Kleckner, N. (1981). "Transposable elements in prokaryotes." *Annu. Rev. Genet.* 15:341–404.

10. McClintock, B. (1950). "The origin and behavior of mutable loci in maize." *Proc. Natl. Acad. Sci.,* USA. 36:344–355.

11. McClinktock, (1953). "Induction of instability at selected loci in maize." *Genetics.* 38:579–599.

12. Campbell, A. (1981). "Evolutionary significance of accessory DNA elements in bacteria." *Ann. Rev. Immunol.* 35:55–83.

13. Finnegan, D.J. (1985). "Transposable elements in eukaryotes." *Int. Rev. Cytol.* 93:281–326.

14. Aldaz, H., Schuster, E. and Baker, T.A. (1996). "The interwoven architecture of the *Mu* transposase couples DNA synthesis to catalysis." *Cell.* 85:257–269.

15. Savilahti, H. and Mizuuchi, K. (1996). "*Mu* transpositional recombination: donor DNA cleavage and strand transfer in *trans* by the Mu transpose." *Cell.* 85:271–280.

16. Bender, J. and Kleckner, N. (1986). "Genetic evidence that Tn10 transposes by a non-replicative mechanism." *Cell.* 45:801–815.

17. Fedorofff, N. (2000). "Transposons and genome evolution in plants." *Proc. Nat. Acad. Sci., USA.* 97:7002–7007.

18. Ros, F. and Kunze, R. (2001). "Regulation of activator/dissociation transposition by replication and DNA methylation." *Genetics.* 157:1723–1733.

19. Singer, T., Yordan, C. and Martienssen, R.A. (2001). "Robertson's mutator transposons in *A. thaliana* are regulated by the chromatin–remodeling gene decrease in DNA methylation (DDM1)." *Genes Dev.* 15:591–602.

20. Engels, W.R. (1983). "The P family of transposable elements in *Drosophila*." *Ann. Rev. Genet.* 17:315–344.

21. Laski, F.A., Rio, D.C., and Rubin, G.M. (1986). "Tissue specificity of *Drosophila* P element transposition is regulated at the level of mRNA splicing." *Cell* 44:7–19.

MODEL QUESTIONS

1. What are transposable elements?

2. What are transposons?

3. Discuss the role of transposable elements in prokaryotes.

4. How do retrotransposons affect human genome?

5. Comment on *Mu* bacteriophage.

6. What are SINEs and LINEs?

7. What are the transposable elements in yeast and *Drosophila*?

8. Comment on yeast transposable elements.

9. What are the different roles of transposons?

HUMAN GENETICS

16

16.1 INTRODUCTION

There are 50,000 genes on the 46 chromosomes. Any of these can be mutated and cause disease in human. Mutations are generally classified into three categories; **genome mutation** (mutation that affects the number of chromosomes), **chromosome mutation** (mutation that alters structure of the chromosome) and **gene mutation** (mutation that alters individual genes). The chromosome mutation is structural alterations such as duplications, deletions, inversions and translocations, which can occur spontaneously or as a result of abnormal segregation of translocated chromosomes during meiosis. Single nucleotide changes or alterations in a group of base pairs can cause gene mutations. If any of these three kinds of mutations can be passed on to future generations, they are called as germ-line mutations. Some mutations occur by chance only in a subset of cells or tissues, and cannot be inherited. These kinds of mutations are called somatic mutations.

Human disease is caused not only due to the mutation of gene but may also be due to environmental influence. Now the diseases are recognized by three main types; single gene disorder, chromosome disorder or multifactorial disorder. Single gene defects are caused by presence of mutation in a gene (example, cystic fibrosis). The mutation may be present on a single allele or on both the alleles. Similarly mutations in the mitochondrial gene can also cause polycystic kidney, cardiomyopathy, etc. Single gene disorders usually exhibit pedigree patterns. Multifactorial inheritance results in many congenital malformations. Here the disease is the result of a combination of variations in genes in concert with environmental factors. This kind of inheritance shows no characteristic pedigree patterns like single gene defect. In chromosome disorder, the defect is due to increase or decrease in whole chromosome or a segment of chromosome.

16.2   GENES IN PEDIGREE

Expression of any human character is likely to require a gene or large number of genes and environmental factors. When gene expressions are extreme, this impairs the health, fitness and reproductive capacity of the individual. Transmission of characters occurs through single gene in any one of the following modes: autosomal dominant inheritance, autosomal recessive inheritance, X-linked recessive inheritance, X-linked dominant inheritance and Y-linked inheritance.

16.2.1   Autosomal Dominant Inheritance

- An affected person usually has an affected parent.
- Affects both male and female.
- Transmitted by both the sexes.
- An affected × unaffected mating has a 50% chance of progeny being affected.

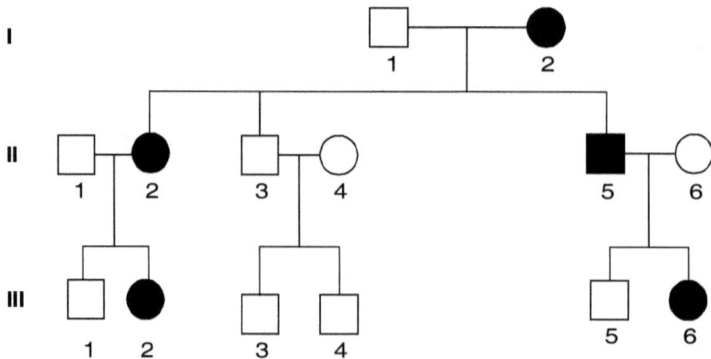

Best examples are

1. Huntington's chorea
2. Polycystic kidney disease
3. Familial hypercholesterolaemia
4. Myotonic dystrophy

16.2.2   Autosomal Recessive Inheritance

*   Autosomal recessive.
*   Parents of affected are usually asymptomatic carriers.
*   Affects both male and female.
*   After the birth of an affected child, each subsequent child has a 25% chance of being affected.

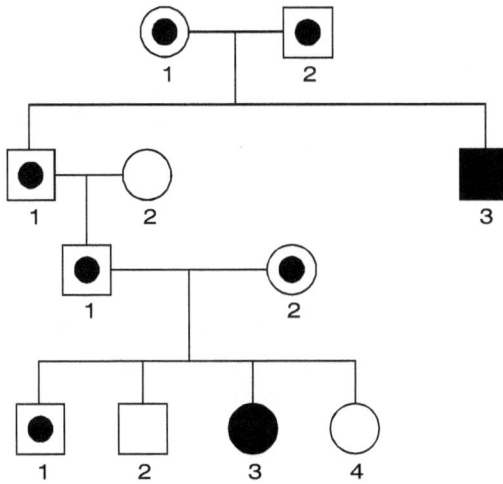

Best examples are

1. Cystic fibrosis
2. Sickle cell anaemia
3. β-thalassemia
4. Phenylketonuria
5. Hurler's syndrome

16.2.3   X-linked Recessive Inheritance

*   Affects almost exclusively males.

- Affected males are usually born to unaffected parents, the mother is normally an asymptomatic carrier and may have affected male relatives.

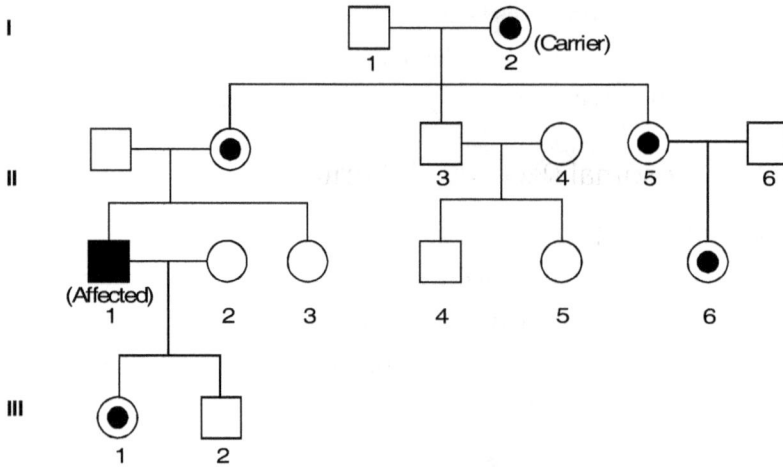

Best examples are

1. Ocular albinism
2. Fragile X syndrome
3. Haemophilia A
4. Haemophilia B
5. Hunter's syndrome

There is no male-to-male transmission in the pedigree (but matings of an affected male and carrier female can give the appearance of male-to-male transmission. This also occurs as a result of non-random X-inactivation.

16.2.4   X-linked Dominant Inheritance

- Affects either sex, but more females than males.
- Females are often more mildly and more variably affected than males.
- The child of an affected female, regardless of its sex, has a 50% chance of being affected.
- For an affected male, all his daughters but none of his sons are affected.

Best examples are

1. Incontinentia pigmenti
2. Orofaciodigital syndrome

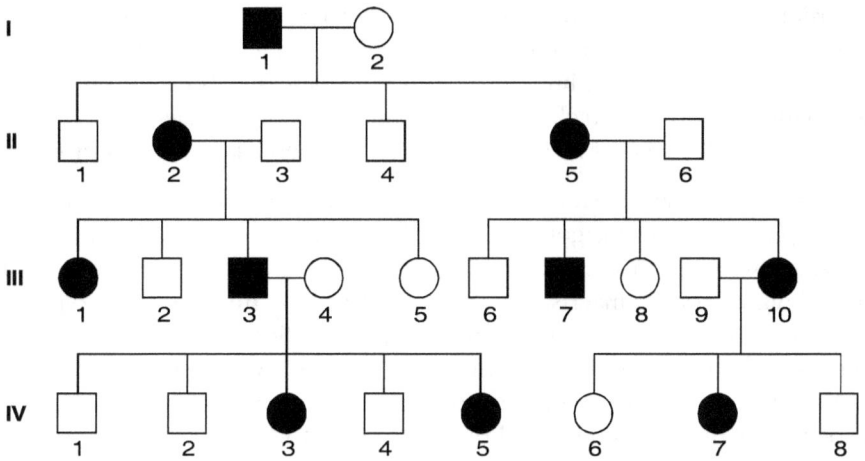

16.2.5   Y-linked Inheritance

- Affects only males.
- Affected males always have an affected father.

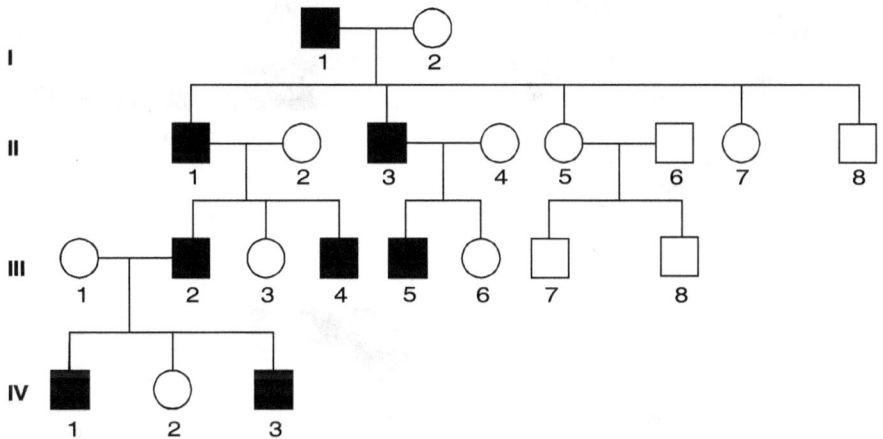

Best examples are

1. Porcupine skin
2. Webbed toes

16.3   HUMAN GENETIC DISEASES

In this chapter, we present a more detailed account of several common genetic disorders and their causes and consequences. Chromosomal disorders mainly happen due to abnormal chromosomal numbers or due to the structural aberration of chromosome. When a nucleus has a single copy of each chromosome (in humans, 23 chromosomes), it is called **euploidy**. When there is variation in chromosomal number (either increase as in Down's syndrome or decrease as in Turner's syndrome) then it is called **aneuploidy**.

Aneuploidy may be caused due to non-disjunction, premature disjunction or anaphase lag during germ-cell division. The frequency of such events increases with increase in maternal age. Mosaicism (different cell lines from single zygote) or chemacrism (different cell line by fusion of two zygotes) also possibly cause chromosomal disorder. Some of the important chromosomal disorders are discussed in this chapter. Variability in expression is also frequent with variance in the degree of abnormality. Some of the important clinical features are shown in Figure 16.1.

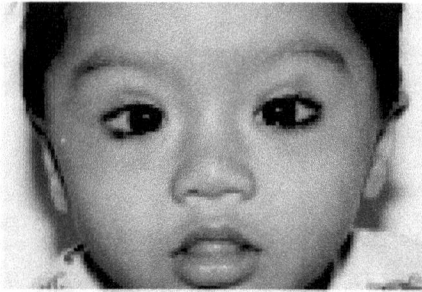

Opening of the mouth crease, protrusion of the tongue, ptosis, blunt nasal bridge

Ear anomaly

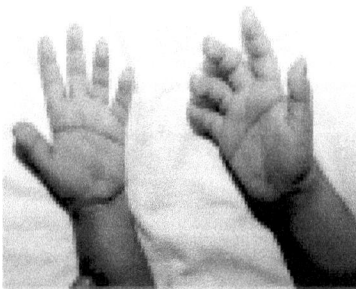

Simian crease

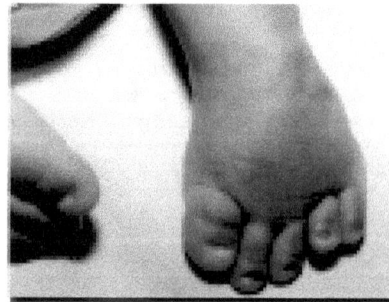

Edema

Figure 16.1   Clinical features of some of the syndromes

(Continues)

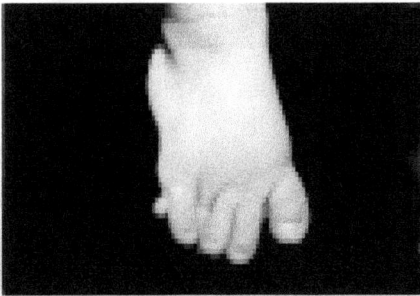

Polydactyly
Clinodactyly
Hypoplasia of nails

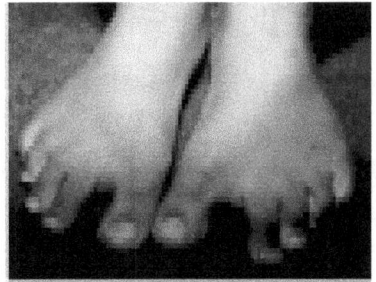

Gap in the toe

Webbed neck

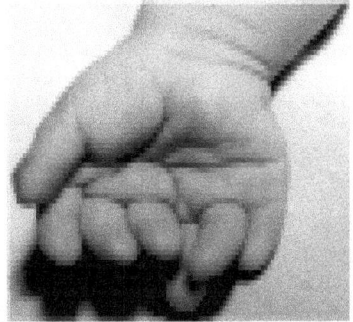

Arachnodactyly

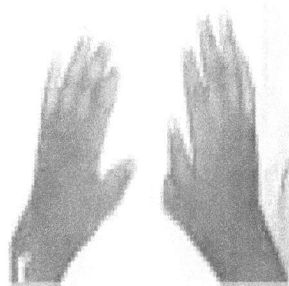

Polydactyly

Figure 16.1   Clinical features of some of the syndromes

16.3.1   Down's Syndrome

Etiology	21 Trisomy (94%), Mosaicism (2.4%), Translocation D group/G group and G group/G group (3.3%).
Abnormalities	Hypotonia, flat facies, slanted palpebral fissures and small ears are most common.
CNS	Mental deficiency.
Cranofacial	Brachycephaly, short hard palate, small noses with low nasal bridge, inner epicanthal folds.
Eyes	Speckling of iris (Brushfield spots), refractive error, mostly myopia.
Ears	Small or absent earlobes.
Dentition	Hypoplasia, irregular placement.
Hands	Hypoplasia of mid-phalanx of 5th finger, clinodactyly, single crease, Simian crease.
Feet	Wide gap between 1st and 2nd toes.
Pelvis	Hypoplasia.
Cardiac	Endocardial cushion defect, ventricular septal defect, auricular septal defect, patent ductus arteriosus.
Skin	Loose folds in posterior neck (infancy).
Genitalia	Relatively small penis, primary gonadal deficiency.
Occasional abnormalities	Gastrointestinal anomalies including duodenal atresia, annular pancreas, hischprung disease, imperforate anus, hip abnormalities, thyroid disorders.

16.3.2   Edward's Syndrome

Etiology	Trisomy 18, 47, XX, +18 or 47, XY, +18, mosaicism and partial trisomy.
Abnormalities	Clenched hand, short sternum, low arch, dermal ridge, patterning in fingertips.
General	Feeble foetal activity, weak cry, growth deficiency, hypoplasia of skeletal muscle, mental deficiency, polyhydraminos.

Cranofacial	Prominent occiput, low set, malformed auricles, small oral opening, micrognathea, wide frontal, microcephaly, inner epicentral folds, cleft lip, cleft palate or both, hypertelorism, cataract.
Hands and feet	Clenched hand, tendency for overlapping of index finger over third, fifth over fourth hypoplasia of nails of fifth finger and toes, low arch, dermal ridge, pattern on six or more fingertips, rocker bottom feet, syndactyly.
Thorax	Short sternum, small nipples.
Pelvis and Hip	Small pelvis, limited hip abduction.
Genitalia	Cryptorchidism in male, hypoplasia of labia majora in females.
Cardiac	Ventricular septal defect, auricular septal defect, patent ductus arteriosus, pulmonary stenosis, coarctation of aorta, anomalous coronary artery, tetralogy of fallot.
Lung	Malsegmentation to absence of right lung.
Endocrine	Thyroid or adrenal hypoplasia.

16.3.3 Klinefelter's Syndrome

Etiology	47, XXY; 47, XXY/XY
Abnormalities	Hypogenitalism, hypogonadism, with or without long legs, dull mentality and/or behavioural problems.
Performance	IQ above average (Mean 85–90).
	Behavioural problems, immaturity, insecurity shyness, poor judgement.
Growth	Long limbs, low upper to lower segment ratio, relatively tall and thin stature.
Hypogonadism and Hypogenitalism	Relatively small penis and testis, testosterone production is inadequate, infertility, gynecomastis.
Others	Mild elbow dysplasia, fifth finger clinodactyly, cryptorchidism, mild to moderate ataxia.

16.3.4 Turner's Syndrome

Etiology	45, XO 45, XO/XX 45, XO/XY

Abnormalities	Short female, broad chest with wide spacing of nipples, congenital lymphedema, gonadal dysgenesis.
Growth	Small stature, tendency to become obese.
Performance	Mean IQ-90, delays in motor skills clusiness, low self esteem, and depression in teenagers.
Gonads	Ovarian dysgenesis with hypoplasia.
Lymph vessels	Congenital lymphedema.
Thorax	Broad chest with widely spaced nipples, often mild pectus excavatum.
Auricles	Anomalous auricles.
Facies	Narrow maxilla, relatively small mandible.
Neck	Low posterior hairline, appearance of short neck, webbed posterior neck.
Other skeletal	Bone dysplasia, dislocation of hip.
Skin	Excessive pigmented nevi, loose skin.
Renal	Horseshoe kidney, double or cleft renal pelvis.
Cardiac	Cardiac defects of bicupid aortic valve, coarctation of aorta, valvular aortic stenosis, mitral valve prolapse.
Central nervous system	Perceptive hearing impairment, mental retardation.

16.3.5   Patau's Syndrome

Etiology	47, XY, + 13 or 47, XX, + 13 Trisomy 13 mosaicism Partial trisomy (13 pter à q14) Partial trisomy (13 q14 à qter)
Abnormalities	Defects of eye, nose and lip, holoprosencephaly of forebrain, polydactyly, narrow, hyperconvex fingernails, skin defects.
Central nervous system	Holoprosencephaly, incomplete development of forebrain, severe mental defect, hyper/hypotoxia, hydrocephalia, cerebellar hypoplasia.
Hearing	Apparent deafness.

Eyes	Microphthalmic, retinal dysplasia.
	Absent eyebrows, shallow supraorbital ridges, hyper/hypotelorism.
Mouth	Cleft lip, cleft palate or both.
Skin	Capillary hemangiomata, loose skin.
Auricles	Abnormal helices with low set ears.
Hands and Feet	Simian crease, hyperconvex narrow fingernails, polydactyly of hands and sometimes feet, posterior prominence of heel.
	Retroflexible thumb, syndactyly, cleft between first and second toes, radial aplasia.
Other skeletal	Thin posterior ribs with or without missing ribs, hypoplasia of pelvis.
Cardiac	Ventricular septal defect, patent ductus arteriosus, auricular septal defect and dextroposition.
	Anomalous venous return, overriding of aorta, pulmonary stenosis, aneurysm of ascending aorta.
Kidney	Polycystic kidney, hydronephrosis, duplicated ureters.
Genitalia	Male—Cryptorchidism, abnormal scrotum, hypospadias.
	Female—Bicornuate uterus, duplication or anomalous insertion of fallopian tubes, uterine cysts.

16.3.6 Shprintzen Syndrome (Velo-Cardio-Facial Syndrome)

* Autosomal dominant mode of transmission
* Interstitial deletion of chromosome 22q11.21–q11.23

Performance	Hearing disabilities, mild intellectual impairment, IQ ranges from 70–90.
Growth	Postnatal onset, short stature.
Ears and Hearing	Conductive hearing loss, secondary to cleft palate.
Craniofacial	Cleft of the secondary palate, velopharyngeal incompetence, narrow palpebral fissures, abundant scalp hair, microcephaly.

Limbs	Slender and hypotonic with hyperextensible hands and fingers.
Cardiac	Ventricular septal defect, right aortic arch tetralogy of fallot.
Nails	Hypoplastic.
Teeth	Neonatal teeth, partial anodontic, small teeth.
Mouth	Short upper lip.
Cardiac	Half of the patients have cardiac defect, common atrial septal defect with single atrium.

16.3.7   Cardio-Facio-Cutaneous (CFC) Syndrome

- Sporadic, autosomal or dominant

Abnormalities	Congenital heart defects, ectodermal anomalies, frontal bossing.
Neurological	Mild and moderate mental retardation, hypotonical, nystagmus, strabismus, cortical atrophy, hypoplasia of the frontal lobes.
Growth	Postnatal growth deficiency.
Craniofacies	Relative macrocephaly with large prominent forehead, bitemporal narrowing, shallow orbital ridges, down-slanting palpebral fissures, hypertelorism, prominent philtrum.
Cardiac	Atrial septal defects, pulmonic stenosis.
Skin and hair	Sparse, curly, slow growing hair, lack of eyebrows and eyelashes, abnormalities of the skin.

16.3.8   Marfan's Syndrome

- Autosomal dominant
- Mutation in fibrillin (FBN1) gene on 15q21.1

Abnormalities	Aracheodactyl with hyperextensibility, lens subluxation, aortic dilation.
Skeletal	Long stature with long slim limbs, little subcutaneous fat, muscle hypotensin, arachnodactyly, narrow facies with narrow palate.

Eyes	Lens ausluxation usually upward, increased axial globe length, myopia.
Cardiac	Ventricular and atrial septal defect.
Dentition	Partial anodontia, enamel hypoplasia.
Musculoskeletal	Joint laxity with scoliosis.
Urinary	Renal anomalies, nephrocalcinosis, bladder diverticula.

16.3.9 Noonan Syndrome

- Gene mapped to 12q22- qter
- Autosomal dominant inheritance
- Genetic heterogeneity seen

Abnormalities	Webbing of the neck, pectus excavatum, cryptorchidism, pulmonic stenosis.
Growth	Short stature of postnatal onset, mental retardation (25%).
Facies	Epicanthal folds, ptosis of eyelids, low nasal bridge, down-slanting palpebral fissures, myopia, increased width of the mouth.
Neck	Low posterior hairline, short or webbed neck.
Thorax	Shield chest and pectus excavatum.
Skeletal	Cubitus valgus.
Heart	Pulmonary valve stenosis, left ventricular hypertrophy, septal defects, patent ductus arteriosus, branch stenosis of pulmonary arteries.
Genitalia	Small penis, cryptorchidism.
Bleeding diathesis	A variety of defects in coagulation and platelet systems, von Willebrand disease, thrombocytopenia.

16.3.10 Ellis-Van-Creveld Syndrome

- Autosomal recessive
- EVCSQ gene mapped to 5q region

Abnormalities	Short distal extremities, polydactyl nail hypoplasia.

Growth	Small stature of prenatal onset.
Skeletal	Disproportionate, short extremities, polydactyly, malformed carpals, narrow thorax with short poorly developed ribs, hypoplasia of upper lateral tibia.
Cat Eye Syndrome (also called as coloboma of iris anal atresia syndrome)	Extra chromosome derived from two identical segments of chromosome 22, consists of a satellite, the entire short arm, the centromere, a tiny piece of long arm (22pter->q11).
Abnormalities	Coloboma of iris, down-slanting palpebral fissures, anal atresia.
Performance	Mild mental deficiency, some normal intelligence but emotionally retarded.
Growth	Normal.
Craniofacial	Mild hypertelorism, down-slanting palpebral fissure, inferior coloboma of iris.
Cardiac	Seen in more than one third cases. Total anomalous pulmonary venous return, persistence of left superior vena cava.
Anus	Anal atresia.

16.4   DISORDERS INVOLVING GENES

16.4.1   Williams Syndrome

Etiology	Deletion in 7q11.23–Elastin allele.
Abnormalities	Prominent lips, hoarse voice, cardiovascular anomalies.
Growth	Mild prenatal growth deficiency, postnatal 75% normal growth rate.
Performance	Average IQ-56, friendly loquacious personality, hoarse voice, hypersensitivity to sound, mild neurological dysfunction, hyperactive deep tendon reflexes and poor coordination.
Facies	Medial eyebrow flace, short palpebral fissures, depressed nasal bridge, epicanthal folds, blue eyes, prominent lips with open mouth.

Limb	Hypoplastic nails, hallus valvus.
Cardiovascular	Supravascular aortic stenosis, peripheral pulmonary artery stenosis, pulmonic valvular stenosis.

16.4.2 Hurler's Syndrome

- Autosomal recessive
- Absence or defect in lysosomal hydroxylase
- L-L- Idurmidas (IDVA)
- IDVA gene maps to 4p16.3

Abnormalities	Coarse facies, stiff joints, mental deficiency, cloudy cornea by 1–2 years.
Growth	Deceleration of growth 6 and 18 months.
Performance	Grossly retarded progress.
Craniofacial and Eyes	Scaphocephalic macrocephaly, coarse facies with full lips, flared nostrils, inner eipcanthal folds, retinal pigmentation.
Mouth	Hypertrophied alveolar ridge, enlarged tongue.
Skeletal and Joints	Diaphyseal broadening, flaring of ribcage, thoraco-lumbar gibbus, secondary to anterior vertebral wedging, short neck, widening of medial end of clavicle.
Cardiac	Murmurs, cardiac failure, internal thickening in coronary vessels or valves.
Urinary excretion	Dermatan and heparan sulphate.

16.4.3 Ehlers-Danlos Syndrome Type I

- Autosomal dominant inheritance
- Heterogeneity

Abnormalities	Hyperextensibility of joints and skin, poor wound healing.
Face	Narrow maxilla.
Auricles	Hypermobils, tendency towards lopears.
Skin	Velvety, hyperextensible and fragile, poor wound healing.

Joints	Hyperextensibility, liability towards dislocation of hip, shoulder, elbow, knee.

16.4.4   Holt-Oram Syndrome

- Autosomal recessive
- Severe defective cholesterol biosynthesis with low plasma cholesterol levels

Abnormalities	Upper limb defect, cardiac anomaly and narrow shoulders.
Skeletal	All gradation defect in upper limb and shoulder girdle, thumbs absent, hypoplastic, syndactyly, defects of ulna, humerus, clavicle, scapula and sternum, carpal anomalies.
Cardiovascular	Ostium secundum, atrial septal defects, sometimes arrhythmias, ventricular septal defect and other types of congenital defects.

16.4.5   Smith–Lemli–Opitz Syndrome

- Autosomal recessive
- Severe defective cholesterol biosynthesis with low plasma cholesterol levels

Abnormalities	Anteverted nostrils and/or ptosis of eyelids, syndactyly of second and third toes and cryptorchidism in male.
Growth	Moderately small at birth, failure to thrive.
Performance	Moderate to severe mental deficiency.
Craniofacial	Microcephaly with narrow frontal area, auricles slanted, ptosis of eyelids, inner epicanthal folds, strabismus, broad maxillary secondary alveolar ridges.
Limb	Simian crease, high frequency of digital whole dermal ridge patterning, syndactyly of second and third toes.
Genito-urinary	Genital abnormalities (70% cases), cryptorchidism, micropenis, hypoplastic scrotum, upper tract anomalies, renal cystic dysplasia, renal duplication.
Blood vessels	Easy bruisability.

Cardiac	Mitral valve prolapse, aortic root and orsinus of valsalva dilation.
Occasional eyes	Wide nasal bridge, blue sclerae.
Skeletal	Small stature, long neck, slim skeleton.
Dentition	Small, irregular placement.
Mental deficiency	Autism.

16.4.6 Deletion 5p Syndrome

- Partial deletion of 5p
- Chromosome region maps to 5p 15.3

Abnormalities	Cat-like cry in infancy, microcephaly, downward slant of palpebral fissures.
General	Low birth weight, slow growth, cat-like cry.
Performance	Mental deficiency, hypotoxin.
Craniofacial	Microcephaly, round face, hypertelorism, epicanthal folds, downward slanting of palpebral fissures, strabismus (divergent).
Cardiac	Congenital heart disease.
Hands	Simian crease, distal axial tsiradius.
Occasional	Cleft lip and palate, absence of kidney and spleen.

16.4.7 Digeorge Syndrome

- Microdeletion of 22q11.2

Abnormalities	Primary defect—Fourth branchial arch and derivatives of third and fourth pharyngeal pouches.
Thymus	Hypoplasia to aplasia, deficit of cellular immunity.
Parathyroid	Hypoplasia to absence, severe hypocalcaemia and seizures.
Cardiovascular	Aortic arch anomalies including right aortic arch, interrupted aorta, conotruncal anomalies at truncus arteriosus, ventricular septal defect, patent ductus arteriosus and tetralogy of fallot.

Facial	Lateral displacement of inner canthi with short palpebral fissures, short plitrum micrognathia, ear anomalies.

16.4.8   Cystic Fibrosis

- Autosomal recessive
- High Incidence in Caucasian – 1/2500
- Mutation of CFTR gene on locus 7q31
  (CFTR—Cystic fibrosis transmembrane conductance regulator)
- A chloride ion channel

Abnormalities	Increased sweat concentrations of Na^+ and Cl^-.
	Chronic obstructive lung disease.
	Thick mucous secretion in lungs.
	Recurrent infections.
	Deficiencies of pancreatic enzymes (preventing normal digestion).
	Death due to pulmonary failure.
	Females—Congential bilateral absence of vas deferens.
	Occasional idiopathic chronic pancreatitis.

16.4.9   Haemophilia A and B (F8C and F9 mutation)

- α–linked
- Haemophilia A common
- Haemophilia A—F8C mutation
- Haemophilia B—F9 mutation

Abnormalities	Bleeding into soft tissues, muscles and weight-bearing joints.
	Continuous bleeding upon injury.
	Newborns—excessive cephalohaematomas.
	Excessive haematomas and haemarthroses.
	Haemophilia A-Factor 8C—cofactor in coagulation system.
	F9-protease.
	Diagnosis—Reduced activity of F8C and F9.

16.4.10   Sickle Cell Anaemia
(β-Globin Val 6 Glu Mutation)

- Autosomal recessive
- Missence mutation of β-Globin chain
- Val-6- Glu mutation
- Mutation–decrease in solubility of deoxygenated haemoglobin, making haemoglobin stiff and leads to cell distortion
- Sickling of RBC's

Abnormalities First 2 years of life—Anaemia, failure to thrive, splenomegaly, repeated infections and dactylitus.

Vaso-occlusive infections, stroke.

Acute chest syndrome.

Renal papillary necrosis.

Autosplenectomy.

Leg ulcers, priapism, bone aseptic necrosis.

Visual loss.

Bony vaso-occlusion.

Functional asplenic- increase susceptibility to bacterial infections.

Later stages of life—aplastic anaemia and haemolytic anaemia.

Patient conferred some resistance to malaria.

16.4.11   Coronary Artery Disease

- Multifactorial
- Highest in morbidity and mortality
- Prime cause is atherosclerosis
- Numbers of modifiable risk factors described include, smoking, age, diabetes
- Lipid and lipoprotein levels
- Many stages in evolution of the atherosclerotic plaque—fatty streak, fibrous tissue, intinal plaques become vascular, bleed, ulcerate, calcify and cause significant vessel narrowing—leads to occlusion and myocardial infarction

Promotion of Atherosclerosis dependent on number of genes

Serum lipid transport and metabolism genes like ApoE, ApoCIII, LDL receptor, lipoproteins (a).

Cholesterol levels.

Vasoactivity as angiotensin converting enzyme.

Blood coagulation, fibrinolysis as plasminogen activator inhibitor-1.

Platelet surface glycoprotein Ib and IIIa.

A number of genetic and nongenetic predisposing factors.

Hesels to Angina, myocardial infarction.

16.4.12   Diabetes Mellitus

- Multifactorial
- Two types – Types I – Insulin-dependent
              Types II – Non-insulin-dependent
- Familial aggregation visible

Type I   MHC locus major genetic factor in type I.

HLA DR3/DR4 heterozygotes are susceptible to type I.

Abnormality   β-cell distribution due to autoimmune process.

Decreased glucose tolerance.

Glucosuria.

Type II   Associated with obesity and raised levels of plasma-free fatty acids.

Poor glucose tolerance.

Non-insulin-dependent diabetes mellitus—derangement of insulin secretion and resistance to insulin action

Persistent hyperglycaemia.

Caused due to persistent levels of high glucose loads leads to insulin resistance.

Uncontrolled glucose levels—ketoacidosis.

Cause atherosclerosis, peripheral neuropathy, renal disease and cataracts and retinopathy.

16.5 DEGREES OF RELATIONSHIP

Figure 16.1 shows the level of relationship of proband (proposita) to various family members.

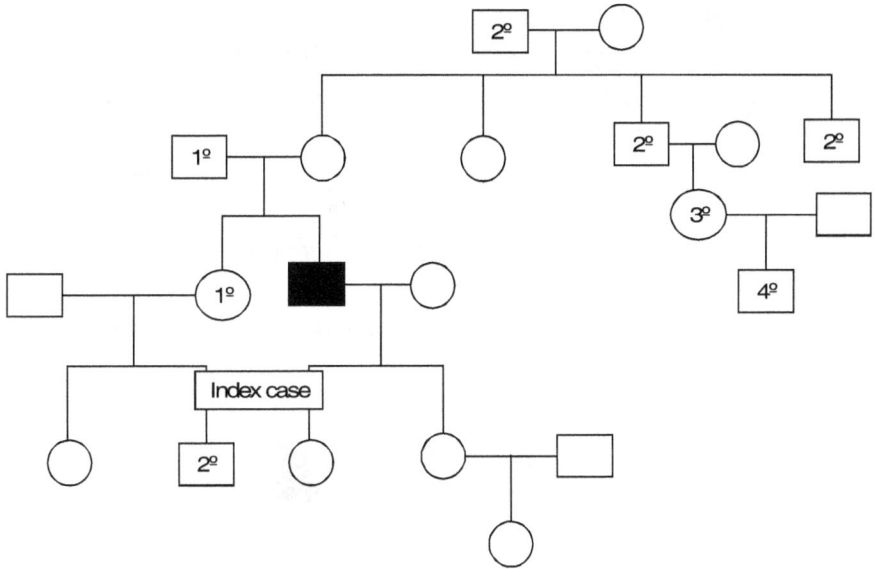

Figure 16.1   Degree of relationship

Degree	Relationship	Genes shared
1°	Parents, sibs, offspring	1/2
2°	Grandparents, parent sibs, Grandchildren niece, nephew	1/4
3°	Cousins	1/8
4°	First cousins once removed	1/16

FOR ADDITIONAL READING

1. Griffiths, A.J.F., Gelbert, W.M., Lewontin, R.C. and Miller, J.H. (2002). *Modern Genetic Analysis*. W.H. Freeman and Company, New York.

2. Griffiths, A.J.F., Gelbert, W.M., Lewontin, R.C., Miller, J.H. and Suzuki, D.T. (2002). *An Introduction to Genetic Analysis*. W.H. Freeman and Company, New York.

3. McKusick, V.A. (1998). *Mendelian Inheritance in Man*, 12th edn. Johns Hopkins University Press, Baltimore.

4. Orel, V. (1996). *Gregor Mendel: The first Geneticist.* Oxford University Press, Oxford, England.

5. Orel, V. and Hartl, D.L. (1994). "Controversies in the interpretation of Mendel's discovery." *History and Philosophy of the Life Sciences.* 16:423.

6. Snustad, D.P. and Simmons, M.J. (2002). *Principles of Genetics.* John Wiley & Sons Inc., New York.

7. Stern, C. and Sherwood, E. (1966). *The origins of Genetics: A Mendel Source Book.* W.H. Freeman and Company, New York.

8. Tom Strachan *et al.* (1996). *Human Molecular Genetics.* BIOS Scientific Publishers Ltd., UK.

9. Vogel, F. and Motulsky, A.G. (1986). *Human Genetics: Problems and Approaches,* 2nd edn. Springer, New York.

MODEL QUESTIONS

1. Draw and explain the degree of relationships.

2. Describe in detail the patterns of inheritance.

3. Discuss in detail various clinical features and etiology of Down's syndrome.

4. Discuss in detail various clinical features and etiology of Turner's syndrome.

5. Discuss in detail various clinical features and etiology of Klinefelter's syndrome.

6. Discuss in detail various clinical features and etiology of William's syndrome.

7. Discuss in detail various clinical features and etiology of Marfan syndrome and Noonan syndrome.

CANCER GENETICS

17.1   INTRODUCTION

Our somatic cells are programmed to divide, grow, differentiate and die in response to various signals that arise inside and outside the body. Cancer results due to a clone of cells, which have uncontrolled cell growth and are freed of the above programming. This leads to a mass of cells called tumour or neoplasm. The formation of tumour is known as tumorigenesis.

Neoplasm has two stages; malignant neoplasm and benign neoplasm. In malignant neoplasm the cancer cells invade nearby tissues and metastasize (spread) to distant sites in the body. But in benign neoplasm the cancer cells do not invade or metastasize nearby tissue.

The tumours are classified based on the tissue type in which they arise;

Carcinomas—epithelial tissue cancer

Sarcomas—connective tissue cancer

Lymphomas—lymphatic tissue cancer from mesoderm

Gliomas—glial cells (central nervous system)

Leukemias—cancer of the haematopoietic organs

Some of the characteristic features of cancer cells are

- Changes in cell morphology
- Loss of capacity for growth arrest
- Loss of dependence on anchorage for growth
- Increased saturation density
- Loss of contact inhibition
- Alterations in the cell surface of glycoproteins and glycolipids
- Easier agglutination by lectins
- Increased glucose transport
- Loss of surface fibronectin and actin microfilament
- Continuous release of transforming growth factors

The tumour cells are derived from a single ancestral cell, that is, they are monoclonal. The basic cause of tumour is damage to the genetic material. The cause of damage to the genetic material can be due to genetic conditions or environmental conditions, or both.

The basic cause of cancer is damage to specific genes. Usually, mutations in the genes accumulate in somatic cells over the years and there will be no risk to the whole organism, until a cell loses a critical number of growth-control mechanisms and initiates a tumour. If damage occurs in cells of the germ line however, an altered form of one of these genes can be transmitted to progeny and predispose them to cancer. The greatly increased risk of cancer in such individuals is due to the fact that each of their cells now carries the first step in a multi-step cancer pathway. The cause of cancer can be due to genetic or environmental conditions, or both.

17.2   GENETIC CONSIDERATION

Sudden or prolonged genetic alterations of cell regulatory systems are the primary basis of carcinogenesis. Cancer can be created in animal models by damaging specific genes. Contrastingly it is also possible in cell culture systems to reverse a cancer phenotype by introducing normal copies of the damaged genes into the cell. Most of the genetic events that cause cancer occur in somatic tissues during the lifetime of the individual. But the frequency of these events can be altered by exposure to mutagens, thus establishing a link to environmental cancer-causing agents (carcinogens). Since these genetic events occur in somatic cells, errors are not transmitted to future generations and they are not inherited.

Recent study shows that it is also possible for cancer-predisposing mutations to occur in germ-line cells. This results in the inheritance of cancer-causing genes from one generation to the next and produces families that have a high frequency of specific cancers. In such "cancer families", even though a single mutant allele is inherited, that is sufficient to cause specific form of cancer. All individuals who inherit such mutant allele will develop a tumour. This is due to cells that carry altered gene that has already taken the first step down the cancer pathway. The congenital retinoblastoma is a good example. A person who inherits a mutant version of the retinoblastoma gene has approximately a 90% chance of developing one or more retinoblastoma tumours.

Though the inheritance of cancer as a single-gene disorder is not so common, there are reports for more frequent clustering of cancer types in families. Families with a breast and colon cancer at first-degree relatives show a several-fold increase in the risk of developing the cancer in the subsequent relatives. This kind of familial clustering of cancer is remaining unspecified. But genetic transmission of altered forms of specific genes may be held responsible.

The extent to which inheritance of germ-line mutations versus somatic cell mutations contributes to human cancer is still obscure. More intensive screening of defined high-risk populations or cancer families could result in early detection and intervention. In turn it leads to better prognosis for individuals and brings down morbidity and mortality in the population.

17.3 ENVIRONMENTAL CONSIDERATION

Environmental factors play a major role in the progression of cancer. But individual's overall risk for cancer depends on a combination of inherited factors and environmental components.

Tumour cells are produced when mutations occur in genes that are responsible for regulating the cell's growth. But the frequency and consequences of these mutations may vary by a large number of environmental factors. For example, many chemicals (carcinogens) that cause mutation in experimental animals also cause cancer. Similarly, certain other environmental agents can enhance the growth of genetically altered cells without directly causing new mutations. Thus, it is often the interaction of genes with the environment that determines carcinogenesis.

One example for the idea that exposure to environmental agents can significantly alter an individual's risk of cancer is that of cigarette smoking. Cigarette smoking can cause lung and other types of cancer. Similarly uranium dust can cause lung cancer among miners and asbestos exposures causing lung cancer among miners are also well documented.

Alternatively, breast cancer, is most prevalent among northern Europeans and Americans but rare among women in developing countries. It is difficult to understand such dissimilarities; it is not clear whether it reflects the differences in lifestyle or in gene frequencies.

The regulation of cell growth is accomplished by substances that include

1. growth factors that send signals between cells
2. specific receptors for the growth factors
3. signal transduction that activates a cascade of phosphorylating reactions within the cell
4. nuclear transcription factors.

The cell integrates and interprets the signals received by the host from its environment. Decisions whether or not to divide, grow and differentiate depends on the result from processing of the signals.

17.4   MAJOR CLASSES OF CANCER GENES

Cancer genes can be classified into three categories;

1. Those that normally inhibit cellular proliferation (tumour suppressors),
2. Those that activate proliferation (oncogenes) and
3. Those that participate in DNA repair.

17.4.1   Tumour Suppressor (TS) Genes

The biochemical action of tumour suppressor (TS) gene products has proved harder to unravel than the function of oncogene products. They have roles similar to cell adhesion molecule and also control the cell cycle progression, as negative regulators (Table 17.1). The two most important genes are RB and P^{53}. The products of these two genes play an important role in the control of cell cycle progression. Simultaneous inactivation of both the gene products is frequently observed in tumour cells, and their functions partially overlap.

The RB1 gene is a 110-kDa nuclear protein (pRb) and widely expressed. It is believed to play a key role in controlling cell proliferation. At least part of this role is to bind and inactivate cellular transcription factors called E2F, which are required for cell cycle progression. In normal cells, the pRb is inactivated by phosphorylation reaction and simultaneously activated by dephosphorylation. Two hours before a cell enters into the S phase of the cell cycle, pRb is phosphorylated. Phosphorylation of pRb releases the inhibition of E2F and allows the cell to proceed to S phase. Phosphorylation is governed by a whole series of cyclins, cyclin-dependent kinases and cyclin kinase inhibitors. This seems to constitute the most crucial single checkpoint in the cell cycle.

The product of the MDM2 oncogene is mostly amplified in sarcomas. This binds to pRb and inhibits. This in turn favours cell cycle progression. In cancer cells pRb can be inactivated in several other ways also. Several viral oncoproteins such as adenovirus E1A, SV40 T-antigen and human papillomavirus E7 protein can bind to pRb and degrade it. This may be directly inactivated by loss-of-function mutations in the RB1 gene and result in retinoblastoma and osteosarcomas.

The other most important tumour suppressor gene is P^{53}. It is found in more than 60% of all human cancers which either have lost P^{53} protein or have mutations. P^{53} is a nuclear phosphoprotein and found in SV40 transformed cells, where it is associated with the T antigen. TP^{53} maps to 17 p 12 and this is one of the commonest regions of tumours. Tumours which have not lost TP^{53} very often have mutated versions of it. To complete the picture of TP^{53} as a TS gene, constitutional mutations in TP^{53} are found in families with the dominantly inherited Li-Fraumeni syndrome. Affected family members suffer multiple primary tumours, typically including soft tissue sarcomas, osteosarcomas, tumours of the breast, brain and adrenal cortex, and leukaemia.

Loss or mutation of TP^{53} is the most common single genetic change in cancer. The biochemical function of P^{53} is to act as a transcription factor. Tetramers of P^{53} bind to DNA and activate transcription of reporter genes placed in the downstream of a P^{53} binding site. However, P^{53} is believed to have a broader role in the cell such as "the guardian of the genome". It stops replication of cells with damaged DNA. P^{53} is also involved in a checkpoint at the G1/S stage of the cell cycle. Normal cells with damaged DNA arrest at this point until the damage is repaired, but cells that lack P^{53} or contain a mutant form do not arrest at G1. Replication of damaged DNA leads to random genetic changes, some of which are oncogenic, similar to cells with a defective mismatch repair system.

Table 17.1 Tumour suppressor genes and cancer

Gene (related genes in parentheses)	Chromosome location	Function of gene product	Disease caused by germ-line mutations
RB1 (p107, p130)	13q14	Cell cycle brake; binds to E2F	Retinoblastoma; osteosarcoma
APC	5q21	Interacts with β-catenin in Wnt signalling pathway	Familial adenomatous polyposis
NF1	17q11	Downregulates Ras protein	Neurofibromatosis type1
NF2	22q12	Possible link between cell membrane proteins and cytoskeleton	Neurofibromatosis type2
p53	17p13	Transcription factor; induces cell cycle arrest or apoptosis	Li-Fraumeni syndrome
VHL	3p25	Regulates transcriptional elongation	Von-Hippel Lindau disease (renal cancer)
WT1	11p13	Transcription factor	Wilms tumour
p16(p15)	9p21	CDK inhibitor	Familial melanoma
BRCA1	17q21	Interacts with RAD51 DNA repair protein	Familial breast/ovarian cancer
BRCA2	13q12	Interacts with RAD51 DNA repair protein	Familial breast cancer
PTEN	10q23	Phosphatase	Cowden disease (breast and thyroid cancer)
AT	11q22	Cell cycle regulator; responds to DNA damage	Ataxia telangiectasia

17.4.2   Oncogenes

The next important category of genes that can cause cancer is termed oncogenes. Most oncogenes originate from proto-oncogenes, which are genes that are involved in the four basic regulators of normal cell growth (growth factors, growth factor receptors, signal transduction molecules, and nuclear transcription factors) (Table 17.2). When a mutation occurs in a proto-oncogene, it can become an oncogene. When a cell proceeds from regulated to unregulated growth, the cell is said to have been transformed.

Unlike tumour suppressor genes, oncogenes are usually dominant at the cellular level. A single copy of a mutated oncogene is enough to contribute to the multi-step process of tumour progression.

The major difference between Ts and oncogenes are

- The tumour suppressors are typically disabled by loss-of-function mutations whereas oncogenes are typically activated by gain-of-function mutations.

- Most tumour suppressor genes are known to exhibit germ-line mutations that cause inherited cancer syndromes (retinoblastoma, Li-Fraumeni syndrome). In contrast, although oncogenes are commonly found in sporadic tumours, germ-line oncogene mutations that cause inherited cancer syndromes are uncommon.

Retroviruses are capable of inserting oncogenes into the host genome and transform the host cell into a tumour-producing cell. The study of such retroviral transmission has identified a number of specific oncogenes. Retroviruses are RNA viruses capable of using reverse transcriptase to transcribe RNA into DNA, which is inserted into a chromosome of a human cell (HIV). These retroviruses carry altered versions of certain growth-promoting genes into cells. In an earlier cycle of infection, a retrovirus may have incorporated a mutant oncogene from the genome of its host. When the retrovirus invades a new cell, it can transfer the oncogene into the genome of the new host, thus transforming the cell.

A number of gene products that receive and interpret extracellular signals for growth or differentiation have been identified through retrovirus oncogenes. For example the *sis* oncogene is carried by the simian sarcoma virus. It has been identified as an altered version of the human gene that encodes platelet-derived growth factor (PDGF). Similarly retrovirus studies identified the gene encoding the receptor molecule for another of the growth factors, epidermal growth factor (EGF), through the *ErbB* oncogene. The *ras* (rat sarcoma) oncogene was first identified through transforming retroviruses. Transforming transcription factor genes, *myc*, *jun*, and *fos*, were identified as other molecular components capable of initiating cell transformation.

Table 17.2   Oncogenes and cancer

Oncogene	Loci	Proposed function	Associated tumour
Growth factors			
hst	11q13	Fibroblast growth factor	Stomach carcinoma
sis	22q12	B subunit of platelet-derived growth factor	Glioma (brain tumour)
Growth factor receptors			
RET	10q	Receptor tyrosine kinase	Multiple endocrine neoplasia
ErbB		Epidermal growth factor	Glioblastoma (brain tumour) breast cancer
ErbA	17q11	Receptor thyroid hormone receptor	Promyelocytic leukemia
neu		Receptor protein kinase	Neuroblastoma
Signal transduction proteins			
H-ras	11p15	GTPase	Carcinoma of colon, lung, pancreas
K-ras	12p12	GTPase	Melanoma, thyroid carcinoma AML
abl	9q34	Protein kinase	Chronic myelogenous leukemia; acute lymphocytic leukemia
Transcription factors			
N-myc	2p24	DNA-binding protein	Neuroblastoma; lung carcinoma
myb	6q22	DNA-binding protein	Malignant melanoma; lymphoma; leukemia osteosarcoma
fos	14q24	Interacts with *jun* oncogene to regulate transcription	

17.4.3 DNA Repair Genes

The last class of genes concerned with cancer are called "DNA repair genes". DNA repair genes code for proteins which are being used to correct errors that arise when cells multiply their DNA before cell division. Mutations in DNA repair genes can lead to a failure in DNA repair (Table 17.3). This in turn allows subsequent mutations in tumour suppressor genes and proto-oncogenes to accumulate. The best example is a person with a condition called Xeroderma pigmentosum, who has an inherited defect in a DNA repair gene. As a result, they cannot effectively repair the DNA damage that normally occurs when skin cells are exposed to sunlight. This in turn leads to skin cancer. Certain forms of hereditary colon cancer also involve defects in DNA repair.

Table 17.3 Inherited syndromes due to defects in DNA repair

Syndrome or Genes	Chromosome location	Function of gene product	Disease caused by germ-line mutations
MSH2, 3, 6, *MLH1*, *PMS2*	2p21, 3p22.3, 7p22.1	Mismatch repair	Colon cancer
Xeroderma pigmentosum (XP) groups A-G	ERCC2 19q13	Nucleotide excision-repair	Skin cancer, cellular UV sensitivity, neurological abnormalities
Ataxia-telangiectasia (AT)	11q22.3	ATM protein, a protein kinase activated by double strand breaks	Leukemia, lymphoma, cellular γ-ray sensitivity, genome instability
BRCA-2	13q13.1	Repair by homologous recombination	Breast and ovarian cancer
Werner syndrome (WRN gene)	8p12	Accessory 3´-exonuclease and DNA helicase	Premature aging, cancer at several sites, genome instability
Bloom syndrome (BLM gene)	15q26.1	Accessory DNA helicase for replication	Cancer at several sites, stunted growth, genome instability
Fanconi anaemia groups A-G	16q24.3, xp22.31, 9q22.32, 3p25.3, 6p21.31, 11p14.3, 9p13.3	DNA interstrand cross-link repair	Congenital abnormalities, leukemia, genome instability

FOR ADDITIONAL READING

1. Lynn, B., Jorde, John, C. Carey *et al.* (1997). *Medical Genetics*, 2nd edn. Mosby Inc., Missouri.

2. Benjamin Lewin. (2000). *Genes VII*. Oxford University Press Inc., New York.

3. De Vita, V.T., Hellman, S. and Rosenberg, S.A. (eds.) (2000). *Cancer: Principles and Practice of Oncology*, 6th edn. Lippincott, Williams and Wilkins, Philadelphia.

4. Hahn, S.A., Schutte, M., Hoque, A.T. *et al.* (1996). "DPC4, a candidate tumour suppressor gene at human chromosome 18q21.1." *Science*. 271:350–353.

5. Hilgers, W. and Kern, S.E. (1999). "Molecular genetic basis of pancreatic adenocarcinoma." *Genes Chromosomes Cancer*. 27:1–12.

6. Varmus, H. and Weinberg, Ra. (1993). *Genes and the Biology of Cancer*. New York Scientific American Library.

7. Vogelstein, B. and Kinzler, K.W. (eds.) (1998). *The Genetic basis of human cancer*. McGraw-Hill, New York.

8. Wallrapp, C., Muller-Pillasch, F., Micha, A. *et al.* (1999). "Strategies for the detection of disease genes in pancreatic cancer:" *Ann. NY Acad. Sci.* 880:122–146.

9. Artandi, S.E. and DePinho, (2000). "Mice without telomerase: What can they teach us about human cancer?" *Nat. Med.* 6:852–855.

10. Chambers, A.F., Naumov, G.N., Vantyghem, S. and Tuck, A.B. (2000). "Molecular biology of breast cancer metastasis: clinical implications of experimental studies on metastatic inefficiency." *Breast Cancer Res.* 2:400–407.

11. Edwards, P.A.W. (1999). "The impact of developmental biology on cancer research: an overview." *Cancer Metastasis Rev.* 18:175–180.

12. Hanahan, D. and Weinberg, R.A. (2000). "The hallmarks of cancer." *Cell.* 100:57–70.

13. Karran, P. (1996). "Microsatellite instability and DNA mismatch repair in human cancer." *Semin. Cancer Biol.* 7:15–24.

14. Kinzler, K.W. and Vogelstein, B. (1996). "Lessons from hereditary colorectal cancer." *Cell.* 87:159–170.

15. Lowe, S.W. and Lin, A.W. (2000). "Apoptosis in cancer." *Carcinogenesis.* 21:485-495.

16. Macleod, K.F. and Jacks, T. (1999). "Insights into cancer from transgenic mouse models." *J. Pathol.* 187:43–60.

17. McCormic, F. (1999). "Signalling networks that cause cancer." *Trends Cell Biol.* 9:M53–56.

18. Mendelsohn, J., Howley, P.M., Israel, Ma., Liotta, L.A. (eds.) (2001). *Molecular basis of cancer*, 2nd edn. Saunders, Philadelphia.

19. Ridley, A. (2000). "Molecular switches in metastasis." *Nature*. 406:466–467.

20. Vogelstein, B., Lane, D.P. and Levine, A.J. (2000). "Surfing the p53 network." *Nature*. 408:307–310.

21. Weinberg, R.A. (1995). "The retinoblastoma protein and cell-cycle control." *Cell*. 81:323–330.

22. Sidransky, D. (1997). "Nucleic acid based methods for the detection of cancer." *Science*. 278:1054–1059.

23. Waldman, T., Khang, Y., Dillehay, L. et al. (1997). "Cell-cycle arrest versus cell death in cancer therapy." *Nat. Med.* 3:1034–1036.

MODEL QUESTIONS

1. What is the role of environment in carcinogenesis?

2. Compare the oncogenes and tumour suppressor genes.

3. What are the various kinds of genetic instabilities in cancer?

4. Comment on DNA repair genes.

5. List various oncogenes and their proposed functions.

6. What are the various germ-line mutations caused by tumour suppressor genes?

7. Why is the chance of members of Li-Fraumeni syndrome families developing cancer relatively small?

GENETICS OF CARDIOVASCULAR DISEASES

18

18.1 INTRODUCTION

The cardiovascular system is the first functional organ to develop and function in the embryonic life. During the first and second week, ovulation, fertilization, blastulation, implantation, gastrulation and development of placenta are the principal events taking place.

On the 15th day, the lateral plate mesoderm gets split into parietal mesoderm and visceral mesoderm. On the 16th day, visceral mesoderm at the anterior edge of the embryonic disc (in the cardiogenic area) differentiates to form small masses of angioblastic tissues. The angioblastic tissues give rise to endothelium and blood vessels. These blood vessels unite in this area to form two distinct heart tubes. These tubes become specialized with contractile filaments and probably beat begins at this stage. Soon these two straight tubes fuse with one another and show a series of dilations, viz. bulbus cordis, ventricle, atrium and sinus venosus. The bulbus cordis represents the atrial end of the heart and it has three sections. The proximal region is the primitive right ventricle, the middle region is conus cordis and distal region is the truncus arteriosus, which later forms the aorta and pulmonary artery.

The sinus venosus represents venous end of the heart and it has two horns. Each horn has one vitelline vein from yolk sac, one umbilical vein from the placenta and one cardinal vein from the body wall. The development of straight heart tube gets completed by the 20th day. On the 21st day, as an intrinsic ability, the straight heart tube forms cardiac loop called Bulbo–ventricular loop. They normally form on the right side (D-loop) or abnormally on the left side (L-loop).

On the 22–24th days the first pair of aortic arches are formed and the development of left ventricle is more advanced than that of the right. On the 26–28th days the 'in-series circulation' begins and at this time blood flows from the morphological right atrium to the left atrium, to the left ventricle, to the right ventricle and to the truncus arteriosus. On the 29th day, the left ventricle, the right ventricle and interventricular septum continue to grow and develop. The conus merges with the primitive ventricle, the dilated lower part communicates posteriorly with the atria through the right and left atrio-ventricular canals. The atrio-ventricular valves grow from the endocardial cushions and ventricular muscles. From the floor of the bulbo–ventricular cavity, the inter ventricular septum grows upward, divides the lower part into right ventricle and left ventricle and partially fuses with atrio-ventricular cushions. By the 33rd day, the right atrium, the left atrium and the atrial septum are well developed. The sinus venosus and the atrial chamber have open communication with each other. Later they become partially separated by grooves that appear on the lateral wall of the heart tube.

The blood now enters the atrium through the right horn of the sinus venosus. The left horn becomes much reduced in size. The septum primum and septum secundum grow as an interatricular septum to close the communication between left atrium and right atrium. However, foramen ovale closes only after birth by the fusion of flaps.

18.2 ANATOMY AND PHYSIOLOGY OF THE CARDIOVASCULAR SYSTEM

The cardiovascular system is also called blood vascular system or the circulatory system. It consists of the heart, which is a muscular pumping device, and a closed system of vessels called arteries, veins and capillaries.

The heart is a conical, hollow muscular organ (Figure 18.1) situated in the middle of mediastinum. It is enclosed within the pericardium. It pumps blood to various parts of the body to meet their nutritive requirements. The heart weighs about 300 g in males and 250 g in females. The interior of the heart is divided into four chambers, two upper and two lower. The upper chambers are named atria and the lower ones as ventricles. Of these, the ventricles are considerably larger and thick-walled than the atria because they carry a heavier pumping burden. The heart has three surfaces such as anterior, inferior and left surfaces and four borders, viz. upper, lower, right and left borders. The atria are separated from the ventricles by an atrio-ventricular or coronary sulcus. The anterior interventricular groove is nearer to the left margin of the heart and runs downwards to the left. The lower end of the groove separates the apex from the rest of the inferior border of the heart. The posterior interventricular groove is situated on the diaphragmatic surface of the heart. The two interventricular grooves meet at the inferior border near the apex.

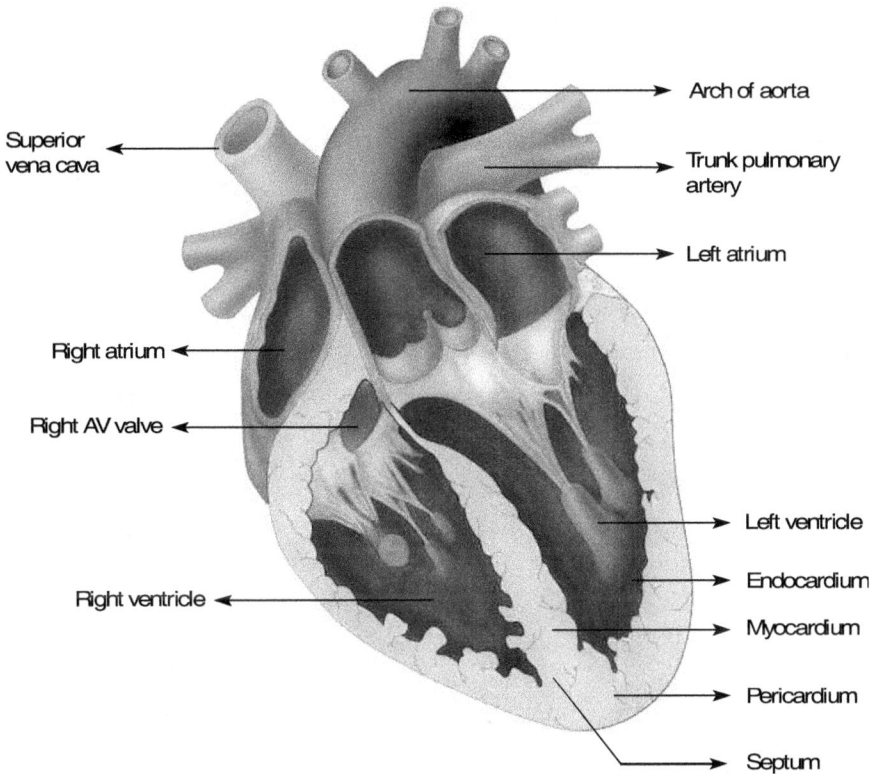

Figure 18.1   Section of the heart showing various parts

18.2.1   Wall of the Heart

There are three distinct layers of tissue that make up the heart wall in both the atria and the ventricles. The outer layer of the heart wall is called the epicardium. The bulk of the heart wall is the thick, contractile, middle layer of specially constructed and arranged cardiac muscle cells called the myocardium. The lining of the interior of the myocardial wall is a delicate layer of endothelial tissues known as the endocardium.

18.2.2   Valves of the Heart

The heart valves are mechanical devices that permit the flow of blood in one direction only. Four sets of valves are of importance to the normal functioning of the heart. Two of these, the atrio-ventricular valves are located in the heart, guarding the openings between the atria and ventricles. The other two, the semilunar valves, are located inside the pulmonary artery and the great aorta just as they arise from the right and left ventricles respectively. The right atrioventricular orifice consists of three flaps (tricuspid) of endocardium anchored to the papillary muscles of the right ventricle by several cord-like structures called chordae tendineae. The valve that guards the left atrio-ventricular orifice is similar in structure to the tricuspid except that it has only two flaps (biscuspid).

18.2.3   Circulation

The heart acts as a pump keeping the blood moving through this circuit of vessels—arteries, arterioles, capillaries, venules and veins. In the heart, right atrium receives deoxygenated blood from the whole body through the superior and inferior vena cavae and the coronary sinus. It controls and sends the blood through the right atrio-ventricular orifice to the right ventricle. The right ventricle contracts and propels the blood into the pulmonary trunk, the pulmonary arteries and finally to the lungs where the blood is oxygenated.

The oxygenated blood returns to the heart through the 4 pulmonary veins and enters the left atrium. The left atrium contracts and sends its blood through the left atrio-ventricular orifice into the left ventricle, which in turn contracts and drives the blood into the ascending aorta to meet the nutritive requirements of the body.

18.2.4   Conduction System

The conducting system of the heart is made up of specialized cells for initiation and conduction of the cardiac impulse. Sinuatrial node is known as the 'Pacemaker' of the heart located in the superior lateral wall of the right atrium. The impulse travels through the atrial wall to reach the atrioventricular node. Then the impulse travels to reach the ventricles through the Bundle of His and bundle branches.

The resting membrane potential of individual mammalian cardiac muscle cells is about -90 mV. Stimulation of action potential is responsible for initiating contraction. The initial rapid depolarization and the overshoot are due to rapid increase in Na^+. The initial rapid repolarization is due to closure of Na^+ channels. The influx of Cl^- and Ca^{++} are also initiated subsequently. Final repolarization is due to closure of Ca^{++} channels and efflux of K^+.

18.2.5   Contraction and Relaxation

Cardiac muscle is the major biochemical transducer that converts potential energy into kinetic energy. They are striated and involuntary in function. Each muscle fibre contains several thousand myofibrils. Each myofibril in turn has myosin filaments and actin filaments which are large polymerized protein molecules that are responsible for muscle contraction. During repolarization, calcium enters the cell and triggers sarcoplasmic reticulum for the release of intracellular stores. The net result is an increase in free calcium concentration. Now calcium binds to a subunit of troponin.

The calcium/troponin/tropomyosin complex then undergoes a conformational change which facilitates myosin interaction with actin filament (sliding of filament) and hence contraction occurs. Relaxation of the myocardium is accomplished by lowering cytosolic calcium concentration. Coronary arteries supply blood to cardiac muscle to meet their nutritive requirements during contraction and relaxation.

The Frank–Starling law states that the force of cardiac contraction increases in proportion to the degree of stretching of the cardiac muscle fibres during diastole. The law thus states a fundamental property of cardiac muscle that can be understood by reference to the sliding filament hypothesis described above.

18.2.6   Nerve Connection

Parasympathetic nerves reach the heart via the vagus. These are cardio-inhibitory, that is, on stimulation they slow down the heart rate. Sympathetic nerves are derived from the upper 3–5 thoracic segments of the spinal cord. These are cardio-acceleratory.

18.3   CONGENITAL HEART DISEASES

Congenital cardiac anomalies are major forms of heart disease in the first decade of life. Some may be incompatible with intrauterine survival; others permit live birth but are soon fatal or produce manifestation early in life; still others become evident only in adult life. There are two types of congential heart defects, viz. cyanotic (blueness of the extremities) and acyanotic.

Septal defect   Any opening in the atrial septum, other than a competent foramen ovale is described as an atrial septal defect. The ostium secundum defect, ostium primum defect and endocardial cushion defect are some of atrial defects. The essential feature of atrial septal defect is a left-to-right shunt of pulmonary venous return into the right atrium that produces a high tricuspid flow and increased blood flow to the lungs, with a normal pulmonary arterial pressure.

Based on developmental pathology the congenital defects are divided into four categories;

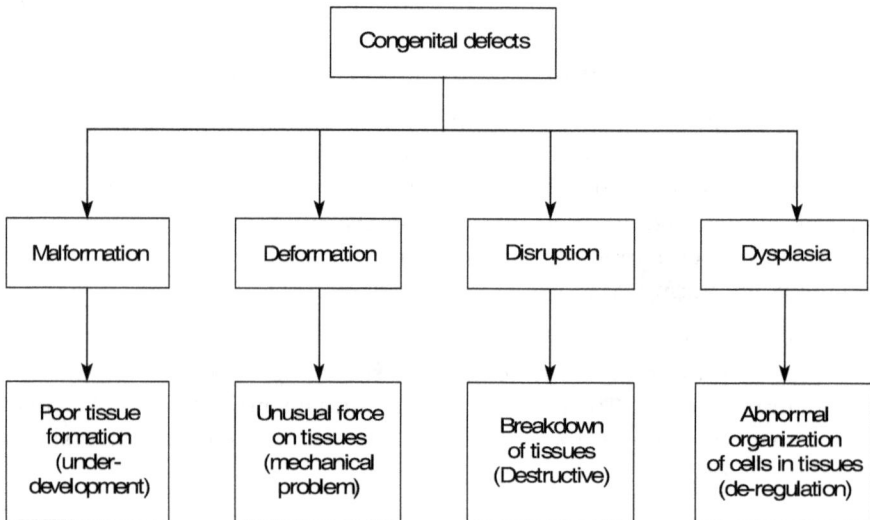

```
                        Congenital defects

Malformation      Deformation      Disruption      Dysplasia

Poor tissue       Unusual force    Breakdown       Abnormal
formation         on tissues       of tissues      organization
(under-           (mechanical      (Destructive)   of cells in tissues
development)      problem)                         (de-regulation)
```

The ventricular septal defect (VSD) describes an opening in the ventricular septum. It may be located anywhere in the ventricular septum, may be single or multiple, and may be of variable size and shape. VSD may be manifested in the isolated condition or with other defects such as atrial defect, patent ductus arteriosus and some valvular diseases. VSD are classified by their location in the septum. A partial list includes membranes, subaortic, subarterial, subpulmonary, infundibular, supracristal, intracristal and subcristal.

Patent ductus arteriosus (PDA)   Functional patency of the ductus arteriosus is abnormal if it persists more than a few hours after birth. The ductus arteriosus is derived from the left sixth embryonic arch and connects the origin of the left main pulmonary artery to the aorta, just below the left subclavian artery. During foetal life, the ductus is large and directs blood from pulmonary artery to the descending aorta; within hours of birth, the ductus closes.

Coarctation of aorta (CoA)   Coarctation of the aorta is an obstruction in the descending aorta located almost invariably at the insertion of the ductus arteriosus. It commonly becomes more obstructive as the child grows. Coarctation affects the heart by causing left ventricular hypertension and hypertrophy.

Transposition of great arteries (TGA)   In transposition of the great arteries, the aorta arises from the right ventricle and the pulmonary artery from the left ventricle. As a consequence, there are separate systemic and pulmonary circulations; life cannot be sustained unless there is some communication between them. Usually there is one or more of the following: a patent foramen ovale, an atrial septal defect, a ventricular septal defect or a persistent ductus arteriosus.

Tetralogy of fallot (TOF)   The four features of the disorder are; VSD, dextaposed aorta, obstruction to the right ventricular outflow and right ventricular hypertrophy. The VSD is usually large. The severity of the clinical manifestations is directly related to the degree of obstruction to right ventricular outflow, producing the right side hypertension, right ventricular hypertrophy and right-to-left flow.

Ebsteins anomaly   In this disorder, the posteromedial part of the tricuspid valve ring is displaced towards the apex of the right ventricle. An atrial septal defect is usually present.

Truncus arteriosus (TA)   In truncus arteriosus, there is a single semilunar valve, because the embryonic truncus has not divided into a pulmonary trunk and aorta. The vessels to the lungs thus arise from the truncus. A ventricular septal defect is always part of the lesion.

In chapter 16, already we have understood about various syndromes and their association with heart diseases. Some of the other important genes which are found to be associated with congenital heart disease, coronary artery disease, cardiomyopathy and arrhythmia are discussed in detail in the following section. Table 18.1 shows various genes and their association with congenital cardiac diseases.

18.3.1   *NKX*2.5 Gene

The human transcriptional factor *NKX*2.5 gene maps to chromosome 5q34. It has an intron sequence spanning between two exons. Both exons together code for 324 amino acids. *NKX*2.5 gene has three conserved regions such as TN domain, NX2-specific domain (NKX2-SD) and homeodomain (HD). The HD that lies within exon 2 consists of three α helices. Helix 3 is important for DNA binding specificity. Many studies indicate that occurrence of mutations in the *NKX*2.5 gene (especially in the highly conserved region of exon 2) can cause septal defects.

Table 18.1   Genes and markers associated with congenital heart defects

Gene name	Gene	Loci	Exons	Defect
Elastin	*ELN*	7q11	35	Aortic stenosis
Jagged 1 precursor	*JAG1*	20p12.2	26	Tetralogy of fallot
Heart and neural crest derivatives-expressed protein 1/eHand	*HAND1*	5q33	2	Aortic stenosis
Heart and neural crest derivatives-expressed protein 2/dHand	*HAND2*	4q33	4	Aortic stenosis
Fibrillin 1 precursor	*FBN1*	15q21	66	Ascending aorta
Iroquois homeobox gene 4	*IRX4*	5p15.3	5	Sinus venosus
GATA4 Transcription factor	*GATA4*	8p22-23	7	Atrial septal defect, ventricular septal defects, atrio-ventricular septal defects, pulmonary valve thickenings, impaired looping of the heart
GATA5 transcription factor	*GATA5*	20q13.3	7	Impaired looping of the heart
T-box transcription factor 1	*TBX1*	20q11.2	11	Digeorge syndrome

(Contd.)

Table 18.1 (Contd.)

Gene name	Gene	Loci	Exons	Defect
T-box transcription factor 5	*TBX5*	12q24.1	11	Tetralogy of fallot
Fibroblast growth factor 8 precursor	*FGF8*	10q24.32	6	Double outlet right ventricle
Homeobox protein NKX-2.5	*NKX2.5*	5q34	2	Impaired looping of the heart, atrial septal defects, ventricular septal defects, double outlet right ventricle, tetralogy of fallot
Cbp/p300 interacting transactivator 2	*CITED2*	6q24.1	2	Ventricular septal defects, double outlet right ventricle
Tyrosine–protein phosphatase, nonreceptor type 11	*PTPN11*	12q24.1	16	Pulmonary stenosis
Transcription factor activating enhancer-binding protein	*TFAP2B*	6p12	8	Patent ductus arteriosus
Ubiquitin fusion degradation protein 1 homolog	*UFD1L*	22q11.2	12	Digeorge syndrome
CDC45–related protein	*CDC45L*	22q11.2	19	Digeorge syndrome
Bone morphogenetic protein 4 precursor	*BMP4*		4	Atrio-ventricular septal defects
NP_690870.2	*DGCR*	22q11	10	Tetralogy of fallot
Tyrosine-protein kinase receptor	*TEK*	9	24	Venous malformations

18.3.2 Genes of Coronary Artery Disease

Coronary artery disease is a progressive disease characterized by the accumulation of lipids and fibrous elements in the large arteries. The early lesions of atherosclerosis consists of subendothelial accumulations of cholesterol-engorged macrophages called foam cells. This stimulates the overlying endothelial cells to produce a number of pro-inflammatory molecules, adhesion molecules and growth factors. Table 18.2 lists various genes that are associated in the augmentation of plaque formation during atherosclerosis.

Table 18.2   Genes and genetic markers involved in coronary artery disease

Gene name	Gene	Locus	Exons	Disease
Angiotensin I converting enzyme (peptidyl-dipeptidase A) 1	*ACE, (Isoforms 1, 2 and 3)*	17q23, 44.8 kb	26	Blood pressure regulation
Angiotensinogen	*AGT*	1q42-q43, 15.5 kb	5	Blood pressure regulation
Angiotensin receptor I	*AGTR I*	3q21-q25, 2.28kb	3	Blood pressure regulation
Apolipoprotein A-I	*APOA1*	11q23-q24, 1.9kb	4	Lipid metabolism
Apolipoprotein B (apoB-100, apoB-48)	*APOB*	2p24-p23, 43 kb	29	Lipid metabolism
Apolipoprotein B48 receptor	*APOB48R*	3.27kb	5	Lipid metabolism
Apolipoprotein C-I	*APOC1*	19q13.2, 4.7 Kb	4	Lipid metabolism
Apolipoprotein E	*APOE*	19q13.2, 3.6 kb	4	Lipid metabolism
Cholesteryl ester transfer protein, plasma	*CETP*	16q21, 22 kb	16	Lipid metabolism
C-reactive protein, pentraxin-related	*CRP*	1q21-q23, 1.9kb	2	Vascular inflammatory pathway
Nitric oxide synthetase-3, (endothelial)	*e-NOS – 3*	7q36, 25.1 Kb	26	Vascular inflammatory pathway
Estrogen receptor 1	*ESR 1*	6q25.1, 2.99 Mb	8	
Ghrelin	*GHRL*	3p26-p25,	4	Lipid absorption

Intercellular adhesion molecule 1 (CD54), human rhinovirus receptor	*ICAM1*	19p13.3-p13.2, 15.5 kb	7	Vascular accumulation of lipids
Interleukin 6	*IL-6*	7p21, 1 kb	5	Vascular inflammatory pathway
Lecithin-cholesterol acyltransferase	*LCAT*	16q22.1, 4.3 kb	6	Lipid metabolism
Low density lipoprotein receptor	*LDLR*	19p13.3, 44.4 Kb	18	Lipid metabolism
Lipoprotein, Lp (a), apolipoprotein (a)	*LPA*	6q26-q27, 132.8 Kb	41	Lipid metabolism
Lipoprotein lipase	*LPL*	8p22, 28 kb	10	Lipid metabolism
Myocyte enhancer family 2A	*MEF2A*	15q26, 5 kb	12	Vascular inflammatory pathway
Paraoxonase 1	*PON1*	7q21.3, 26.9 kb	9	Inflammatory pathway
Renin	*REN*	1q32	9	Blood pressure regulation
Thrombomodulin	*THBD*	20p12-cen, 8.5kb	1	Vascular cell adhesion
Thrombospondin	*THBS-1*	13q14.3, 2kb	5	Vascular cell adhesion
Tumour necrosis factor α	*TNFα*	6p21.1-21.3		Inflammatory pathway

18.3.3   Genetics of Cardiomyopathy

Cardiomyopathy means "diseases of the heart muscle" which leads to heart failure or sudden death. There are 3 main types of cardiomyopathy; **dilated** cardiomyopathy or DCM (heart cavity is enlarged and stretched), **hypertrophic** cardiomyopathy or HCM (muscle mass of the left ventricle enlarges) or **restrictive** cardiomyopathy RCM (ventricular muscle of the heart becomes exclusively rigid). Cardiomyopathy progresses from childhood and the onset of the disease varies according to the family history. Although transplantation may be an effective strategy in these patients, its implementation is hindered by inavailability of donor as well as numerous ethical, social, economic and legal issues.

The molecular etiology is not known in many cases of cardiomyopathies affecting children as well as adults, with an annual incidence of 2–8 out of 100,000 in United States and Europe. Though there are reports on association of mutations in nuclear genome and cardiomyopathy, quite a number of cases do not show any such mutations. As there is a close relationship with the cardiac muscle contraction and energy metabolism, it is quite reasonable also to speculate the role of mitochondrial DNA variations as possible cause of these cases. Various gene mutations that are found to associate with cardiomyopathy are listed below (Table 18.3).

Table 18.3   Genes and their association with cardiomyopathy

Gene name	Gene	Loci	Exons	Disease
Cardiac α-actin	*ACTC*	15q11	6	HCM
Cardiac troponin C	*TNNC1*	3p21	6	HCM
Cardiac troponin I	*TNNI3*	19q13	8	HCM,RCM
Cardiac troponin T	*TNNT2*	1q31	17	HCM, DCM
β-myosin heavy chain	*MYH7*	14q11	41	HCM, DCM
α-myosin heavy chain	*MYH6*	14q12	37	HCM, DCM
Myosin binding protein C	*MYBPC3*	11p11	34	HCM
Myosin essential light chain	*MYL3*	3p21	7	HCM
Myosin regulatory light chain	*MYL2*	12q23	7	HCM

(Contd.)

Table 18.3 (Contd.)

Gene name	Gene	Loci	Exons	Disease
A-Tropomysin	*TMSA*	15q22	15	HCM, DCM
Titin	*TTN*	2q24		HCM, DCM
Z- disk associated proteins				
Muscle LIM protein	*CLP*	11q15.1	4	DCM
Telethonin	*T-cap*	17q12	2	DCM
Sarcolemma cytoskeleton				
Dystropin	*DMD*	Xp21.2	79	DCM
A-Dystrobrevin	*DTNA*	18q12		DCM
Metavinculin	*VCL*	10q22.1-q23	22	DCM
B-sarcoglycon	*SGCB*	4q12	6	DCM
δ-sarcoglycon	*SGCD*	17q21	8	DCM
Intermediate filaments				
Desmin	*DES*	2q35	9	DCM, RCM
Lamin A/C	*LMNA*	1q21.2-3	12	DCM
Signalling enzymes				
AMP-dependent kinase γ2	*PRKAG2*	7q35-36	16	HCM
Tyrosine phosphatase	*PTPN11*	12q24.1	24	HCM
Myosin light chain kinase	*MYLK2*	20q13.3	14	HCM
Calcium regulators				
Phospholmaban	*PLN*	6q22.1	1	HCM
ATPase	*SERCA2*	12q23-q24.1	21	HCM
Tafazzin	*G4.5*	Xq28	11	DCM

18.3.4   Genes of Arrhythmia

Electrophysiological dysfunction resulting in delayed myocardial cell depolarization and repolarization is caused by sequence variation in the proteins encoded by these genes (for example, Long QT syndrome and Brugada syndrome). More recent evidence suggests that mutations in these genes (Table 18.4) can cause enhanced cell repolarization that may result in converse disorders, such as "short-QT syndrome" believed to enhance sudden death risk.

Table 18.4   Genes and markers associated with arrhythmia

Gene name	Gene	Loci	Exons	Defect
Cardiac potassium ion channel	*KCNQ1*	11p15.5	18	Congenital long QT syndrome
Cardiac potassium ion channel	*KCNH2*	7q35-q36	17	Congenital long QT syndrome
Cardiac sodium ion channel	*SCN5A*	3p21	29	Congenital long QT syndrome, Brugada syndrome and progressive cardiac conduction defect
Cardiac potassium ion channel KGN112	*KCNE1*	21q22.1	4	Congenital long QT syndrome
Ryanodine receptor 2	*RyR2*	1q42.1-q43	108	Catecholaminergic polymorphic ventricular tachycardia and arrhythmogenic right ventricular cardiomyopathy

Ion channel gene sequence variations have now been identified as being arrhythmogenic in more than half a dozen rare inherited conditions, including Anderson's disease, one form of catecholaminergic-induced ventricular tachycardias and one form of arrhythmogenic cardiomyopathy.

FOR ADDITIONAL READING

1. Schott, J., *et al*. (1997). "Isolation and characterization of a gene from the Digeorge chromosomal region homologous to the mouse Tbx1 gene." *Genomics.* 43:267–277.

2. Hatcher, C.J., *et al*. (2003). "Transcription factor cascades in congenital heart malformation." *Trends. Mol. Med.* 9:512–515.

3. Bruneau, B.G. (2002). "Transcriptional regulation of vertebrate cardiac morphogenesis." *Circ. Res.* 90:509–519.

4. Gelb, B.D. (2004). "Genetic basis of congenital heart disease." *Curr. Opin. Cardiol.* 19:110–115.

5. Deal, K.K., England, S.K., Tamkun, M.M. (1996). "Molecular physiology of cardiac potassium channels." *Physiol. Rev.* 76:49–67.

6. Tyson, J., Tranebjaerg L., Bellamn S, Wren, C., Taylor, J, Bathen, J., Aslaksen, B., Sorland, S.J., Lund, O., Malcolm, S., Pembrey, M., Bhattacharya, S. and Bitner-Glindzicz, M. (1997). "*Isk* and *KvLQT1*: mutation in either of the two subunits of the slow component of the delayed rectifier potassium channel can cause Jervell and Lange-Nielsen syndrome." *Hum. Mol. Genet.* 6:2179–2185.

7. Splawski, I., Shen, J., Timothy, K.W., Lehmann, M.H., Priori, s., Robinson, J.L., Moss, A.J., Schwartz, P.J., Towbin, J.A., Vincent, G.M., Keating, M.T. (2000). "Spectrum of mutations in long-QT syndrome genes. *KVLQT1, HERG, SCN5A, KCNE1,* and *KCNE2.*" *Circulation.* 102:1178–1185.

8. Keating, M.T., Sanguinetti, M.C. (2001). "Molecular and cellular mechanisms of cardiac arrhythmias." *Cell.* 104:569–580.

9. Brugada, R., Hong, K., Dumaine, R., Cordeiro, J., Gaita, F., Borggrefe, M., Menendez, T.M., Brugada, J., Pollevick, G.D., Wolpert, C., Burashnikov, E., Matsuo, K., Wu, Y.S., Guerchicoff, A., Bianchi, F., Giustetto, C., Schimpf, R., Brugada, P. and Antzelevitch, C. (2004). "Sudden death associated with short-QT syndrome linked to mutations in HERG." *Circulation.* 109:30–35.

10. Priori, S.G., Napolitano, C., Tiso, N., Memmi, M., Vignati, G., Bloise, R., Sorrentino, V.V. and Danieli, G.A. (2001). "Mutations in the cardiac ryanodine receptor gene (hRyR2) underlie catecholaminergic polymorphic ventricular tachycardia." *Circulation.* 103:196–200.

11. Tiso, N., Stephan, D.A., Nava, A., Bagattin, A., Devaney, J.M., Stanchi, F., Larderet, G., Brahmbhatt, B., Brown, K., Bauce, B., Muriago, M., Basso, C., Thiene, G., Danieli, G.A. and Rampazzo, A. (2001). "Identification of mutations in the cardiac ryanodine receptor gene in families affected with arrhythmogenic right ventricular cardiomyopathy type 2 (ARVD2)." *Hum. Mol. Genet.* 10:189–194.

12. Marian, A.J., Roberts, R. (2001). "The molecular genetic basis for hypertrophic cardiomyopathy." *J. Mol. Cell Cardiol.* 33:655–670.

13. Wang, L., Fan, C., Topol, E.J., Wang, Q. (2003). "Mutation of MEF2A in an inherited disorder with features of coronary artery disease." *Science.* 302:1578–1581.

14. Roden, D.M. (2003). "Genetics polymorphisms, drugs, and proarrhythmia." *J. Interv. Card. Electrophysiol.* 9:131–135.

15. Risch, N.J. (2000). "Searching for genetic determinants in the new millennium." *Nature.* 405:847–856.

16. Halushka, M.K., Fan, J.B., Bentley, K., Hsie, L., Shen, N., Weder, A., Cooper, R., Lipshutz, R. and Chakravarti, A. (1999). "Patterns of single-nucleotide polymorphisms in candidate genes for blood-pressure homeostasis." *Nat. Genet.* 22:239–247.

17. Roberts, R. and Brugada, R. (2000). "Genetic aspects of arrhythmias." *Am. J. Med. Genet.* 97:310–318.

18. Watkins, H., Conner, D., Thierfelder, L., Jarcho, J.A., MacRae, C., McKenna, W.J., Maron, B.J., Seidman, J.G. and Seidman, C.E. (1995). "Mutations in the cardiac myosin-binding protein-C gene on chromosome 11 cause familial hypertrophic cardiomyopathy." *Nat. Genet.* 11:434–437.

19. Marian, A.J., Kelly, D., Mares, A., Jr, Fitzibbons, J., Caira, T., Qun, T., Hill, R., Perryman, M.B. and Roberts, R. (1994). "A missense mutation in the β myosin heavy chain gene is a predictor of premature sudden death in patients with hypertrophic cardiomyopathy." *J. Sports Med. Phys. Fitness.* 34:1–10.

20. Kimura, A., Harada, H., Park, J.E., Nishi, H., Satoh, M., Takashi, M., Hiroi, S., Sasaoka, T., Ohbuchi, N., Nakamura, T., Koyanagi, T., Hwang, T.H., Choo, J.A., Chung, K.S., Hasegawa, A., Nagai, R., Okazaki, O., Nakamura, H., Matsuzaki, M., Sakamoto, T., Toshmia, H., Koga, Y., Imaizumi, T. and Sasazuki, T. (1997). "Mutations in the cardiac troponin I gene associated with hypertrophic cardiomyopathy." *Nat. Genet.* 16:379–382.

21. Fananapazir, L., Dalakas, M.c., Cyran, F., Cohn, G. and Epstein, N.D. (1993). "Missense mutations in the β-myosin heavy chain gene cause central core disease in hypertrophic cardiomyopathy." *Proc. Natl. Acad. Sci.,* USA. 90:3993–3997.

22. Dausse, E., Komajda, M., Felter, L., Dubourg, O., Dufour, C., Carrier, L., Wisnewsky, C., Bercovici, J., Hengstenberg, C. and Al-Mahdawi, S., *et al.* (1993). "Familial hypertrophic cardiomyopathy: microsatellite haplotyping and identification of a hot spot for mutations in the β-myosin heavy chain gene." *J. Clin. Invest.* 92:2807–2813.

MODEL QUESTIONS

1. Describe in detail various chromosomal mutations associated with cardiac diseases.

2. List and explain various genes and their involvement in developing congenital cardiac anomalies.

3. What are various kinds of congenital heart diseases?

4. Comment on transcriptional regulatory genes and their involvement in congenital heart diseases.

5. Describe gene mutations and their role in the augmentation of atherosclerosis.

6. Heart muscle disease is due to gene mutation. Justify your answer.

7. Explain with an example how arrhythmia is caused due to genetic factors.

8. Explain with a neat diagram, the physiological function of the heart?

9. With a neat diagram, explain the molecular structure of the heart muscle proteins.

10. Explain various environmental risk factors and their influence on genes in developing developmental as well as acquired heart disease?

NEUROGENETICS

19

19.1   INTRODUCTION

The nervous system is a cellular communication network that controls and coordinates all bodily functions. It is divided into two kinds; central nervous system (CNS) and peripheral nervous system (PNS). The central nervous system consists of the brain and spinal cord. All the nerves extending from and going to the CNS comprise the PNS. The external and internal stimuli are received by PNS and transmitted to CNS. The CNS processes the input impulse and interprets them accordingly. Response signals are transferred to organs or specific cells through nerves. These nerves are made up of many nerve cells, called neurons. Each neuron is a highly specialized cell that transmits electrical impulses and has a variety of shapes, sizes and roles.

A neuron has three basic structural elements—soma (cell body), an axon and dendrites. Figures 19.1 and 19.2 show the structure of a neuron and the different kinds of neurons respectively. The cell body is an enlarged portion of the neuron containing the nucleus. Dendrites are branched elements extending from cell body. These dendrites receive stimuli and transmit to the cell body. Each neuron has only one axon extending from cell body. The tip of each axon branch is a specialized structure called the synaptic bulb. This synaptic bulb has contact with dendrite, cell body or gland cells or with axon of other cells. This junction forms a synapse, where impulses are transmitted.

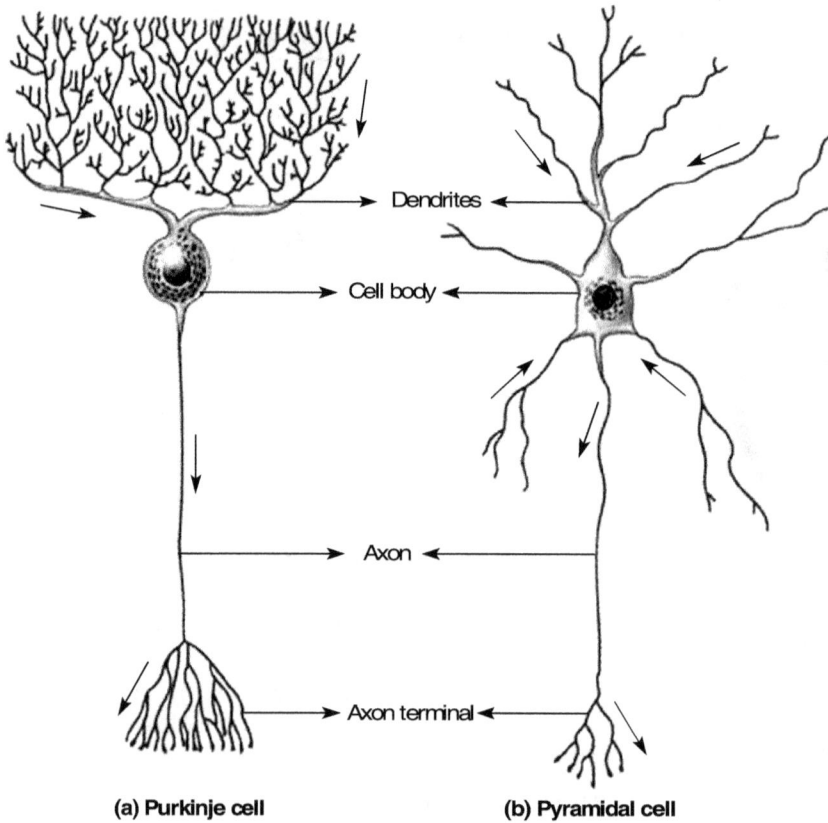

Dendrites

Cell body

Axon

Axon terminal

(a) Purkinje cell   **(b) Pyramidal cell**

Figure 19.1   Structure of a neuron

The synaptic vesicles in a neuron usually contain any one of the neurotransmitters such as acetylcholine, dopamine, serotonin, gamma aminobutyric acid (GABA), somatostatin, glutamic acid or endorphin. Each of these molecules are synthesized in the cell body and transported to the synaptic bulb through axon. They are essential for the transmission of nerve impulses from one neuron to another.

Neurons are classified into 3 types; afferent neurons, efferent neurons, and interneurons. The afferent neurons do not have dendrites and transmit impulses from specialized cells to the CNS. Efferent neurons conduct electrical signals from the CNS to muscles or gland cells. The interneurons reside entirely within the CNS.

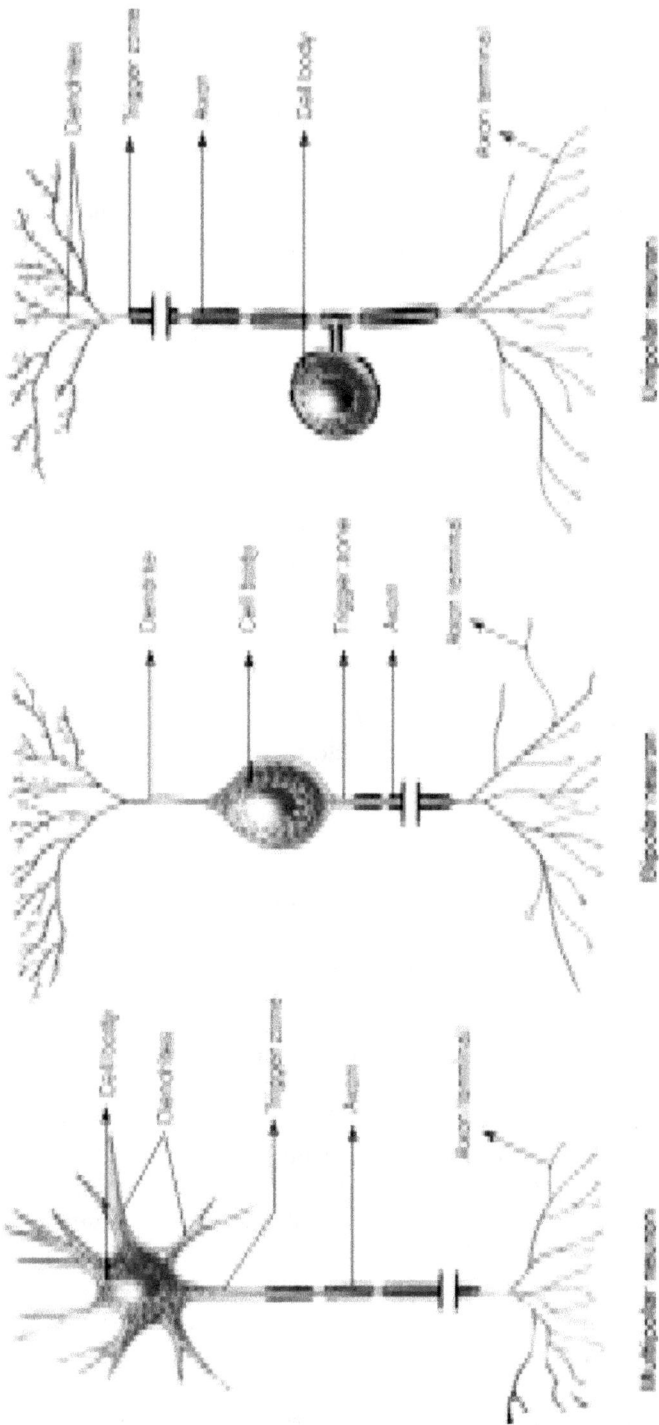

Figure 19.2   Different kinds of neurons

Response to internal and external changes are generated and propagated by neuronal cell membrane. It conveys the impulses in the form of electrical signals when the neuron is at rest. The outside of the neuronal cell membrane is positively charged and the inside of the cell membrane is negatively charged. Now the cell membrane is said to be polarized. The difference in the charge at this point across the cell membrane is called the resting potential of the membrane. In this stage, the extracellular concentration of sodium ions (Na^+) is greater than intracellular concentration but intracellular concentration of potassium ions (K^+) is greater than extracellular concentration. These gradients are maintained by Na^+–K^+ ATPase. This pumps 3 sodium ions out of the cell and brings 2 potassium ions inside the cell. Thus it maintains the polarity and prevents negatively charged proteins going out.

Other neuronal membrane proteins help in entry and exit of ions. These proteins are termed as channel proteins (ion channels). Some channels are always open. But, many of them are open and closed according to the intra- and extracellular signals. This kind of channel is called as gated channel. If channels are opened and closed based on changes, they are called voltage-gated channels; when Na^+ channel is open, sodium ions enter the cell. When K^+ channel is open, potassium ions exit the cell.

A nerve impulse is initiated when there is a signal or excitation. This leads to a change in resting potential of a neuronal cell membrane. When the channel protein opens, Na^+ rushes into the cell and K^+ rushes out. Now the cell membrane is depolarized. This alteration spreads to the adjacent neurons.

Soon after 1 to 2 milliseconds, the Na^+ and K^+ channels close and Na^+/K^+ ATPase starts pumping K^+ and Na^+ in and out of the cell, respectively. This restores the resting membrane potential, that is, the cell membrane is repolarized. If the initial stimulus is strong, then the impulse is propagated along the cell membrane.

The laboratory-determined transmembrane potential is expressed in millivolts (mV). The resting membrane potential for neuron is −70 mV and the threshold potential is about −50 mV. The maximum response for opening the voltage gated Na^+/K^+ channel is +30 mV and this is called action potential. When the action potential reaches a region of the cell membrane near the synaptic bulb, the depolarization opens voltage-gated calcium ion (Ca^{2+}) axon and Ca^{2+} rushes into the synaptic bulb. This causes dumping of neurotransmitter molecules (GABA) into the synaptic space (chemical synapses). This stimulates the receptor protein to open the ion channels. Now Na^+ rushes into the postsynaptic neuron. The membrane is depolarized and an impulse is generated. In addition to chemical synapses, neurons in the heart muscles can transmit impulse by electrical coupling. In this case, the space between two cells is bridged by barrel-shaped hexameric transmembrane protein complexes called connexons. One connexon complex in the membrane of a presynaptic cell joins with a connexon complex in the membrane of a post-synaptic cell. This facilitates uninterrupted flow of ions from presynaptic cell into the post-synaptic cell during action potential. Electrical synapses increase the synchrony of action of heart muscle and smooth muscle of the intestine. This electrical synaptic transmission is rarely used in the CNS.

Brain consists of the following major parts; cerebrum, diencephalon (thalamus, epithalamus, hypothalamus and pineal gland), cerebellum, brain stem (midbrain, pons, medulla oblongata). The brain stem continues with the spinal cord. Cerebrum is the largest part of the brain (Figure 19.3). Blood flows to the brain through the internal carotid artery and vertebral arteries. The internal jugular vein returns blood from the head to the heart. The brain represents only 2% of total body weight and consumes 20% of the oxygen and glucose at resting condition.

The cerebrum is the "seat of intelligence". It has right and left halves, called cerebral hemispheres. The sensory areas interpret sensory impulses and motor areas control muscular movements. Each cerebral hemisphere is further subdivided into four lobes namely frontal, parietal, temporal and occipital lobes (named after bones). The cerebral cortex contains many convolutions, fissures and sulci. The basal ganglia in cerebral hemisphere helps to control automatic movements of skeletal muscles and also helps in regulating muscle tone.

The diencephalon extends from the brain stem to the cerebrum. It consists of the thalamus, hypothalamus, epithalamus and subthalamus. The thalamus is the major relay station for most sensory impulses. It allows appreciation of pain, heat and pressure, and also mediates some motor activities. Below the thalamus, hypothalamus is found. It mainly controls autonomic nervous system. It also controls growth hormone secretion, aggression and body temperature and establishes circadian rhythms. The epithalamus is a small region superior and posterior to thalamus. It consists of pineal gland and habenular nuclei. The pineal gland is part of the endocrine system. It secretes the hormone melatonin to promote sleepiness and to set the body's biological clock. The habenular nuclei are involved in olfaction and specified emotional response to odours. The subthalamus contains subthalamic nuclei and portions of the red nuclei and also the substantia nigra. These regions communicate with the basal ganglia to help control body movements.

The cerebellum is the second largest part of the brain. A deep groove known as the transverse fissure and the tentorium cerebelli support the posterior region of the cerebrum, but separate the cerebrum from the cerebellum. The cerebellum is the brain region that regulates posture and balance. The presence of reciprocal connections between the cerebrum and cerebellum suggests that the cerebellum may also have cognition function (acquisition of knowledge) and language processing.

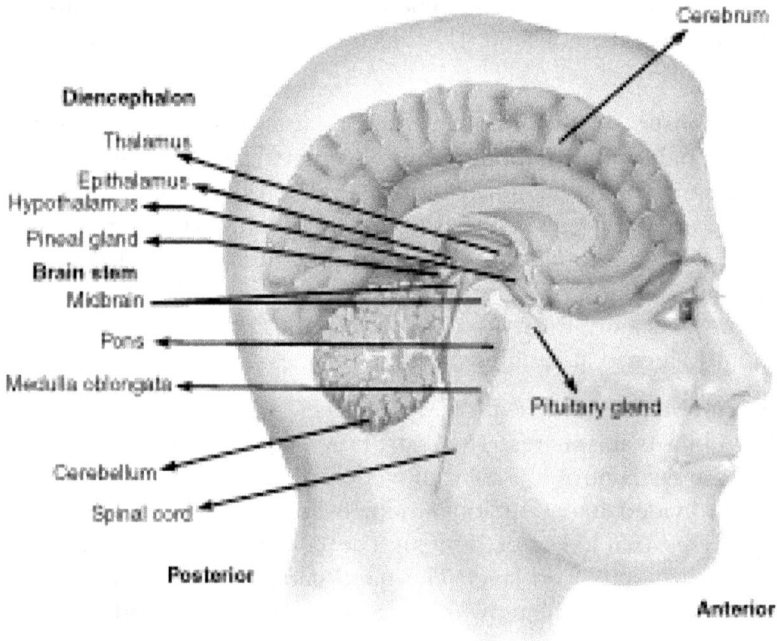

Figure 19.3   Various parts of the brain

The brain stem is the part between the spinal cord and diencephalon. It consists of three structurally and functionally connected regions—medulla oblongata, pons and midbrain. The medulla oblongata is continued as the spinal cord. It begins at the foramen magnum. The medulla regulates the rate and force of the heart beat and diameter of the blood vessel. It also adjusts the basic rhythm of breathing. Other nuclei in the medulla control reflexes for vomiting, coughing, sneezing and hiccupping. Also, the medulla contains nuclei that receive sensory input and motor output to cranial nerves VIII (vestibulocochlear), IX (glossopharyngeal), X (vagus), XI (accessory) and XII (hypoglossal). The pons consists of both nuclei and tracts. It contains nuclei associated with cranial nerves V (trigeminal), VI (abducens), VII (facial) and VIII (vestibulocochlear). The midbrain is also called mesencephalon. It transmits motor impulses from cerebrum to the spinal cord through cerebellum. The sensory impulses are transmitted from the spinal cord to the thalamus. It also contains nuclei associated with cranial nerves III (occulomotor nerve) and IV (trochlear nerve). A large portion of the brain stem consists of the small areas of gray matter and white matter called the reticular formation. It helps to maintain consciousness, causes awakening from sleep and contributes to regulating muscular tone.

19.2 SCHIZOPHRENIA

Schizophrenia is a psychiatric disorder. It leads to a significant loss of perception of reality. Schizophrenia is probably the result of errors during brain development. Schizophrenics are unable to finish work task and reluctant to associate with others. They lose interest in all their personal needs. Their conversation is generally incoherent and chaotic.

Schizophrenia has a significant genetic component (Table 19.1). A number of chromosome regions with possible schizophrenia genes have been subjected to higher resolution analysis. Dystrobrevin-binding protein 1 or dysbindin (*DTNBP1*) gene at 6p22.3 was found to be associated with schizophrenia. Dysbindin is present in the brain and all other tissues, and binds to β-dystrobrevin to form dystrophic protein complex (DPC). The DPC maintains the integrity of neuromuscular synapses. In addition to this, DPC also facilitates neural activity in parts of the brain.

Another gene that is found to be associated with schizophrenia is neuregulin1 (*NRG1*) at 8012-p21. The neuregulin-1 protein regulates neurotransmitter receptor N-methyl –D-aspartate (NMDA), acetylcholine and γ aminobutyric acid. It also participates in the myelination process.

A balanced translocation was found to be associated with schizophrenia. This balanced translocation was between chromosome 1 and chromosome 11,t(1;11)(q42.1;q21). Molecular genetic analysis showed chromosome 1 (1q42.1) region codes for many genes. *DISC 1* and *DISC 2* were the two genes disrupted in schizophrenia. The *DISC 1* gene encodes an 854 amino acid protein and the *DISC 2* gene encodes RNA that is complementary to *DISC 1* sequence. The *DISC 1* gene is transcribed during the development of the cerebral cortex. The *DISC 1* protein is multifunctional and interacts with a number of brain proteins that are present during formation of cerebral cortex.

Alleles of other genes (Table 19.1) that have been implicated in increasing the risk for schizophrenia include D-amino acid oxidase activator (DAOA) at 13q33.2, D-amino acid oxidase (DAO) at 12q24 and regulator of G protein signalling 4 (RGS4) at 1q23.2. The D-amino acid oxidase activator binds with D-amino acid oxidase (a detoxifying enzyme). It can also affect neurotransmission and intraneuronal signalling.

Table 19.1   Various gene loci and schizophrenia

Chromosome loci	Gene name	Protein name	Clinical feature
6p22.3	*DTNBP1*	Dysbindin (Dystrobrevin-binding protein 1)	Schizophrenia
8p12-p21	*NRG1*	Neuregulin -1	Schizophrenia
1q42.1	*DISC 1* *DISC 2*	–	Schizophrenia (protein for cerebral cortex development)
13q33.2	D-amino acid oxidase activator		
12q24	D-amino acid oxidase		Schizophrenia
1q23.2	Regulator of G-protein signalling 4 (RGS4)		

19.3   NEURONAL CHANNELOPATHIES

Channel proteins facilitate and regulate ion movement in and out of the cell. The channels may be gating (opening and closing of channels) or ion channels (handling ions to shuttle) or voltage-gated (electrical potential across the membrane). The voltage-gated channels have an extensive family of more than 400 related proteins. Disorders caused by mutations of channel-encoding genes are called channelopathies.

The neuronal channelopathies tend to feature intermittent loss of brain function (seizures, convulsions, epilepsy), uncontrolled muscle movement (ataxia) and migraine (severe headache with nausea and sensitivity to light). The genes for different types of neuronal channelopathies are listed in Table 19.2. There is an abnormal ion current in each of these cases. Mutations of the *KCNQ2* and *KCNQ3* genes which are of K^+ channel protein genes are responsible for benign familial neonatal convulsion. Similarly, mutations of the *SCNA2* Na^+ channel gene are responsible for benign familial neonatal–infantile seizures. Most of the individuals do not experience seizures or convulsion beyond 12 months of age.

Table 19.2 Different types of neuronal channelopathies

Chromosome location	Gene locus	Channel proteins	Protein function	Disease pattern
12p13	KCNA1	Potassium	Repolarization of axons	Episodic ataxia; neuromyotonia (decreased)
20q13.3	KCNQ2		Low threshold current modulated by muscarinic receptors	Benign familial neonatal convulsions (decreased)
8q24	KCNQ3		Low threshold current modulated by muscarinic receptors	Benign familial neonatal convulsions (decreased)
19p13	CACNA1A	Calcium	Current in Purkinje and granule cells; presynaptic terminals	Unilateral migraine, episodic ataxia; spinocerebellar ataxia (decreased)
2q24	SCN1A	Sodium	Somatodendritic Na^+ influx	Seizures with high fever; infant epilepsy (increased)
2q23-q24.3	SCN2A		Axonal fast Na^+ influx	Benign familial neonatal–infantile seizures (increased)
19q13.1	SCN1B		Accessory protein	Seizures with high fever (increased)
5q31.1-q33.1	GABRG2	Aminobutyric acid (GABA) receptors	Fast inhibition	Seizures with high fever (decreased Cl^-)
20q13.2-q13.3	CHRNA4	Nicotinic acetylcholine (ACh) receptors	Presynaptic transmitter release	Nocturnal frontal lobe epilepsy (increased Na^+)
1p21	CHRNB2		Presynaptic transmitter release	Nocturnal frontal lobe epilepsy (increased Na^+)
5q32	GLRA1	Glycine receptors	Fast inhibition	Extreme startle reponse, i.e., hyperekplexia (decreased Cl^-)

19.4   ALZHEIMER'S DISEASE

The clinical and neuropathological features of Alzheimer's disease was described by the German neurologist Alois Alzheimer in 1907. Alzheimer's disease accounts for two-third of all diagnosed cases of dementia. Inability to create new memories and the loss of short-term memory are the prime clinical features of Alzheimer's disease. The three distinctive neuropathological features are the following.

1. Loss of synapses and neurons within the hippocampus and the entorhinal cortex.

2. Dense spherical structures called senile plaques surrounded by cellular debris from disintegrated axons and dendrites. This structure is prevalent outside the neurons of the hippocampus and other parts of brain.

3. Aggregation of neurofibrillary tangles (NFT) within the cell bodies and dendrites of the neurons of the hippocampus, the neocortex, the amygdale and other parts of the brain.

The core of a senile plaque is a densely packed fibrous structure. It is called an amyloid body. The production of amyloid bodies is called amyloidosis. The protein of Alzheimer's disease amyloid bodies is a 4-kilodalton peptide termed AB protein. This AB protein consists of many isoforms (amino acid ranges from 39 to 43). The two main isoforms are containing 40 (AB-40) and 42 (AB-42) amino acids.

Once they form the core of an amyloid body, it becomes fibrillogenic. The AB proteins are primarily derived from amyloid precursor protein (APP). This APP is located in 21q21.2. Mutation or a typical processing of ATP leads to an excess of the AB42 isoform. This is probably responsible for the destruction of synapses during the early stages of Alzheimer's disease, the formation of senile plaques and further to this degradation of neurons.

The NFTs are numerous and form paired helical filaments. APHF consists of two protein strands intertwined. A number of *tau* (I) protein molecules are found in the PHF to form a *tau* filament. The *MAPT* gene encodes the *tau* protein. The *MAPT* gene is located at chromosome site 17q21. The *tau* proteins bind to tubulin and help in the formation of microtubules. It also maintains the stability of microtubules. The dissolution of the neuronal microtubule system would prevent axonal transport, which in turn would cause the loss of synapses. This further leads to degeneration and formation of synaptic bulbs. Various chromosomal sites were associated with Alzheimer disease. The loci 14q24.3 (presenilin gene 1) and 1q32-42 (presenilin gene 2) are the two loci showing linkage with early onset of

Alzheimer's disease. But linkage between Alzheimer's disease and 19q13.2 site was observed in families with late onset.

19.5 HUNTINGTON'S DISEASE

Huntington's disease (HD) is a catastrophic neurological disorder. In the initial stage, muscle coordination is slightly impaired associated with forgetfulness. Further to this, cognitive disorganization and personality changes are evident. Eventually speech is slurred and body movements become uncontrolled with jerking and writhing. Autopsied brains of HD patients show neuronal degeneration throughout the cerebral cortex, caudate nucleus and substantia nigra. However a combination of granular and neuronal intranuclear inclusions accumulate in the nuclei. This results in dementia due to the loss of neurons within the cerebral cortex.

Huntington's disease is inherited in an autosomal dominant fashion. The first disease loci were localized to a chromosome location 4p16.3. The disease gene was isolated in 1993. It has 67 exons and covers about 200 kb of genomic DNA. An interesting finding was that of an uninterrupted stretch of CAG/GTC repeats in the first exon. Normal individuals have between 6 and 35 CAG/GTC repeats, whereas those with Huntington's disease have more than 40 repeats in one Huntington disease allele. This CAG codes for glutamine and during translation this repeat leads to polyglutamine production. A set of contiguous trinucleotides is called a trinucleotide repeat.

FOR ADDITIONAL READING

1. Berkovic, S.F. *et al.* (2004). "Benign familial neonatal–infantile seizures: characterization of a new sodium channelopathy." *Ann. Neurol.* 55:550–557.

2. Bertram, L. and Tanzi, R.E. (2004). "Alzheimer's disease: one disorder, too many genes?" *Hum. Mol. Genet.* 13:R135–141.

3. Carola, R.J.P., Harley, and Noback, C.R. (1992). *Human Anatomy and Physiology,* 2nd edn. McGraw-Hill, New York.

4. Cooper, E.C. and Jan, L.Y. (1999). "Ion channel genes and human neurological disease: recent progress, prospects, and challenges." *Proc. Natl. Acad. Sci., USA.* 96:4759–4766.

5. Finckh, U. (2003). "The future of genetic association studies in Alzheimer disease." *J. Neural. Transm.* 110:253–266.

6. Harris, D.A. (2003). "Trafficking, turnover and membrane topology of PrP." *Br. Med. Bull.* 66:71–85.

7. Hubner, C.A. and Jentsch, T.J. (2002). "Ion channel diseases." *Hum. Mol. Genet.* 11:2435–2445.

8. Murphy, R.M. (2002). "Peptide aggregation in neurodegenerative disease." *Annu. Rev. Biomed. Eng.* 4:155–174.

9. O'Donovan, M.C., Williams, N.M. and Owen, M.J. (2003). "Recent advances in the genetics of schizophrenia." *Hum. Mol. Genet.* 12:125–133.

10. Reichert, H. (1992). *Introduction to Neurobiology.* Oxford University Press, New York.

11. Riesner, D. (2003). "Biochemistry and structure of PrPc and PrPsc." *Br. Med. Bull.* 66: 21–33.

12. Rubinsztein, D.C and Carmichael, J. (2003). "Huntington's disease: Molecular basis of neurodegeneration." *Expert. Rev. Mol. Med.* 5:1–21.

13. Selkoe, D.J. amd Podlisny, M.B. (2002). "Deciphering the genetic basis of Alzheimer's disease." *Annu. Rev. Genomics Hum. Genet.* 3:67–99.

14. Sharp, A.H. and Ross, C.A. (1996). "Neurobiology of Huntington's disease." *Neurobiol. Dis.* 3:3–15.

15. Smith, M.A. (1998). "Alzheimer disease." *Int. Rev. Neurobiol.* 42:1–54.

16. Spillantini, M.G. *et al.* (1998). "Mutation in the *tau* gene in familial multiple system tauopathy with presenile dementia." *Proc. Natl. Acad. Sci.,* USA. 95:7737–7741.

17. Tandon, A. and Fraser, P. (2002). "The presenilins." *Genome Biol.* 3:1–9.

18. Wadsworth, J.D. *et al.* (2003). "Molecular and clinical classification of human prion disease." *Br. Med. Bull.* 66:241–254.

MODEL QUESTIONS

1. What is a neuron, an axon and a synapse?

2. What is an action potential and how is it propagated?

3. What are channel proteins?

4. Discuss the genetic aspects of neuronal channelopathy.

5. What are the functions of various parts of the brain?

6. Write a brief note on the molecular basis of schizophrenia.

7. What is Alzheimer's disease? Comment on the genetic background of AD.

8. List various chromosomal sites linked to neuronal channelopathy, schizophrenia and Alzheimer's disease.

BIOCHEMICAL GENETICS

20

20.1   INTRODUCTION

Based on requirements, cells synthesize, degrade, excrete and also recycle (at times) macromolecules. Each of this metabolic process consists of a sequence of catalytic steps that are mediated by different enzymes. Genes encode for these enzymes. Normally these genes replicate and inherit with high fidelity. Sometimes mutations in any of the enzyme genes can block the metabolic pathway. A block in a metabolic pathway results in the accumulation of intermediary products (substrate or precursor or metabolite) as well as a deficiency of the end product, and this condition is termed as **inborn error of metabolism**.

More than 350 different types of errors of metabolism have been described to date and most of these are rare. Sir Archibald Garrod (1902) was the first to recognize these variants of metabolism in his studies of alkaptonuria. Most metabolic disorders are inherited in an autosomal recessive pattern, that is, only individuals having two mutant alleles are affected. Although a mutant allele in a heterozygous carrier produces reduced enzyme activity, it usually does not alter the health of an individual.

20.2   CLASSIFICATION OF METABOLIC DISORDERS

Metabolic disorders have been classified in many different ways. They are as follows:

1. the pathological effects of the pathway blocked (absence of end product, accumulation of substrate)
2. different functional classes of proteins (receptors, hormones)
3. associated cofactors (metals, vitamins) and
4. pathways affected (glycolysis, citric acid cycle).

The various metabolic disorders are listed in Table 20.1.

Table 20.1   Disorders of metabolism

Name	Mutant gene product	Chromosomal location
Carbohydrate disorders		
Galactosemia	Galactose 1-phosphate uridyl transferase	9p13
Fructose intolerance	Fructose 1,6-biphosphate aldolase	9q13-q32
Fructosuria	Fructokinase	2p23
Hypolactasia	Lactase	2q21
Diabetes mellitus (type I)	Unknown	Polygenic
Diabetes mellitus (type II)	Unknown	Polygenic
Amino acid disorders		
Phenylketonuria	Phenylalanine hydroxylase	12q24
Tyrosinemia	Fumarylacetoacetate hydrolase	15q23-25
Maple syrup urine disease	Branched-chain α-ketoacid decarboxylase	Multiple loci

(Contd.)

Table 20.1   (Contd.)

Name	Mutant gene product	Chromosomal location
Alkaptonuria	Homogentisic acid oxidase	3q2
Homocystinuria	Cystathionine β-synthase	21q2
Oculocutanoeus albinism	Tyrosinase	11q
Cystinosis	Unknown	17p
Cystinuria	SLC3A1(type I)	2p
Lipid disorders		
MCAD deficiency	Medium-chain acyl CoA dehydrogenase	1p31
LCAD deficiency	Long-chain acyl-CoA dehydrogenase	2q34-q35
SCAD deficiency	Short-chain acyl-CoA dehydrogenase	12q22qter
Organic acid disorders		
Methylmalonic acidemia	Methylmalonyl-CoA mutase	6p
Propionic acidemia	Propionyl-CoA carboxylase	13q32;3q
Lysosomal storage disorders		
Hurler/Scheie	α-L-Iduronidase	4p16.3
Hunter	Iduronate sulphatase	Xq 28
Maroteaux-Lamy	Aryl sulphatase B	5p11-q 13
Sly	β-Glucuronidase	5q11
Urea cycle defects		
Ornithine transcarbamylase deficiency	Ornithine carbamyl transferase	Xp21
Carbamyl phosphate synthetase deficiency	Carbamyl phosphate synthetase I	2p
Argininosuccinic acid synthetase deficiency	Argininosuccinic acid synthetase	9q34

(Contd.)

Table 20.1   (Contd.)

Name	Mutant gene product	Chromosomal location
Energy production defects		
Cytochrome C oxidase deficiency	Cytochrome oxidase peptides	Multiple loci
Pyruvate carboxylase deficiency	Pyruvate carboxylase	11q
Pyruvate dehydrogenase complex (E_1) deficiency	Pyruvate decarboxylase E1α	Xp22
NADH-CoQ reductase deficiency	Multiple nuclear genes	Multiple loci
Heavy metal transport defects		
Wilson disease	ATP7B	13q14
Menkes disease	ATPase7A	Xq13
Hemochromatosis	HFE	6p21

20.3   GALACTOSEMIA

Galactosemia is one of the most common metabolic disorders of carbohydrate metabolism. It is caused by mutations in galactose 1-phosphate uridyl transferase gene. It has monogenic pattern of inheritance. This gene is composed of 11 exons distributed across 4 kb of DNA. Approximately 70% of galactosemia-causing alleles contain a single missense mutation in exon 6. As a result, diminished GAL 1-P uridyl transferase activity takes place. This in turn affects the effective conversion of galactose to glucose. Consequently, galactose is alternatively metabolized to galactitol and galactonate.

The common clinical signs of galactosemia are failure to thrive, hepatic insufficiency, cataracts and developmental delay. Long-term disabilities include poor growth, mental retardation and ovarian failure in females. In newborns, measuring plasma GAL 1-P uridyl transferase activity from the blood helps in screening for galactosemia. Early identification helps in eliminating dietary galactose.

Galactosemia can also be caused by mutations in the galactokinase and uridine diphosphate galactose 4-epimerase genes. Galactokinase deficiency can cause cataract but does not cause growth failure, mental retardation, or hepatic disease. UDP-galactose 4-epimerase deficiency can be limited to red blood cells and leukocytes but causing no ill effects.

20.4 GLUCOSE IMBALANCE

Imbalance of glucose level in the human body is also one of the errors in carbohydrate metabolism. The cause of this disorder is heterogeneous due to the involvement of both environmental and genetical factors. The disorder associated with elevated levels of plasma glucose is termed hyperglycaemia. The hyperglycaemia is classified into three categories: they are: diabetes mellitus type 1 (reduced level or absence of plasma insulin, may be from childhood), diabetes mellitus type 2 (insulin resistance and adult onset) and maturity onset diabetes of youth (MODY). These deficient forms of diabetes have a different clinical and genetic basis. Recent studies show that mutations in the insulin receptor gene can be associated with hyperglycaemia. Similarly mutations in the mitochondrial DNA and in the genes encoding insulin and glucokinase have also been associated with elevated levels of plasma glucose. MODY type is associated with mutations in genes called HNF4A (chromosome 12q), insulin promoting factor-1 gene (chromosome 13q12.1) and gene TCF 2 (chromosome 17 Cen q21.3).

20.5 GLYCOGEN

Glycogen is the major storage form of carbohydrates in human muscles and liver. Impaired synthesis or degradation of glycogen is considered as a disorder of carbohydrate metabolism. This disorder causes liver enlargement (hepatomegaly) and low plasma glucose level (hypoglycaemia). In muscles, glycogen storage disorder causes exercise intolerance, progressive weakness, cramping, cardiomyopathy and early death (Figure 20.1).

20.6 HYPERPHENYLALANINEMIAS

Defects in the phenylalanine metabolism are the first and one of the most widely studied metabolic disorders. This disorder is caused by mutations of the genes encoding enzymes of the phenylalanine hydroxylation pathway. High levels of phenylalanine in plasma disturb myelination and protein synthesis in the brain and can also cause mental retardation (Figure 20.2).

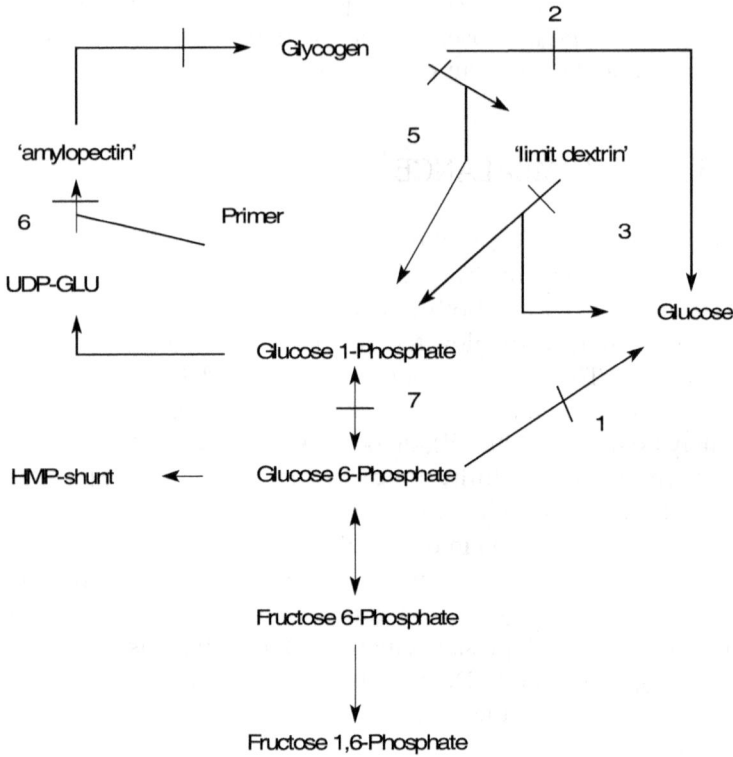

Figure 20.1   Glycogen storage diseases (1) glucose 6-phosphatase
deficiency (von Gierke's); (2) lysosomal α-1, 4-glucosidase
deficiency (Pompe's); (3) amylo-1, 6-glucosidase deficiency
(Forbes'); (4) amylo-(1,4 or 1,6) - transglucosidase deficiency
(Andersen's); (5) muscle and liver phosphorylase deficiency
(McArdle's and Hers', respectively); (6) glycogen synthetase
deficiency; (7) phosphoglucomutase deficiency.

Any mutation (such as nucleotide insertion, deletion or substitution) in
the phenylalanine hydroxylase gene can cause phenylketonuria (that is,
excretion of phenylalanine in the urine). A complete lack of phenylalanine
is fatal. Similarly hyperphenylalaninemia in a pregnant woman can cause
an embryopathy through poor growth, congenital defects, microcephaly and
mental retardation.

Elevated levels of plasma phenylalanine disturb the essential cellular
processes such as myelination and protein synthesis in the brain eventually
producing severe mental retardation. Most cases of hyperphenylalaninemia
are caused by mutations of phenylalanine hydroxylase (PAH) gene and
produce classical phenylketonuria (PKU). So far, more than 100 mutations

have been identified in PAH, including substitutions, insertions and deletions. The prevalence of hyperphenylalaninemia varies widely among ethnic groups, with PKU ranging from 1/10,000 in caucasians to 1/90,000 in Africans. Less commonly, hyperphenylalaninemia is caused by defects in the synthesis of tetrahydrobiopterin, the cofactor necessary for the hydroxylation of phenylalanine, or by a deficiency of dihydropteridine reductase.

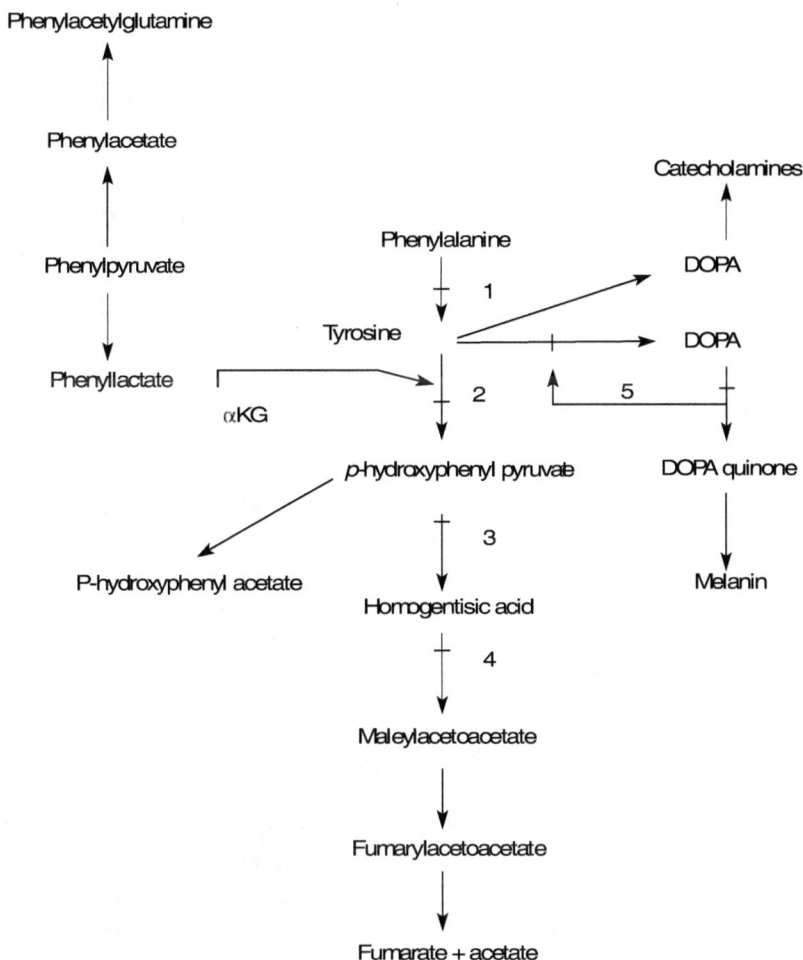

Phenylacetylglutamine

Phenylacetate

Catecholamines

Phenylalanine

Phenylpyruvate   DOPA

1

Tyrosine   DOPA

Phenyllactate

αKG   2   5

p-hydroxyphenyl pyruvate   DOPA quinone

3

P-hydroxyphenyl acetate   Melanin

Homogentisic acid

4

Maleylacetoacetate

Fumarylacetoacetate

Fumarate + acetate

Figure 20.2   Disorders of phenylalanine and tyrosine metabolism (1) phenylalanine hydroxylase deficiency (phenylketonuria); (2) **tyrosine:** α-ketoglutarate amino-transferase deficiency (tyrosinosis); (3) *p*-hydroxyphenyl pyruvate oxidase deficiency (tyrosinaemia); (4) homogentisic oxidase deficiency (alkaptonuria); (5) tyrosinase deficiency (oculocutaneous albinism).

20.7   TYROSINEMIA

Tyrosine is used to synthesize catecholamines, thyroid hormones and melanin pigments. Elevated levels of serum tyrosine as a result of an inborn error of catabolism can cause severe hepatocellular dysfunction. Hereditary tyrosinemia type 1 (HT 1) is caused by a deficiency of fumarylacetoacetate hydrolase (FAH). Accumulation of fumarylacetoacetate, and its precursor, maleylacetoacetate, is thought to be mutagenic and toxic to the liver. HT 1 is clinically characterized by renal tubular dysfunction, peripheral neuropathy, liver cirrhosis, and liver cancer (hepatocellular carcinoma). Liver transplantation can be the only curative therapy.

Tyrosinemia type 2 is another type of metabolic disorder caused by the deficiency of tyrosine aminotransferase. This metabolic defect is characterized by corneal erosions, skin thickening in palms and soles, and mental retardation. Tyrosinemia type 3 is associated with reduced activity of 4-hydroxy-phenylpyruvate dioxygenase and neurological dysfunction.

20.8   MCAD DEFICIENCY

The most common type of inborn error of fatty acid metabolism is due to the deficiency of medium-chain acyl CoA dehydrogenase (MCAD). MCAD deficiency is mainly due to hypoglycaemia, which is often stimulated by fasting. Fasting results in the accumulation of fatty acid intermediates that in turn fail to produce ketones for the tissue glucose demand. Cerebral edema and encephalopathy are the result of fatty acid intermediates in the central nervous system. If glucose is not provided promptly, it leads to death of the person.

20.9   LCHAD AND SCAD DEFICIENCY

The deficiencies of long-chain acyl-CoA dehydrogenase (LCAD) and long-chain 3-hydroxyacyl-CoA dehydrogenase (LCHAD) have been found in fatty acid metabolism. Patients with LCAD deficiency may show clinical symptoms such as fasting-induced coma, hepatomegaly, cardiomyopathy, and hypotonia. Similarly LCHAD deficiency can cause liver disease, muscle weakness, cardiomyopathy and peripheral neuropathy.

20.10   LYSOSOMAL STORAGE DISORDERS

Most of the lysosomal disorders are due to the deficiency of enzymes within lysosomes that catalyse the stepwise degradation of sphingolipids, glycosaminoglycans, glycoproteins and glycolipids. Accumulation of excess substrates results in the dysfunction of cell, tissue and organ system. Lysosomal disorders can also be caused due to the inability to activate an enzyme or failure to transport an enzyme to a subcellular compartment.

The mucopolysaccharidoses (MPS) are a heterogeneous group of disorders caused by a reduced ability to degrade glycosaminoglycans such as dermatan sulphate, heparan sulphate, keratan sulphate and chondroitin sulphate. Deficiencies in ten different enzymes that break the glycosaminoglycans can cause six different MPS disorders. These glycosaminoglycans are degradation products of proteoglycans found in the extracellular matrix. All MPS disorders are characterized by impairment of hearing, vision, joint, and cardiovascular dysfunction. Deficiency of iduronidase (MPS I) can cause Hurler syndrome, Hunter syndrome and Sanfilippo syndromes that are characterized by mental retardation.

Hunter syndrome (MPS II) is caused due to the deficiency of iduronate sulphatase. Normally, the disease onset occurs between the age group of 2 and 4 years. Defective children develop coarse facial features, short stature, limb deformities, joint stiffness, and mental retardation. The gene for iduronate sulphatase is made up of 9 exons spanning 24 kb. Twenty percent of all identified mutations are large deletions, and most of the remainder are missense and nonsense mutations.

Defects in the breakdown of sphingolipids result in their gradual accumulation, which leads to multiorgan dysfunction. Acid β-glucosidase deficiency causes accumulation of glucosylceramide. This leads to Gaucher disease. The clinical features are visceromegaly, multiorgan failure and unbearable skeletal disease.

20.11 UREA CYCLE DISORDERS

Urea cycle is to prevent and eliminate the accumulation of nitrogenous wastes such as nitrogen, urea, uric acid, etc. from the human body. This urea cycle is also responsible for the *de novo* synthesis of arginine. The urea cycle consists of five major biochemical reactions. The defects in each of these steps have been described in humans. Deficiencies of carbamyl phosphate synthetase (CPS), orinithine transcarbamylase (OTC), argininosuccinic acid synthetase (ASA) and argininosuccinase (AS) result in the accumulation of ammonium and glutamine. The clinical presentations of individuals with CPS, OTC, ASA, and AS deficiencies are similar and produce progressive lethargy and coma (Figure 20.3).

Except OTC deficiency that is X-linked, other disorders are inherited in an autosomal recessive pattern. The goal of therapy for each of this disorder is to prevent hyperammonaemia and provide sufficient calories and protein for normal growth.

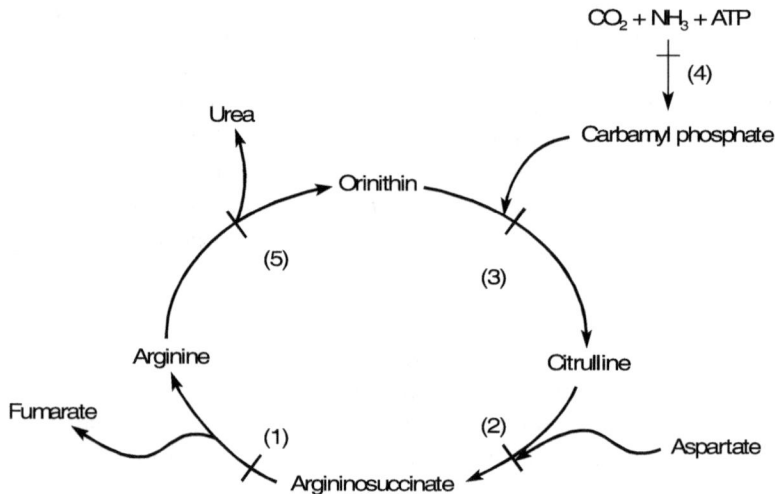

Figure 20.3   Disorders of the urea cycle, (1) argininosuccinase deficiency (argininosuccine aciduria); (2) argininosuccinate synthetase deficiency (citrullinaemia); (3) ornithine transcarbamylase deficiency (hyperammonaemia I); (4) carbamyl phosphate synthetase deficiency (hyperammonaemia II); (5) arginase deficiency (hyperargininaemia).

OTC deficiency is the most prevalent of the urea cycle disorders. The OTC gene has 10 exons spanning approximately 85 kb. This gene is located just proximal to dystrophin on Xp. A variety of exon deletions and missense mutations have been described.

20.12   DISORDERS OF PURINE/PYRIMIDINE METABOLISM

Gout is a human disorder due to the abnormalities of purine or pyrimidine metabolism. The pain, swelling and tenderness of the joints are due to the accumulation of salt or uric acid, which results in the inflammatory response. Lesch-Nyhan disease is an X-linked disorder of purine metabolism. This is due to the deficiency of the enzyme hypoxanthineguanine phosphoribosyl transferase, which results in the increased levels of phosphoribosyl pyrophosphate. The latter is normally a rate-limiting substrate in the synthesis of purines. An excess of phosphoribosyl pyrophosphate leads to an increased rate of purine synthesis, which results in an accumulation of excessive amounts of uric acid and other metabolic precursors. Although allopurinol inhibits uric acid formation and lowers uric acid levels, any satisfactory treatment is yet to be found (Figure 20.4).

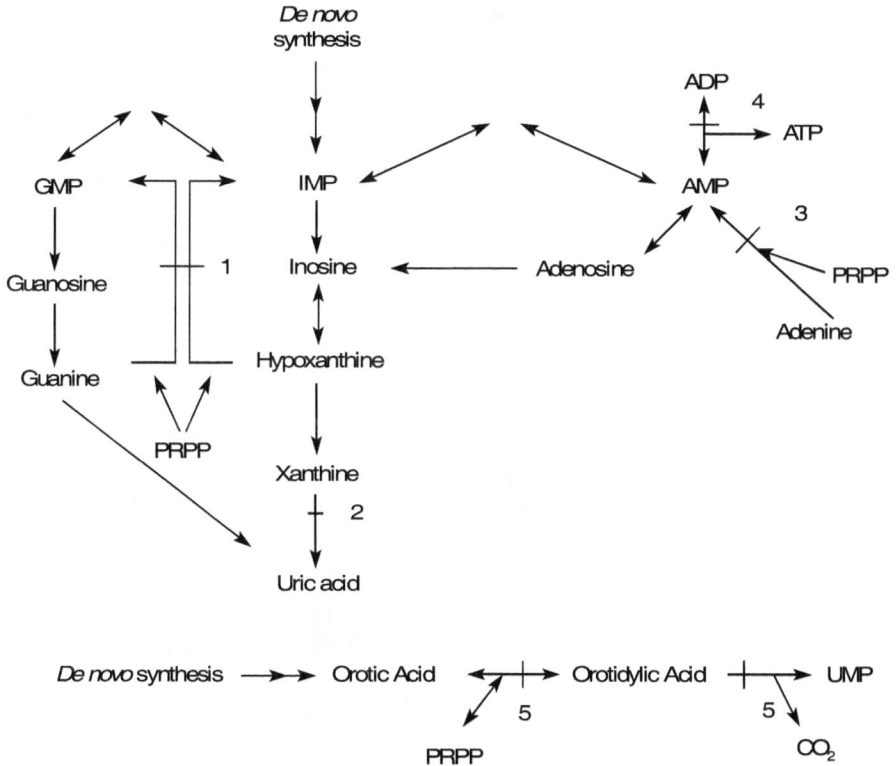

Figure 20.4   Disorders of purine and pyrimidine metabolism deficiencies, (1) hypoxanthine-guanine phosphoribosyl transferase (Lesch-Nyhan); (2) Xanthine oxidase (xanthinuria); (3) adenine phosphoribosyl transferase; (4) adenylate kinase; (5) orotidylic pyrophosphorylase and decarboxylase (oroticaciduria).

FOR ADDITIONAL READING

1. Brock, D.J.H. and Mayo, O. *The Biochemical Genetics of Man.* Academic Press, London.

2. Cooper, D.N. and Krawczak, M. (1993). *Human Gene Mutation.* BIOS Scientific Publishers, Oxford, England.

3. Harris, H. (1980). *The Principles of Human Biochemical Genetics*, 3rd edn. Elsevier, North-Holland, Amsterdam.

4. McKusick, V.A. (1972). *Heritable Disorders of Connective Tissue*, 4th edn. CV Mosby, St. Louis.

5. Scriver, C.R., Beaudet, A.L., Sly, W.S., Valle, D. (eds.) (2001). *The Metabolic and Molecular Bases of Inherited Disease*, 8th edn. Mc-Graw Hill, New York.

6. Scriver, C.R. and Childs, B. (1989). *Garrod's Inborn Factors and Disease*. Oxford University Press, New York.

7. Chillon, M., Casals, T., Mercier, B. *et al.* (1995). "Mutations in the cystic fibrosis gene in patients with congenital absence of the vas deferens." *N. Engl. J. Med.* 332:1475–1480.

8. Chinnery, P.F. and Turnbull, D.M. (1999). "Mitochondrial DNA and disease." *Lancet.* 354:S117–S121.

9. Davis, L., Britten, J.J. and Morgan, M. (1997). "Cholinesterase—its significance in anaesthetic practice." *Anaesthesia.* 52:244–260.

10. Dubowitz, V. (1997). "The muscular dystrophies—clarity or chaos?" *N. Engl. J. Med.* 336:650–651.

11. Goldstein, J.L., Hobbs, H.H. and Brown, M.S. (2001). "Familial hypercholesterolemia." In Scriver CR, Beaudet AL, Sly WS, Valle D (eds). *The Metabolic and Molecular bases of inherited disease*, 8th edn. McGraw-Hill, New York. pp. 2863–2914.

12. Lightowlers, R.N., Chinnery, P.F., Turnbull, D.M. and Howell, N. (1997). "Mammalian mitochondrial genetics: Heredity, heteroplasmy and disease." *Trends Genet.* 13:450–454.

13. Martin, J.B. (1999). "Molecular basis of the neurodegenerative disorders." *New Engl. J. Med.* 340:1970–1980.

14. Nance, M.A. (1997). "Clinical aspects of CAG repeat disease." *Brain Pathol.* 7:881–900.

15. Scriver, C.R. and Kaufman, S. (2001). "The hyperphenylalaninemias: phenylalanine hydroxylase deficiency." In Scriver, C.R., Beaudet, A.L., Sly, W.S., Valle, D. (eds). *The Metabolic and Molecular Bases of Inherited Disease*, 8th edn. McGraw-Hill, New York. pp. 1667–1724.

16. Scriver, C.R. and Waters, P.J. (1999). "Monogenic traits are not simple: Lessons from phenylketonuria." *Trends Genet.* 15:267–272.

17. Shoffner, J.M. (1999). "Oxidative phosphorylation disease diagnosis." *Ann. NY Acad. Sci.* 893:42–60.

18. Worton, R. (2000). "Muscular dystrophies: Diseases of the dystrophin-glycoprotein complex." *Science.* 270:755–756.

19. Zielinski, J. (2000). "Genotype and phenotype in cystic fibrosis." *Respiration.* 67:117–133.

20. Zielinski, J., Corey, M., Rozmahel, R., *et al.* (1999). "Detection of a cystic fibrosis modifier locus for meconium ileus on human chromosome 19q13." *Nat. Genet.* 22:128–129.

MODEL QUESTIONS

1. What is an inborn error of metabolism?

2. How does inborn metabolic error bridge biochemistry and genetics?

3. Discuss briefly the disorders of phenylalanine and tyrosine metabolism.

4. Comment on various inborn metabolic errors in urea cycle.

5. List out various defects of carbohydrate metabolism.

6. Illustrate inborn errors in fatty acid metabolism.

7. Explain various disorders of purine metabolism.

8. Discuss the following;

   i.   Alkaptonuria   ii.   Albinism
   iii. Pompe's disease   iv.   Hyperammonaemia

9. Illustrate pyrimidine metabolism and disorders.

10. A 5-year-old girl develops hyperammonaemia and is critically ill. Liver biopsy confirms that she has OTC deficiency. Which genetic test is to be conducted next and why?

11. Explain why OXPHOS System is associated with elevated blood lactic acid levels.

12. Alkaptonuria is found to be more common in the offspring of consanguineous matings. Why?

PHARMACOGENETICS

21

21.1 INTRODUCTION

Individual and racial differences in people's response to drugs have become a challenge to the medical practitioners and pharmaceutical industries. Pharmacogenetics is a new discipline that identifies gene-based genetic variations in response to a drug among individuals across a population. The term pharmacogenetics was introduced by Vogel in 1959. The metabolism of many drugs involves conjugation with another molecule, which usually takes place in the liver. This biochemical transformation facilitates the excretion of the drug. The ways in which many drugs are metabolized vary from person to person and can be genetically determined.

21.2 DRUG METABOLISM

The sequence of events which is involved when a drug is metabolized is shown below.

```
Intake ──┐
         ▼
   Absorption ──┐
               ▼
      Distribution ──┐
                    ▼
         Drug-cell interaction ──┐
                                ▼
                 Breakdown ──┐
                            ▼
                     Excretion
```

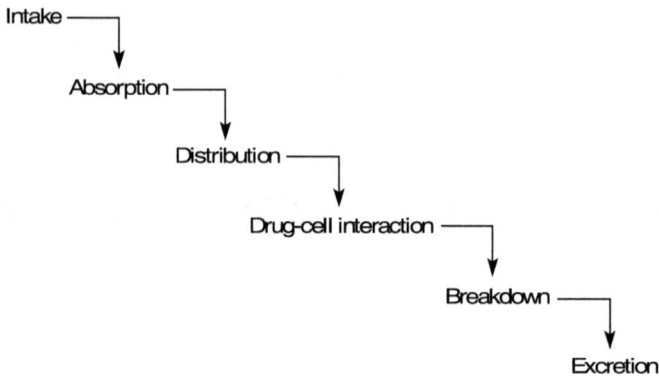

When a drug is taken by mouth, it is first absorbed from the gut and passes into the bloodstream. This in turn gets distributed in the various tissues and tissue fluids. Only a small proportion of the total dose interacts with cells to produce a specific pharmacological effect. Most drugs are either broken down or excreted unchanged.

The breakdown process takes place mainly in the liver and this also may vary with different drugs. Some drugs are completely oxidized to carbon dioxide and exhaled. Others are either metabolized in the kidneys and excreted through urine, or metabolized in the liver and thence the faeces. Many drugs undergo various biochemical transformations, which increase their solubility so that they are more easily and readily excreted.

One important biochemical transformation of morphine and codeine is that of conjugation with the carbohydrate glucuronic acid (glucuronide conjugation), which occurs primarily in the liver. Isoniazid is an important drug used in the treatment of tuberculosis; it is 'acetylated' before it is excreted. Similarly sulphonamides are often acetylated before they are excreted.

21.3   PHARMACOGENETICS OF ASTHMA

Asthma is a common respiratory syndrome characterized by partially reversible airflow obstruction, airway hyperresponsiveness, and airway inflammation. The disease is caused by a complex interaction between genetic and environmental factors that result in a spectrum of biological and clinical features. Among the spectrum of features, a highly variable response to asthma therapeutics is one. Three major classes of asthma therapeutics are available:

1. beta 2-agonists (e.g. albuterol, salmeterol, fenoterol)
2. glucocorticoids (e.g. beclomethasone, triamcinolone, prednisone) and

3. inhibitors of the cysteinyl–leukotriene pathway (e.g. montelukast, zafirlukast, zileuton).

Asthmatic patients vary greatly in their response to all three classes of drugs. It is estimated that up to 60–80% of this variability may have a pharmacogenetic basis.

The variations of therapeutic response may be due to

1. Genetic variants associated with altered uptake, distribution, or metabolism of drug (pharmacokinetic),

2. Genetic variants resulting in an unintended action of a drug outside of its therapeutic indication (idiosyncratic) or

3. Genetic variation in the drug target leading to altered drug efficacy (pharmacodynamic).

21.4 PHARMACOGENETICS OF β-AGONISTS

Albuterol is a β-agonist that binds to β_2-adrenergic receptors (β_2AR) of a cell-surface G-protein-coupled receptor. β_2-adrenergic receptor is attached to the stimulatory guanine-nucleotide-binding protein, G_s. G_s is a heteromeric G protein, which upon activation by receptor dissociates into α and βγ subunits. $G_{\alpha s}$ stimulates the enzyme adenylyl cyclase (AC) and catalyses the conversion of ATP to cAMP. The βγ subunits carry out signal transduction by altering the activities of adenylyl cyclase, phospholipase C and mitogen-activated protein kinase. The bronchodilating action of β_2 adrenergic receptor is mainly by $G_{\alpha s}$ activation of adenylyl cyclase in bronchial smooth muscle and this results in the increase of intracellular cAMP. cAMP activates protein kinase A, which phosphorylates myosin light chain kinase, potassium channels, sodium potassium ATPase, phosphorlamban, and one or more pumps that lead the sarcoplasmic reticulum to calcium uptake (Figure 21.1). This also inhibits inositol phosphates. The net effect is low levels of intracellular calcium and phosphorylation of contractile proteins. This in turn leads to smooth muscle relaxation. β_2-adrenergic receptor function is altered by its phosphorylation by the β-adrenergic receptor kinase and the binding of β-arrestin.

β_2-adrenergic receptor as well as other cell-surface G-protein-coupled receptors, can actively combine in the absence of agonist between the activated (R_A, also called R*) and inactivated (R_I, also called R) state. In the absence of agonist, the majority of receptors are in the R_I state. In the presence of agonist, R_A is stabilized and accumulates such that this state predominates. Inverse agonists stabilize the R_I state thereby lowering basal adenylyl cyclase activity.

Partial agonists stabilize the R_A conformation but not as efficiently as full agonists, such that maximal partial-agonist-stimulated adenylyl cyclase activity is less than that found with full agonists. Neutral antagonists stabilize both R_A and R_I. Agonist binding also initiates a series of events leading to a loss of β_2-adrenergic receptor function over time. This waning of the response despite continuous stimulation is termed desensitization. β_2-adrenergic receptor and many other G-protein-coupled receptors can undergo agonist-specific desensitization (termed homologous) or heterologous desensitization due to the action of ligands at other receptors and their resultant intracellular signalling. Clinically, desensitization is typically termed tachyphylaxis, which refers to a loss of clinical effectiveness of a therapeutic agent upon repetitive administration. In asthma, chronic β-agonist administration has been associated with tachyphylaxis in some but clearly not in all studies.

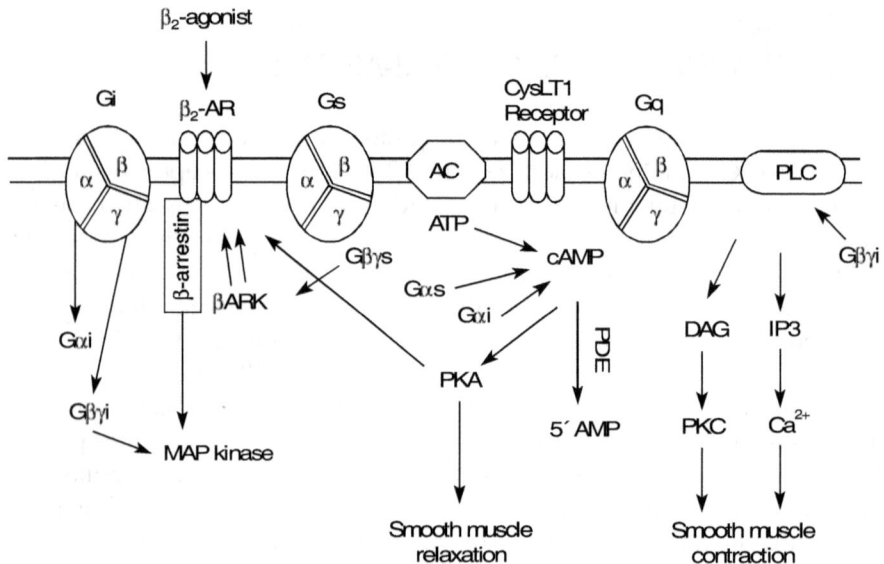

Figure 21.1 β_2 adrenergic pathway. β_2-AR—β_2-adrenergic receptor; βARK—βAR kinase; CysLT1—cysteinyl leukotriene receptor 1; Gs—heterotrimeric guanine nucleotide-binding protein with α, β and γ subunits; AC—adenylyl cyclase; PKA—protein kinase A; PLC—phospholipase C; MAP-kinase—mitogen-activated protein kinase; SR—sarcoplasmic reticulum.

In addition, excessive use of β-agonist has been associated with increased asthma mortality and morbidity. These associations with excess use of β-agonist use are not universally found, though, and it is not known whether tachyphylaxis *per se* is the basis of such observations.

21.5 PHARMACOGENETICS OF GLUCOCORTICOIDS

Glucocorticoids penetrate the plasma membrane and enter the target cell. Then they bind to a specific cytoplasmic glucocorticoid receptor (GR). The GR has three domains:

1. Ligand-binding domain at carboxylic terminal

2. DNA-binding domain with zinc finger structure and

3. Transcription regulatory domain at amino terminal

There is only one glucocorticoid receptor gene located on chromosome 5q31. The unliganded glucocorticoid receptor is a heterohexamer. It contains ligand-binding and DNA-binding domains, two molecules of heat-shock protein 90 (hsp90) and one molecule each of hsp70, hsp56, hsp26 and other proteins.

Administration of glucocorticoids to asthma patient causes binding of glucocorticoid to the receptor. This results in the dissociation of the heat-shock proteins from receptor and the DNA binding site is exposed (Figure 21.1). Now the activated receptor regulates transcription by binding to specific inverted repeat sequences (palindrome) of DNA. This results in anti-inflammatory effects due to their ability to alter the expression of many proinflammatory genes. Although most asthma patients respond to glucocorticoid therapy, a subset has poor control. They are referred to as steroid-resistant or steroid-insensitive asthmatics.

21.6 PHARMACOGENETICS OF LEUKOTRIENES

The leukotrienes are derived from arachidonic acid metabolism and exhibit a wide range of pharmacological and physiological actions (Figure 21.1). Leukotrienes cause protracted contraction of smooth muscles, constrict airways and play a role in cell function. There are three enzymes involved in the production of cysteinyl leukotrienes. They are arachidonate 5-lipoxygenase (ALOX5), leukotriene C4 (LTC4) synthase and leukotriene A4 epoxide hydrolase (LTA4H). Inhibition of the action of ALOX5 or antagonism of the action of the cysteinyl leukotrienes at their receptor level is helpful in treating asthma. The regulation of ALOX5 can be worked out at various levels such as at the transcription or translation level. The ALOX5 gene promoter contains numerous consensus binding sites for transcription factors (SP1, SP3, Egr-1, NF-KB, GATA, etc.). Mutation or deletion in the promoter sequence causes lack of binding of one or more transcription factors which in turn leads to failure in gene expression.

21.7   MULTI-DRUG RESISTANCE GENE POLYMORPHISMS

The products of multi-drug resistance gene belong to ATP-binding cassette (ABC) transporter protein family. The multi-drug resistance proteins (MDR) are composed of two transmembranous units. These units are made up of six membrane-spanning helices and cytoplasmatic nucleotide-binding domains. This domain binds with ATP which hydrolyses them to generate energy for the transport process.

Multi-drug resistance protein in cancer cells causes serious effects in cancer chemotherapy. The MDR proteins function as efflux pumps for xenobiotics and provide a barrier to entry of drugs and cellular metabolites, which lead to therapeutic failure.

MDR1 and other ABC transporters play an important role in absorption, distribution and elimination of many drugs and xenobiotics. But P-glycoprotein (PGP) serves as a barrier against entry of drugs/ligands into the body tissues. Another example is the uptake of digoxin (for heart disease) which is influenced by intestinal PGP expression. Single nucleotide polymorphism in exon 26 (C3435T) of the MDR1 gene can cause elevated levels of plasma digoxin. This increased level of plasma digoxin may cause adverse effects. Moreover the genetic variability in the PGP activity can influence the drug penetration into blood and tissue, and also mode of action. These kinds of inter-individual variation in drug action due to MDR proteins and drug-metabolizing enzymes, will help us to individualize drug therapy. This in turn will minimize the adverse effect of the drug on humans and maximize their therapeutic benefits.

21.8   PEPTIDE NUCLEIC ACID

Peptide Nucleic Acid (PNA) is a synthetic molecule that readily links with single-stranded nucleic acids. It has a polyamide structure instead of sugar phosphate backbone. DNA is a nonchiral and electrostatically neutral molecule.

Since nucleases or peptidases do not degrade DNA, it is both chemically and biologically stable. Purine and pyrimidine bases of DNA permit an efficient hybridization to complementary nucleic acids such as DNA or mRNA.

DNA-based drugs are used in treating bacterial diseases caused by staphylococci, streptococci, and coli bacillus which are multi-drug-resistant in nature (Figure 21.2).

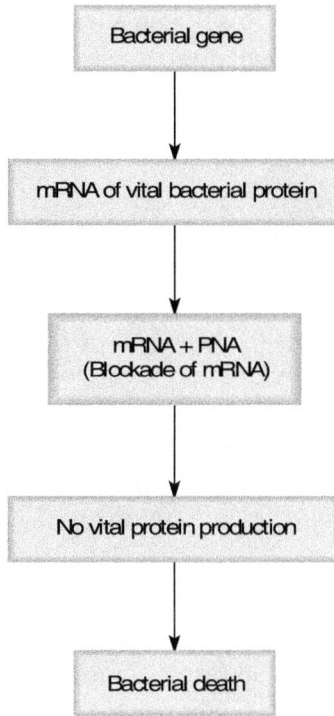

Figure 21.2   The antisense principle

21.9   GENETIC VARIATIONS DUE TO THE EFFECTS OF DRUGS

21.9.1   Acatalasia

In 1946, Takahara, a Japanese otorhinolaryngologist, treated an 11-year-old girl for a gangrenous lesion of her mouth. The infected tissue was excised and hydrogen peroxide was poured on the wound for sterilization. Normally with this treatment the blood oozing from the wound remains bright red and there is frothing. But Takahara observed that the blood which came in contact with the peroxide turned brownish black and no bubbles formed. Takahara suggested that the patient's red cells might be deficient in the enzyme catalase which breaks down hydrogen peroxide into water and oxygen.

Subsequent studies showed that this condition is in fact due to lack of catalase and it has therefore been called acatalasia. Investigations of this girl's family and other families have shown that actalasia is a rare recessive trait. Measurements of blood catalase activity have distinguished three classes of persons: those homozygous for the normal gene with normal levels of enzyme; those homozygous for the acatalasia gene with no enzyme in their blood, and heterozygous individuals with intermediate levels of enzyme.

21.9.2   Isoniazid Inactivation

Isoniazid is one of the most important drugs used in the treatment of tuberculosis. With regard to the metabolism of isoniazid, two classes of persons can be clearly distinguished: rapid and slow inactivators. Family studies have shown that slow inactivators of isoniazid are homozygous for an autosomal recessive gene for the liver enzyme N-acetyl transferase.

The implication in these studies is that the rapid inactivator produces an enzyme which inactivates isoniazid but this enzyme is absent in slow inactivators.

Phenelzine, a drug used in the treatment of depressive illness, has a molecular configuration similar to isoniazid. Studies suggest that slow inactivators of isoniazid respond better to phenelzine than fast inactivators. Similarly with hydralazine, an antihypertensive, and sulphasalazine, a sulphonamide derivative used to treat Crohn's disease, toxic side effects are more likely in slow inactivators of isoniazid.

21.9.3   Glucose 6-phosphate Dehydrogenase Variants

Quinine was the drug of choice in the treatment of malaria but although it has been very effective in acute attacks it is not very effective in preventing relapses.

Family studies have shown that G6PD deficiency is inherited as an X-linked recessive trait. Red-cell G6PD deficiency is much more common in negroes than caucasians but in affected negroes the enzyme activity in their white cells is normal whereas it is greatly reduced in most affected males of Mediterranean origin.

21.9.4   Coumarin Metabolism

The coumarin anticoagulant drugs are used in the treatment of myocardial infarction to prevent the blood from clotting. Observations have suggested that there is discontinuous variation in the response of patients taking these drugs. In this case there is no increased sensitivity but an increased resistance to the effects of the drug. For example, a patient has been described who required 20 times the usual dose in order to maintain adequate anticoagulation. This resistance appears to be transmitted as an autosomal dominant trait.

21.9.5 Organophosphate Metabolism

Paraoxonase is an enzyme which catalyses the breakdown of organophosphates. Some individuals have high serum enzyme activity and others low activity which results from a two-allele polymorphic system. It could be that those who are homozygous for the low-activity allele may be predisposed to a particular sensitivity to organophosphates which are widely used insecticides in agriculture.

FOR ADDITIONAL READING

1. Ambudkar, S.V., Dey, S. and Hrycyna, C.A. *et al.* (1999). "Biochemical, Cellular, and pharmacological aspects of the multi-drug transporter." *Annu. Rev. Pharmacol. Toxicol.* 39:361–398.

2. Bamberger, C.M., Bamberger, A.M., de Castro, M. and Chrousos, G.P. (1995). "Glucocorticoid receptor beta, a potential endogenous inhibitor of glucocorticoid action in humans." *J. Clin. Invest.* 95:2435–2441.

3. Barnes, P.J. (1986). "Non-adrenergic non-cholinergic neural control of human airways." *Ach. Int. Pharmacodyn. Ther.* 280:208–228.

4. Bend, Jr, Karmazyn, M. (1996). "Role of eicosanoids in the ischemic and reperfused myocardium." EXS. 76:243–262.

5. Bleeker, E., Postma, D. and Meyers, D. (1997). "Evidence for multiple genetic susceptibility loci for asthma." *Am. J. Resp. Crit. Care Med.* 156:S113–116.

6. Bodwell, J.E., Hu, J.M., Hu, L.M. and Munck, A. (1996). "Glucocorticoid receptors: ATP and cell cycle dependence, phosphorylation, and hormone resistance." *Am. J. Respir. Crit. Care. Med.* 154:S2–S6.

7. Borst, P., Evers, R., Kool, M. and Wijnholds, J. (1999). "The multidrug resistance protein family." *Biochim. Biophys. Acta.* 1461:347–357.

8. Cole, S.P., Bhardwaj, G., Gerlach, J.H. *et al.* (1992). "Overexpression of a transporter gene in a multidrug-resistant human lung cancer cell line." *Science.* 258:1650–1654.

9. Day, A.B. (1996). "TH2-type cytokines in asthma." *Ann. NY. Acad. Sci.* 796:1–8.

10. Denzlinger, C. (1996). "Biology and pathophysiology of leukotrines." *Crit. Rev. Oncol. Hematol.* 23:167–223.

11. Drazen, J.M., Silverman, E.K. and Lee, T.H. "Heterogeneity of therapeutic responses in asthma." *Br. Med. Bull.*

12. Drazen, J.M. and Silverman, E.K. (1997). "Genetics of asthma: conference summary." *Am. J. Resp. Crit. Care Med.* 156:S69–71.

13. Drazen, J.M., Yandava, C., Dube, L.M., Szczerback, N., Hippensteel, R., Boodhoo, T.L. *et al.* (1999). "Pharmacogentic association between ALOX5 promoter genotype and the response to anti-asthma treatment." *Nat. Genet.* 22:168–170.

14. Encio, I.J. and Detera-Wadleigh, S.D. (1991). "The genomic structure of the human glucocorticoid receptor." *J. Biol. Chem.* 266:7182–7188.

15. Green, S., Turki, J., Innis, M. and Liggett, S.B. (1994). "Amino terminal polymorphisms of the human β_2-adrenergic receptor impart distinct agonist-promoted regulatory properties." *Biochemistry.* 33:9414–9419.

16. Green, S.A., Cole, G., Jacinto, M., Innis, M. and Ligett, S.B. (1993). "A polymorphism of the human β_2-adrenergic domain alters ligand binding and functional properties of the receptor." *J. Biol. Chem.* 268:23116–23121.

17. Green, S.A., Turki, J., Bejarano, I.P., Hall, I.P. and Ligett, S.B. (1995). "Influence of β_2-adrenergic receptor genotypes on signal transduction in human airway smooth muscle cells." *Am. J. Resp. Cell. Mol. Biol.* 13:25–33.

18. Hakonarson, H. and Grunstein, M. (1998). "Regulation of second messengers associated with airway smooth muscle contraction and relaxation." *Am. J. Respir. Crit. Care. Med.* 158:S115–S122.

19. Hancox, R.J., Sears, Mr., Taylor, D.R. (1998). "Polymorphism of the β_2 adrenoceptor and the response to long-term β_2-agonist therapy in asthma." *Eur. Resp. J.* 11:589–593.

20. Hogg, J.C. (1997). "The pathology of asthma." *APMIS.* 105:735–745.

21. Holgate, S. (1993). "Mediator and cytokine mechanisms in asthma. *Thorax.* 48:103–109.

22. Holgate, S.T. (1997). "Asthma: a dynamic disease of inflammation and repair." *Rising Trends in Asthma.* 2:6–34.

23. Hoshiko, S., Radmark, O. and Samuelsson, B. (1990). "Characterization of the human 5-lipoxygenase gene promoter." *Proc. Natl. Acad. Sci., USA.* 87:9073–9077.

24. Juliano, R.L. and Ling, V. (1976). A surface glycoprotein modulating drug permeability in Chinese hamster ovary cell mutants." *Biochim. Biophys. Acta.* 455:152–162.

25. Kam, L.K., Mansur, T., Britton, J., Williams, C., Pavord, I. and Richards, K. *et al.* (1999). "Association between 308 tumor necrosis promoter polymorphism and bronchial hyperreactivity in asthma." *Clin. Exp. Allergy.* 29:1204–1208.

26. Martinez, F., Graves, P., Baldini, M., Solomon, S. and Erickson, R. (1997). "Association between genetic polymorphisms of the β_2-adrenoceptor and response to albuterol in children with and without a history of wheezing." *J. Clin. Invest.* 100:3184–3188.

27. Mcfadden, E.R. and Gilbert, I.A. (1992). "Medical progress – asthma." *N. Engl. J. Med.* 327:1928–1937.

28. McGraw, D.W., Forbes, S.L., Kramer, L.A. and Liggett, S.B. (1998). "Polymorphisms of the 5′ leader cistron of the human β_2-adrenergic receptor regulate receptor expression." *J. Cn. Invest.* 102:1927–1932.

29. Muller, M. and Renkawitz, R. (1991). "The glucocorticoid receptor." *Biochim. Biophys. Acta.* 1088:171–182.

30. Oakley, R.H., Sar, M. and Cidlowski, J.A. (1996). "The human glucocorticoid receptor beta isoform. Expression, biochemical properties and putative function." *J. Biol. Chem.* 271:9550–9559.

31. Panhuysen, C.I.M., Meyers, D.A., Postma, D.S. and Bleecker, E.R. (1995). "The genetics of asthma and atopy." *Allergy.* 50:863–869.

32. Sala, A., Voelkel, N., Maclouf, J., Murphy, R.C. and Leukotriene. (1990). "E4 elimination and metabolism in normal human subjects." *J. Biol. Chem.* 265:21771–21778.

33. Samama, P., Cotecchia, S., Costa, T. and Lefkowitz, R.J. (1993). "A mutation induced activated state of the β_2-adrenergic receptor." *J. Biol. Chem.* 1; 268:4625–4636.

34. Turki, J., Pak, J., Green, S., Martin, R. and Liggett, S.B. (1995). "Genetic polymorphisms of the β_2-adrenergic receptor in nocturnal and non-nocturnal asthma: evidence that Gly 16 correlates with the noctural phenotype." *J. Clin. Invest.* 95:1635–1641.

35. Urbatsch, I.L., Sankaran, B., Bhagat, S. and Senior, A.E. (1995). "Both P-glycoprotein nucleotide-binding sites are catalytically active." *J. Bio. Chem.* 270:26956–26961.

36. Weber, W.W. (1997). *Pharmacogenetics.* Oxford University Press, USA.

MODEL QUESTIONS

1. Define pharmacogenetics and the expression of pharmacogenomics.

2. What are the potential impacts of pharmacogenetics on medicine?

3. What is the potential impact of pharmacogenomics on medicine?

4. Describe pharmacogenetics of β-agonist in asthma.

5. Describe pharmacogenetics of glucocorticoids in asthma.

6. Describe pharmacogenetics of leukotrienes in asthma.

7. List various examples to find the effect of drug and genetic variations.

8. Polymorphism of the *MDR1* gene affects pharmacokinetics of the drug. Explain with a suitable example.

9. When will genetic medicine come to the drugstore?

10. Comment on various types of drug-metabolizing enzymes and their variants.

IMMUNOGENETICS

22

22.1   INTRODUCTION

Our environment contains a large variety of infectious agents (also called antigens). Any of these can enter into the human being (host), multiply rapidly, cause pathological damage and also kill us, if unchecked. It is evident that in normal individuals the infections are found for a limited duration with very little pathological damage. This is due to the individual's immune system, which fights against the infectious agents.

22.2   INFECTION

Infection is the lodgment and multiplication of the infectious agent in or on the tissue of the host. The infection may be acquired by contact, inhalation, ingestion, or inoculation, or from insects, or may be iatrogenic or congenital. It can be classified into **primary infection** (initial infection with an infectious agent in a host), **reinfection** (subsequent infection by the same infectious agents in the host and **secondary infection** (a new agent sets up an infection when pre-existing infectious disease lowers the resistance).

Depending upon the source, infection is classified into **exogenous** where infection is from the external sources and **endogenous** when infection is from the host's own body.

22.3   IMMUNITY

Immunity is the defence mechanism that protects an individual (host) from infectious diseases. They are of two types, viz. **innate immunity** and **acquired immunity**. Innate immunity is the inherited immunity. It acts as a first line of defence against infectious agents and does not exhibit any specificity. Innate immunity may be considered at the level of a species, race or individual.

The resistance that an individual acquires during life is known as acquired immunity. This type of immunity exhibits specificity. A particular infectious agent induces lymphocytes to proliferate, mature, secrete and "remember" that particular agent (primary immune response). Subsequent infection by the same parasite produces increased resistance and is called secondary immune response.

Acquired immunity has four essential features. They are as follows:

- an induction phase,
- recognition,
- specificity and
- immunological memory.

Acquired immunity is mediated by two interrelated and interdependent mechanisms—**humoral immunity** and **cell mediated immunity**. In humoral immunity the active protein component, immunoglobulins (antibodies), are present in the cell-free portion of the blood (plasma or serum). These immunoglobulins are specific for the infectious agents (antigens). They are derived from bone marrow lymphocytes or B-cells. Thymus-derived T-lymphocytes or T-cells reside in the peripheral blood and lymphoid tissues, which comprise the cellular component of the immune system. The infectious agents or their products stimulate the proliferation and differentiation of the T-cell and its progeny for the defence action.

22.4   ANTIGENS

Immune responses arise as a result of exposure to foreign 'stimuli'. The compound that evokes the response is referred to either as **antigen** or

immunogen. An immunogen is any infectious agent capable of inducing an immune response. The antigen is any agent capable of binding specifically to lymphocytes/antibodies. All immunogens are antigens but not all antigens are immunogens.

There are three characteristic features that a compound must possess to be immunogenic. They are foreignness, chemical complexity and high molecular weight. Epitopes are the sites either on or within the antigen with which the antibody reacts. They are also called as determinant groups or antigenic determinants. Haptens are low molecular weight compounds. They are antigenic and react with immune lymphocytes or antibodies, but they are not immunogenic, e.g. allergic response of some persons to penicillin. Penicillin is a hapten; it can couple with body protein and elicit an immune response.

The binding of antigen with antibodies or immunocompetent cells does not involve covalent bonds. The binding may involve electrostatic interactions, hydrophobic interactions, hydrogen bonds and van der Waal's forces. The strength of the attraction between antigen and antibody is referred to as the affinity. Avidity refers to the strength of interaction between multivalent antigens and antibodies.

22.5   ANTIBODIES

Antibodies are otherwise called immunoglobulins. They are made up of glycoproteins and found in the gamma globulin fraction of the serum. They are produced by B-cells (B lymphocytes) or plasma cells in response to exposure to an antigen. Humoral immunity is mediated by serum antibodies. The structural unit of an immunoglobulin molecule is called monomer. Immunoglobulin consists of four unbranched polypeptide chains linked covalently by disulphide bonds. The four-chain monomeric immunoglobulin structure is composed of two identical heavy chains (H-chain) and two identical light chains (L-chain). Heavy chain has 400 amino acids and has a molecular weight of 50 to 75 kDa. Amino acid differences in the carboxy terminal portion of the H-chains identify five antigenically distinct H-chain isotypes (IgG, IgM, IgA, IgD and IgE). The heavy chain subclasses correspond to immunoglobulin subclasses, e.g. IgG1 to IgA1. Light chain is composed of about 200 amino acids and has a molecular weight of ~23 kDa. Light chains are of two types Kappa (κ) and Lambda (λ). Immunoglobulin molecule may contain either two identical kappa chains or two identical lambda chains. Intrachain bonds occur within individual chains and are stronger than interchain bonds (light chains have two; IgG, IgA and IgD heavy chains have four; IgM and IgE chains have five). Each immunoglobulin chain consists of a series of globular regions called domains. Each heavy chain has four or five domains. Each light chain has two domains. Each domain consists of about 110 amino acids.

22.6   GENETICS OF IMMUNE RESPONSE

To produce different immunoglobulin molecules but without requiring excessive numbers of genes, a special genetic mechanism is being used. It is called DNA rearrangement or RNA splicing. Each immunoglobulin chain consists of a distinct variable (V) and constant (C) region. For each type of immunoglobulin chain either kappa light chain (κL) or lambda light chain (λL), and any one of the 5 heavy chains (αH, δH, γH, εH and μH), are found.

There is a separate pool of immunoglobulin gene segments located on different chromosomes (Figure 22.1). Each pool contains a set of different V gene segments widely separated from the D gene (diversity, seen only in H chains), J gene (joining), and C gene segments. In the synthesis of a heavy chain (H chain), for example, a particular V region is brought close to a D segment, several J segments, and a C region. These genes are transcribed into mRNA, and all but one of the J segments are removed by splicing the RNA. During B cell differentiation, the first translocation brings a V_{11} gene near a C_μ gene, leading to the formation of IgM as the first antibody produced in a primary response.

The V region of each light chain is encoded by 2 gene segments ($V + J$). The variable (V) region of each heavy chain is encoded by 3 gene segments ($V + D + J$). These various segments are united into one functional V gene by DNA rearrangement or gene splicing. Each of these assembled V genes is then transcribed with the appropriate C genes and spliced to produce mRNA that codes for the complete peptide chain. Light and heavy chains are synthesized separately on polysomes and then assembled in the cytoplasm by means of disulphide bonds. Finally, an oligosaccharide is added to the constant region of the heavy chain and the immunoglobulin molecule is released from the cell.

The gene organization mechanism outlined above permits the assembly of a very large number of different molecules. Antibody diversity depends on

- multiple gene segments
- their rearrangement into different sequences
- the combining of different L and H chains in the assembly of immunoglobulin molecules and
- mutations

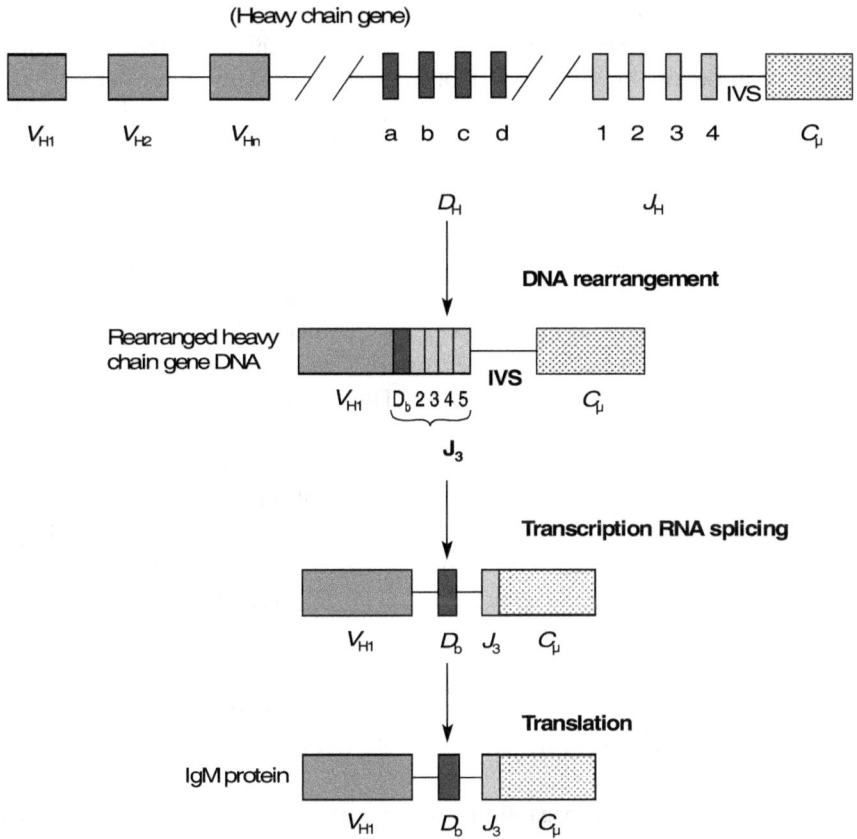

Figure 22.1 Gene arrangement to produce a μ H Chain (IVS—intervening sequence; V—variable; C$_\mu$—constant chain; J—joining; D—diversity)

A fifth mechanism called functional diversity applies primarily to the antibody heavy chain. Functional diversity occurs by the addition of new nucleotides at the splice junctions between the *V-D* and *D-J* gene segments. The diversity of the T cell antigen receptor is also dependent on the joining of *V, D* and *J* gene segments and the combining of different alpha and beta polypeptide chains. However, unlike antibodies, mutations do not play a significant role in the diversity of the T-cell receptor.

22.7   GENETIC BASIS OF STRUCTURE AND DIVERSITY

During initial infection, all B cells will produce IgM antibodies in response to exposure to that antigen. Thereafter gene rearrangement permits the elaboration of antibodies of the same antigenic specificity but of different immunoglobulin classes. Always the antigenic specificity remains the same for the lifetime of the B cell and plasma cell. This happens because the specificity is determined by the variable region genes (V, D and J genes on the heavy chain and V and J genes on the light chain) no matter which heavy chain constant region is being utilized.

In class switching, the assembled V_H genes can sequentially associate with different C_H genes and produce different classes of immunoglobulins (IgG, IgA or IgE), and have different biological characteristics. Once a B cell has "class" switched it can no longer make that class of H chain because the intervening DNA is excised and discarded. Class switching occurs only in heavy chains and not in light chains. The control of class switching is dependent on interleukins and CD40 protein. For example the interleukin 4 (IL-4) always enhances the production of IgE whereas IL-5 increases IgA. The interaction of the CD40 protein on the B cell with CD40 ligand protein on the helper T cell helps the B cell to switch to the production of IgG, IgA or IgE. A single B cell expresses either the paternal or maternal set of only one L-chain (either or) and one H-chain allele. This is called allelic exclusion.

22.8   GENOMIC ORGANIZATION OF MAJOR HISTOCOMPATIBILITY COMPLEX

The success of tissue and organ transplants depends on the human leukocyte antigens (HLA) encoded by the HLA genes. These human leukocyte antigens are proteins. They differ among members of the same species. If the HLA proteins on the donor's cells differ from those on the recipient's cells, an immune response occurs in the recipient. The genes for the HLA proteins are clustered in the major histocompatibility complex (MHC) which is located on the short arm of chromosome 6. MHC is classified into class 1 MHC and class 2 MHC. There are three genes (*HLA-A*, *HLA-B* and *HLA-C*) coding for the class I MHC proteins. Several *HLA-D* loci (*DP*, *DQ* and *DR*) determine the class II MHC proteins.

Each person has 2 haplotypes, one from the paternal and the other from the maternal chromosome 6. These genes are highly polymorphic (that is, there are many alleles of the class I and class II genes). There are at least 40 *HLA-A* genes, 80 *HLA-B* genes, and 10 *HLA-C* genes. Any individual inherits only a single allele at each locus from each parent. Thus no more than 2 genes are codominant; i.e, the protein encoded by both the paternal and maternal genes is produced. Each person can make as many as 12 HLA proteins—3 at class I loci and 3 at class II loci from both chromosomes.

In addition to the major antigens encoded by the *HLA* genes, there are many minor antigens encoded by genes at sites other than the *HLA* locus. These minor antigens can induce a weak and slow immune response, which can result in slow graft rejection. Sometimes the cumulative effect of minor antigens can lead to a more rapid rejection response. At present, there are no laboratory tests for minor antigens. Between the class I and class II gene loci there is a third locus called class III. This locus contains several immunologically important genes, namely 2 cytokines (tumour necrosis factor and lymphotoxin) and two complement components (C2 and C4).

22.8.1   Human Class I Loci

Various gene positions within the human MHC have been mapped precisely using pulsed field gel electrophoresis (PFGE) technique. This technique relies on restriction enzymes that cut rarely within the genome and an electrophoretic system that resolves large DNA fragments. Southern hybridization analysis using probes for MHC genes allows mapping of large regions of DNA and more recently this procedure has been facilitated by cloning of the whole of the 4 Mbp of MHC DNA in yeast artificial chromosome clones, or YACs.

The human class I region contains three well characterized loci, called *HLA-A, B* and *C*, which encode each of the major transplantation antigens HLA-A, B and C. The *HLA-A, B* and *C* loci reside within class I region spanning about 2 million bases of DNA. The *HLA-A, B* and *C* genes are separated by long stretches of DNA. Another set of class I-related genes, the *MIC* loci, is located around the *B* and *C* loci. *MIC* genes are inducible with increase in temperature.

Other class I-related genes have also been detected telomeric to the *HLA-G* gene. These genes may be the human counterparts of the murine *Q/ T/M* genes. The human *MOG* gene is located at the telomeric end of the human MHC, providing a landmark linking the mouse and human regions.

22.8.2   Human Class II Region

Human class II genes are located in the D region of the HLA system. There are at least 6 alpha and 10 beta chains encoded there. These are organized in three different families of genes called *DR, DQ* and *DP*. Additionally other genes such as *DNA, DOB, DMA* and *DMB* have also been identified. The *DR* family consists of a single gene (*DRA*) and four *DRB* genes (*DRB* 1–4), although the number of *DRB* genes varies depending on the particular haplotype, whereas the *DQ* and *DP* families have two *A* and two *B* genes each. These genes are called *DQA1, DQA2, DQB1, DQB2* and *DPA1, DPA2, DPB1, DPB2*. The *DR, DQ* and *DP* α-chains associate primarily with the β-chains of their own family. The *HLA-D* region spans about 850 kb.

22.9   GENETIC BASIS OF MHC POLYMORPHISM

Most class I and II molecules are highly polymorphic. This polymorphism was originally defined through the use of alloantisera, alloreactive lymphocyte populations. Recently use of monoclonal antibodies and direct DNA sequencing are being used to study the polymorphism. The list of different HLA specificities has grown very large. In some cases the location of the epitopes recognized by the antibodies or T cells used to define these specificities is known.

Class II								Class I		
DPB1	DPA1	DMA	DMB	DQB1	DQA1	DRB1	DRA	B	C	A
66	8	4	4	25	16	137	2	136	38	60

A high degree of polymorphism is also seen in class II molecules. Two-dimensional gel electrophoresis studies, sequence analysis and Southern hybridization analysis have shown that most polymorphism occurs in *DRB, DQB* and *DPB* as well as *DQA* genes. The *DPA* genes are non-polymorphic and *DRA* is invariant. As with class I molecules, allelic variations are not random but occur clustered in particular regions of the molecule around the peptide groove.

22.10   REGULATION OF *MHC* GENE EXPRESSION

A major level of regulation of class I and class II gene expression occurs at the transcriptional stage. For both class I and class II genes, the *cis*-acting DNA elements and *trans*-acting factors that bind to these DNA elements have been identified. This reveals a complex series of positive and negative regulatory events. Another way of controlling class I protein expression on the cell surface is by regulating the supply of peptides, so that low *TAP* gene expression will result in most of the class I molecules remaining in the endoplasmic reticulum, as in T2 or RMA-S mutant cells. Loss of class I expression is a feature of many tumours in order to escape immune recognition and this can take place by loss of β_2m or of TAP expression. Tumour cells can also switch off class I expression by failing to express the factors KBF1 and NF-kB, which bind to an upstream enhancer element. Regulation of expression of both class I and class II is highly complex.

22.10.1   Class I Gene Transcription

Class I genes, the region approximately −200 to −150 bp upstream contains the enhancer A and the interferon response element (IRE). Other regulatory sequences such as promoters, operator, and repressor are placed either side. Repression of class I gene expression takes place along with down-regulating immune responses. This is thought to happen in response to hormones. This proves why hormone treatments are sometimes effective in autoimmune diseases such as rheumatoid arthritis.

22.10.2   Class II Gene Regulation

Interferon g is one of the most potent inducers of class II gene expression and its activity is directed through the S, X1 and Y elements. TNFα, TGFβ, IFNα and β gene products also modulate class II expression and are dependent on the tissue and its state of differentiation.

FOR ADDITIONAL READING

1. Lynn, B., Jorde, John, C. Carey, *et al.* (1997). *Medical Genetics,* 2nd edn. Mosby Inc., Missouri.

2. Alt, F.W., Blackwell, T.K. and Yancopoulos, G.D. (1987). "Development of the primary antibody repertoire." *Science.* 238:1079–1087.

3. Blackwell, T.K. and Alt, F.W. (1989). "Mechanism and developmental program of immunoglobulin gene rearrangement in mammals." *Ann. Rev. Genet.* 23:605–636.

4. Max, E.E., Seidman, J.G. and Leder, P. (1979). "Sequences of five potential recombination sites encoded close to an immunoglobulin k constant region gene." *Proc. Nat. Acad. Sci.,* USA. 76:3450–3454.

5. Lewis, S., Gifford, A. and Baltimore, D. (1985). "DNA elements are asymmetrically joined during the site-specific recombination of kappa immunoglobulin genes." *Science.* 228:677–685.

6. Gellert, M. (1992). "Molecular analysis of VDJ recombination." *Ann. Rev. Genet.* 26:425–446.

7. Jeggo, P.A. (1998). "DNA breakage and repair." *Adv. Genet.* 38:185–218.

8. Honjo, T., Kinoshita, K. and Muramatsu, M. (1992). "Molecular mechanism of class switch recombination: linkage with somatic hypermutation." *Ann. Rev. Immunol.* 20:165–196.

9. Rajwesky, K. (1996). "Clonal selection and learning in the antibody system." *Nature*. 381:751–758.

10. Davis, M.M. (1990). "T-cell receptor gene diversity and selection." *Ann. Rev. Biochem*. 59:475–496.

11. Steinmetz, M. and Hood, L. (1983). "Genes of the MHC complex in mouse and man." *Science*. 222:727–732.

12. Hoffman, J.A., Kafatos, F.C., Janeway, C.A. and Ezekowitz, R.A. (1999). "Phylogenetic perspectives in innate immunity." *Science*. 284:1313–1318.

MODEL QUESTIONS

1. What are antigens, antibodies, immune cells and immune response?

2. Describe various types of immunoglobulins and their structures.

3. How are the genetic arrangements of immunoglobulin light chain made?

4. Discuss the genetic aspects of immunoglobulin heavy chain.

5. Compare the functions of Class I and Class II MHC molecules.

6. Discuss the difference and similarities of T-cell receptors and immunoglobulins.

7. Comment on the importance of MHC.

8. Write a detailed note on MHC polymorphism.

9. During organ transplantation, how is the matching of donor and recipient made? Why do organs get rejected and what could be the genetic background?

GENETIC COUNSELLING

23

23.1 INTRODUCTION

Most of the genetic disorders are serious. A few of the disorders are amenable to satisfactory treatment and many of them are not curable. The recent advancements in science and technology have made it possible to identify genes responsible for most of the genetic disorders; thus they help in prenatal diagnosis and in providing better genetic counselling. This often leads to great savings in cost of hospitalization and lifetime care for children with birth defects. **Genetic counselling** is the process by which patients or their families at the risk of developing a disorder (may be hereditary) are advised on consequences, probability, pattern of inheritance and prevention of the disorder.

23.2 INDICATIONS

- Consanguinity
- Recurrent spontaneous abortions
- Advanced maternal age

- Known hereditary disorder in the family
- A foetus or child with birth defects
- Exposure to teratogens

23.3   THE GOALS

- Comprehending the medical facts
- Understanding the mode of inheritance and the recurrence risks
- Understanding the reproductive options

23.4   COUNSELLING PROCESS

Collecting genetic information is the first step in genetic counselling. It is best achieved by drawing up a family tree. The main symbols used in constructing a **family tree** are given below.

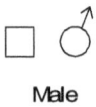

Male

Female

Sex unknown

Affected male

Affected female

Deceased

Individual without offspring

Consanguineous marriage

Illegitimate offspring

Abortion or stillbirth

Twins (Monozygotic)

Heterozygote (Autosomal recessive)

Heterozygote
(X-linked)

Proband/proposites
(Female proposita)

The following practical points deserve emphasis for a **good case history** preparation

- Infant deaths, still births and abortions
- Consanguinity
- Illegitimacy
- Basic details about both sides of the family
- Record date of birth
- Note address of relevant members

Based on this information antenatal and postnatal diagnosis is recommended.

23.5   PEDIGREE CONSTRUCTION

A 5-year-old male patient with Down's syndrome and AV canal defect has three elder brothers and two younger sisters. The parents are normal but mother's father had ventricular septal defect (VSD) at birth, had been surgically treated and lived normal. The patient's father's parents were normal. The pedigree tree looks as shown in Figure 23.1.

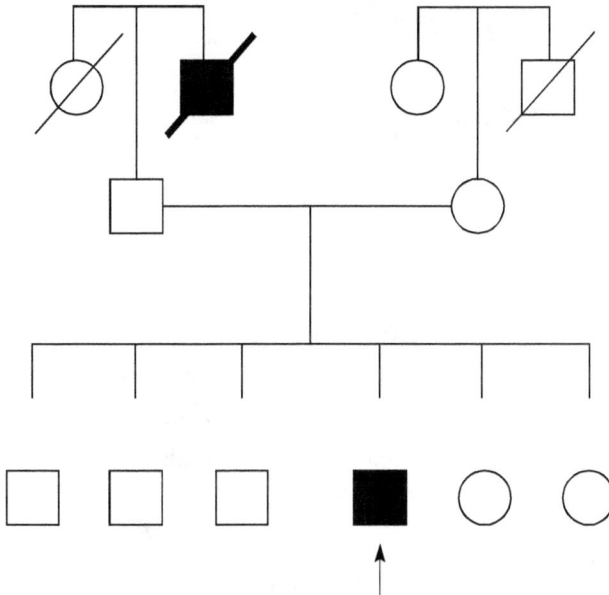

Figure 23.1   Construction of family tree

23.6   ANTENATAL DIAGNOSTIC TECHNIQUES

Some of the techniques used for antenatal diagnosis are shown in Figure 23.2.

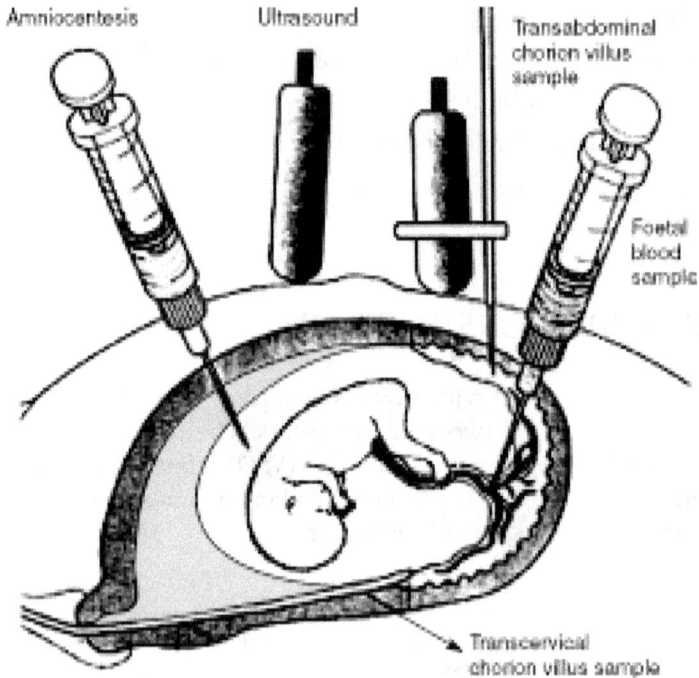

Figure 23.2   Antenatal diagnosis

- Ultrasound
- Scanning
- Cytogenetic study using amniotic fluid or chorionic villus sampling or foetal blood
- Sampling or foetal tissue sampling
- Biochemical genetics—analysis for various enzymes and hormones.

Using the above techniques, the identification of a genetic disorder, genetic predisposition to a disease, or an individual with risk of having a child with birth defects can be done.

23.7   COMPONENTS OF GENETIC COUNSELLING

- Formal interview with patient/parents and relatives

- Family history
- Pedigree preparation and analysis
- Clinical assessment
- Establishment of an accurate diagnosis
- Confirmatory tests and procedures to be done
  - Chromosomal analysis
  - Biochemical tests
  - Molecular analysis
  - Immunological tests
  - X-ray
  - Electrocardiography
  - Echocardiography
  - Dermatoglyphics
  - Biopsy
  - Linkage analysis
- Review of Literature
- Determination of recurrence risk
- Communication of the results and risks
- Discussion of options
- Referrals for prenatal diagnosis and specialists
- Follow-up

23.8 GENETIC COUNSELLING FOR CARDIOVASCULAR DISEASE

Congenital cardiovascular malformation refers to heart defects present at birth, including those discovered later. The severity of the defects varies from patient to patient depending on their stage of development. However, defects in which symptoms are presented particularly at birth are burdensome, as they may cause early death or lifelong disability.

The defects of the heart can occur in many ways:

- as an isolated case
- in conjunction with a chromosomal disorder (e.g. Turner's syndrome, Down's syndrome)

- as part of a Mendelian single gene disorder
  (e.g. Marfan syndrome, Holt-Oram syndrome)
- as part of a syndrome of unclear genetic etiology
  (e.g. VATAR association)
- due to mitochondrial disorder
  (e.g. Kearns Sayre syndrome)
- secondary to maternal infection
  (e.g. Rubella)
- secondary to maternal drug/chemical exposure
  (e.g. Lithium, retinoic acid or alcohol)
- secondary to maternal radiation exposure
  (e.g. X-rays, UV rays)
- secondary to maternal status (such as diabetes, hypertension)

The major significance of identifying inherited types of cardiovascular disease or susceptibility to cardiovascular disease lies in the possibility of prevention or modulation of effects, early detection and diagnosis, early implementation of a treatment plan, education, counselling, and the ability to make life and reproductive plans. In comparison to cancer, genetic risk assessment for cardiovascular disease is less advanced and less widely performed to date. Genetic testing and screening in children or adolescents for conditions such as HCM, FH, and LQT, when indicated, can help to save lives through preventive treatment and therapeutic interventions. Genetic testing at the molecular level for relatives of persons already identified to have heritable cardiac conditions is becoming more and more integral to mainstream primary healthcare. Appropriate genetic counselling must accompany risk assessment, genetic testing, and screening for cardiovascular disease.

The following is information that should be provided by the healthcare provider to the client considering genetic testing for cardiovascular disease:

- reasons that testing is appropriate for this person/family that includes assessment for the likelihood of having a gene mutation for a given disorder;
- what is being tested for;
- what will be analysed;
- what estimation of risk and surveillance recommendation can be done without genetic testing;
- what is the procedure being considered including description and cost;

- what can and cannot be tested. If relevant, this should include the information that although some mutations will be looked for and detected, other rare ones might not and that negative results refer only to that which was being tested for;
- what would positive or negative results mean;
- the accuracy, validity, and reliability of the test, including the likelihood of false negative or false positive results and the suitability of this test for the information the client is seeking;
- the possibility that testing will not yield additional risk information;
- the length of time between the procedure and obtaining of results;
- how the results will be communicated;
- what happens to the sample used for testing; who owns it and what uses are possible;
- a discussion of the possible risks of life and health insurance coverage and/or employment discrimination after testing results are known. Sometimes benefits result, for example, if a person is free of a certain mutation, better insurance rates or coverage might result. However, risks may be more serious.
- the level of confidentiality of results and what this means (who can know and who can find out the results);
- risks of psychological distress and negative impact on family, including stigmatization and altered self-image;
- what disclosure the client might consider for other family members and who (he/she) will tell (if anyone) about the test results; what obligation the health provider might feel to inform other family members and what this means in the context of both positive and negative tests.

As in other genetic testing, a negative test can have several meanings—the individual is truly free of the disease or mutant allele, the result is falsely negative, due to laboratory error, or the person tested possesses alleles that were not among those tested for disease.

23.9 REASONS FOR LACK OF CLEAR DIAGNOSIS

- The affected individual may have lived a considerable time ago when relevant diagnostic investigations were not available.
- The affected individual may have died without essential investigations having been done or without autopsy being performed.

- A firm diagnosis cannot be reached even with the affected individual living
- Magnitude of risk
- Severity of disorder
- Person's experience of the disorder
- Family size
- Parental or cultural and ethical values

FOR ADDITIONAL READING

1. Harper, P.S. (1998). *Practical Genetic Counselling*, 5th edn. Wright Pub., London.

2. Fuhrmann, W. and Vogel, F. (1983). *Genetic Counselling: A Guide for the Practicing Physician.* 2nd edn. Springer-Verlag, Berlin.

3. Kelly, T.E. (1986). *Clinical Genetics and Genetic Counselling.* Medical Publisher Inc. Year Book, Chicago.

4. Batista, D.A.S., Pai, G.S. and Stetton, G. (1994). "Molecular analysis of a complex chromosomal rearrangement and a review of familial cases." *American Journal of Medical Genetics.* 53:255–263.

5. Beaudet, A.L., Feldman, G.L., Fernbach, S.D., Buffone, G.J., and O'Brien, W.E. (1989). "Linkage disequilibrium, cystic fibrosis, and genetic counselling." *American Journal of Human Genetics.* 44:319–326.

6. Bell, J.I. (1993). "Polygenic disease." *Current opinion in Genetics and Development.* 3:466–469.

7. Bonaiti, C. (1978). "Genetic counselling of consanguineous families." *Journal of Medical Genetics.* 15:109–112.

8. Bonaiti-Pellie, C. and Smith, C. (1974). "Risk tables for genetic counselling in some common congenital malformations." *Journal of Medical Genetics.* 11:374–377.

9. Carter, C.O. Evans, K., Coffey, R., Fraser-Roberts, J.A., Buck, A. and Fraser-Roberts, M. (1982). "A three generation family study of cleft lip with or without cleft palate." *Journal of Medical Genetics.* 19:246–261.

10. Curnow, R.N. (1972). "The multifactorial model for the inheritance of liability to disease and its implications for relatives at risk." *Biometrics.* 28:931–946.

11. Dennis, N.R. and Carter, C.O. (1978). "Use of overlapping normal distributions in genetic counselling." *Journal of Medical Genetics.* 15:106–108.

12. Friedman, J.M (1985). "Genetic counselling for autosomal dominant diseases with a negative family history." *Clinical Genetics.* 27:68–71.

13. Hunter, A.G.W. and Cox, D.M. (1979). "Counselling problems when twins are discovered at genetic amniocentesis." *Clinical Genetics.* 16:34–42.

14. Lathrop, G.M. and Lalouel, J.M. (1984). "Easy calculations of lod scores and genetic risks on small computers." *American Journal of Human Genetics.* 36:460–465.

15. McKusick, V.A. (1997). *Mendelian inheritance in man,* 12th edn. Johns Hopkins University Press, Baltimore.

16. Maag, U.R. and Gold, R.J.M. (1975). "A simple combinational method for calculating genetic risks." *Clinical Genetics.* 7:361–367.

MODEL QUESTIONS

1. A 26-year old woman has blood drawn for maternal serum triple screening in her first pregnancy. The result is reported that indicates that her foetus has a significantly increased risk for Down's syndrome. All of the following should be done to follow up his risk **EXCEPT**
   - verifying that the gestational age used in the interpretation is correct
   - checking that the patient's correct age was included in the interpretation
   - undertaking a detailed ultrasound examination for assessment of foetal anatomy
   - offering the patient amniocentesis for foetal karyotyping
   - obtaining blood from the patient for chromosome analysis.

2. What should a genetic counsellor discuss with the couple?

3. What are the indications for genetic counselling?

4. How can molecular genetic techniques be used to improve the genetic counselling?

5. Who are the members of a genetic counselling team?

CLONING

24

24.1 INTRODUCTION

The term **clone** means to make identical copies. Identifying, isolating, preparing and studying small segments of DNA derived from a large chromosome(s) are collectively termed **cloning**. There are six general procedures involved in DNA cloning;

- isolation of gene of interest
- selection of vectors
- restriction digestion of genome at precise locations
- joining of DNA segments of interest into the vector genome
- inserting vectors into the host
- screening host cells for recombinant DNA

24.2 ISOLATION OF GENE OF INTEREST

Two major methods are followed to isolate the gene of interest from the genomic DNA. They are **cell-based cloning** and **selective amplification** method. In the cell-based cloning strategy, the genomic DNA is divided into many small fragments and all fragments are cloned in different bacteria of the same species. Such a collection of clones is called a *library*. This library is then screened to find the clone of interest (gene of interest). The other strategy is to amplify the gene of interest using polymerase chain reaction and the specific gene alone is cloned. There are two strategies applied, viz. functional cloning and positional cloning. In the functional cloning approach, based on information of a protein the corresponding gene is identified. Using contig or genomic clone, the disease gene is mapped to a chromosomal site and mutation analysis is carried out.

24.3   VECTOR

A number of vectors are available for cloning; they are used or they are designed for a specific requirement. These vectors vary in their specificity, insert capacity and expressions. There are six major types of vectors commonly used in the recombinant DNA technology.

- Bacteriophages (Lambda, Charon 40)
- Plasmids (pUC 19, pBR 322, pACYC, ColE1)
- Phagemids (M13)
- Cosmids
- Fosmids
- P1-derived artificial chromosome (PAC)
- Bacterial artificial chromosome (BAC)
- Yeast artificial chromosome (YAC)

Lambda **phage** has a double-stranded DNA and it is about 50 kilobases (kb) in length. The foreign gene (up to 20 kb) is carried as a linear double-stranded molecule with cohesive end. Soon after the entry of phage into the host bacterium, it becomes a circular molecule by using host DNA ligase. **Charon 40** is a replacement vector designed for cloning large fragments of DNA up to 24 kilobases.

The **plasmid** pUC 19 is a circular double-stranded DNA vector used for cloning foreign DNA fragments up to 8 kb in length. It lacks *rop* gene (which controls copy number) near the replication origin region. It has the following important regions, viz. replication origin region, a regulatable segment of the beta galactosidase gene (*lacZ*), ampicillin resistance gene (*amp^r*), a gene for repressor protein (*lacI*), and a short multiple cloning sequence with 14 different cloning sites. Using histochemical staining technique, the recombinants can be identified. **Cosmids** are plasmids that carry a lambda *cos* site and can be 8 kb or less in size. It can carry DNA fragments up to 50 kb. The vector is infective and gets replicated similar to a plasmid. **Fosmids** contain F plasmid origin of replication and a lambda *cos* site. They can maintain DNA insert up to 50 kb.

Bacterial artificial chromosome (BAC) is capable of maintaining human and plant genomic fragments of 300 kb length. It has lambda *cos* N and P1*lox* P sites, two restriction sites for *Hind* III and *Bam* HI and various restriction sites for G+C enzymes. T7 and SP6 promoters are found within the cloning sites. This vector can remain for 100 cycles with a high degree of stability. **P1-derived artificial chromosome** (PAC) has the capacity to insert DNA fragments of about 100 kb. The carrying capacity is twice that of the cosmid. It has the combined features of P1 vectors and BACs. PAC contains packaging site (*pac*) and two *lox*P sites which are recognized by recombinase. This vector has been used to construct genomic libraries of *Drosophila*, mouse, yeast and human DNA. A high copy number can be induced by P1 lytic replicon.

Yeast artificial chromosome (YAC) is capable of maintaining foreign genomic insert of 100 to 2000 kb. These vectors are propagated in *Saccharomyces cerevisiae* and are based on chromosomes. Each chromosome has three important components: the **centromere** (helps in nuclear division), **telomeres** (found at the ends of chromosomal DNA) and **origins of replication** (initiate synthesis of new DNA). Its advantage is the high capacity cloning system in a simple genetic background.

24.4   RESTRICTION ENZYMES

Restriction enzyme is an enzyme that binds to DNA at a specific site of a sequence and cuts it. There are four types of restriction enzymes, viz. type I, type II, type III and type IIs.

The type I and type III bind to the DNA at recognition site and cleave at random site whereas type II binds and cuts at the same position. Contrastingly, type IIs has two different subunits. Recognition sequence is asymmetric. Cleavage occurs about 20 bp away but on one side of recognition sequence. For cloning, type II restriction enzymes are widely used. Some of the type II enzymes are listed in Table 24.1.

Restriction enzymes cut DNA in two different ways (Figure 24.1). They cleave the DNA either with **blunt end** (restriction enzyme cuts exactly at the axis of dyad symmetry) or with **cohesive end** (restriction enzyme cleaves

DNA on the 5′ side of the axis of dyad (asymmetrical) yielding protruding cohesive 5′ termini).

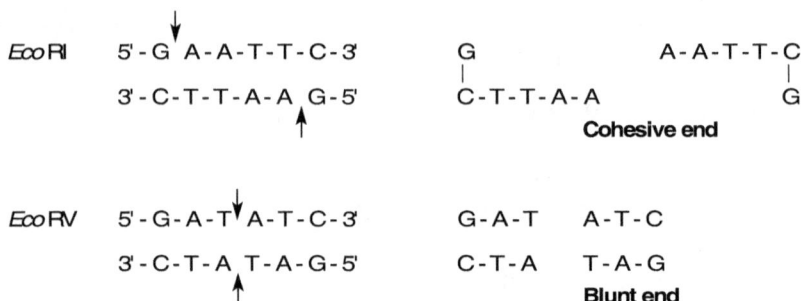

*Eco*RI   5′-G↓A-A-T-T-C-3′   G   A-A-T-T-C
           3′-C-T-T-A-A↑G-5′ | |
                                      C-T-T-A-A   G

                                                 Cohesive end

*Eco*RV   5′-G-A-T↓A-T-C-3′   G-A-T   A-T-C
           3′-C-T-A↑T-A-G-5′   C-T-A   T-A-G

                                                 Blunt end

Figure 24.1   Different kinds of restriction digestion

Restriction enzyme *Eco*RI, derives its designation based on the following principle;

- *E* denotes first letter of the generic name (*Escherichia*).
- *co* denotes first and second letters of species name (*E. coli*).
- R represents strain.
- I represents first type of restriction enzyme isolated from source (Second type of restriction enzyme from the same source is denoted as II).

Table 24.1   Restriction enzymes, their recognition sequence and source

Restriction enzyme	Restriction/cut site	Source
Alu I	5′ - A - G↓C - T - 3′ 3′ - T - C↑G - A - 5′	*Arthrobacter luteus*
Alw44 I	5′ - G↓T - G - C - A - C - 3′ 3′ - C - A - C - G - T↑G - 5′	*Acinetobacter iwoffii*
Apa I	5′ - G - G - G - C - C↓C - 3′ 3′ - C↑C - C - G - G - G - 5′	*Acetobacter pasteurianus*

(Contd.)

Table 24.1   (Continued)

Restriction enzyme	Restriction/cut site	Source
Bam HI	5´ - G G - A - T - C - C - 3´ 3´ - C - C - T - A - G G - 5´	*Bacillus amyloliquefaciens H.*
BssH II	5´ 3´	*Bacillus stearothermophilus*
Cfo I	5´ - G - C - G C - 3´ 3´ - C G - C - G - 5´	*Clostridium formicoaceticum*
Cla I	5´ - A - T C - G - A - T - 3´ 3´ - T - A - G - C T - A - 5´	*Caryophanon latum L.*
Dra I	5´ 3´	*Deinococcus radiophilus*
EclX I	5´ - C G - G - C - C - G - 3´ 3´ - G - C - C - G - G C - 5´	*Enterobacter cloacae*
Eco RI	5´ - G A - A - T - T - C - 3´ 3´ - C - T - T - A - A G - 5´	*Escherichia coli BS5*
Eco RV	5´ 3´	*Escherichia coli*
Hae III	5´ - G - G C - C - 3´ 3´ - C - C G - G - 5´	*Haemophilus aegyptius*
Hind III	5´ - A A - G - C - T - T - 3´ 3´ - T - T - C - G - A A - 5´	*Haemophilus influenzae*

(Contd.)

Table 24.1   (Continued)

Restriction enzyme	Restriction/cut site	Source
Hpa II	5´ - C ↓ C - G - G - 3´ 3´ - G - G - C ↑ C - 5´	*Haemophilus parainfluenzae*
Kpn I	5´ - G - G - T - A - C ↓ C - 3´ 3´ - C ↑ C - A - T - G - G - 5´	*Klebsiella pneumoniae*
Ksp I	5´ - C - C - G - C ↓ G - G - 3´ 3´ - G - G ↑ C - G - C - C - 5´	*Kluyvera sp.*
Mlu I	5´ ↓ 3´ ↑	*Micrococcus luteus*
Nco I	5´ ↓ 3´ ↑	*Nocardia corallina*
Nde I	5´ - C - A ↓ T - A - T - G - 3´ 3´ - G - T - A - T ↑ A - C - 5´	*Neisseria denitrificans*
Nhe I	5´ - G ↓ C - T - A - G - C - 3´ 3´ - C - G - A - T - C ↑ G - 5´	*Neisseria mucosa heidelbergensis*
Not I	5´ ↓ 3´ ↑	*Nocardia otidiscaviarum*
Nru I	5´ ↓ 3´ ↑	*Nocardia rubra*
Nsi I	5´ - A - T - G - C - A ↓ T - 3´ 3´ - T ↑ A - C - G - T - A - 5´	*Neisseria sicca*
Pst I	5´ - C - T - G - C - A ↓ G - 3´ 3´ - G ↑ A - C - G - T - C - 5´	*Providencia stuartii*

(Contd.)

Table 24.1   (Continued)

Restriction enzyme	Restriction/cut site	Source
Pvu I	5´ - C - G - A - T ↓ C - G - 3´ 3´ - G - C ↑ T - A - G - C - 5´	*Proteus vulgaris*
Rsa I	5´ - G - T ↓ A - C - 3´ 3´ - C - A ↑ T - G - 5´	*Rhodopseudomonas sphaeroides*
Sac I	5´ - G - A - G - C - T ↓ C - 3´ 3´ - C ↑ T - C - G - A - G - 5´	*Streptomyces achromogenes*
Sal I	5´ - G ↓ T - C - G - A - C - 3´ 3´ - C - A - G - C - T ↑ G - 5´	*Streptomyces albus G.*
Sau3A I	5´ G - A - T - C - 3´ 3´ - C - T - A - G 5´	*Staphylococcus aureus*
Sca I	5´ - A - G - T ↓ A - C - T - 3´ 3´ - T - C - A ↑ T - G - A - 5´	*Streptomyces caespitosus*
Sma I	5´ ↓ 3´ ↑	*Serratia marcescens*
Spe I	5´ ↓ 3´ ↑	*Sphaerotilus sp.*
Sph I	5´ ↓ 3´ ↑	*Streptomyces phaeochromogenes*
Ssp I	5´ - A - A - T ↓ A - T - T - 3´ 3´ - T - T - A ↑ T - A - A - 5´	*Sphaerotilus sp.*
Stu I	5´ ↓ 3´ ↑	*Streptomyces tubercidicus*
Swa I	5´ ↓ 3´ ↑	*Staphylococcus warneri*
Xba I	5´ ↓ 3´ ↑	*Xanthomonas badrii*

24.5   JOINING DNA

Both vector and donor DNA are treated with *Eco* RI enzyme to produce a number of fragments carrying the same sticky ends. DNA fragments of both are mixed so that fragments from different sources can unite through their sticky ends. This generates a population of chimeric DNA (vector). The sugar–phosphate backbones are sealed by **DNA ligase** by a phosphodiester bond. The recombinant vector is transferred into a bacterium for perpetuation.

24.6   SCREENING THE HOST CELLS FOR RECOMBINANT DNA

It is important to screen the colonies that are having the gene of our interest. This can be achieved by DNA hybridization method. The double-stranded DNA is denatured to single-stranded DNA using heat or alkali treatment. A denatured DNA probe labelled with a radioisotope (or other tagging materials) is added and both are together incubated under hybridization condition. If DNA probe is complementary to a sequence in one of the samples (if our gene of interest is found) then base pairing will occur and this can be identified using autoradiography.

24.7   SYNTHESIS OF HUMAN SOMATOSTATIN IN *E. COLI*

Somatostatin is a human growth hormone. This was the first human protein to be synthesized in *E. coli* (Figure 24.2). It has only 14 amino acids and is ideally suited for artificial gene synthesis. The *lacZ* (β-galactosidase gene) is present in a pBR 322-type vector. The strategy involves insertion of the artificial gene into a *lacZ* vector. This produces a fusion protein and somatostatin is separated using cyanogen bromide.

24.8   REASONS FOR EMBRYO CLONING

The reasons for cloning embryos are as follows:

- To improve the life of future generations
- To find the causes of spontaneous abortions
- To develop the best contraceptive method that would prevent embryos from implanting in the uterus

- To study various functions of cancer cells for better prevention
- To improve cell-based therapy for diseases and disorders (e.g. cardiomyopathy)

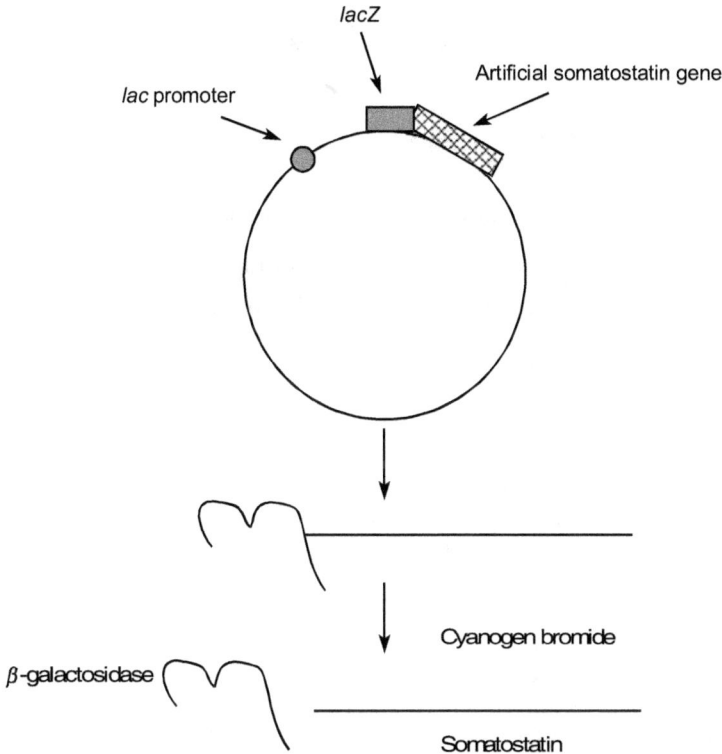

Figure 24.2   Recombinant somatostatin

Human cloning is done for two purposes;

- Reproductive cloning is to produce a clonal embryo, which is implanted in a woman's womb with the intent to create a fully formed living child.
- Therapeutic cloning is to produce clonal embryo for stem cell generation.

The purpose of using clonal embryos to generate stem cells is to allow creation of tissues or organs that the clonal donor can use without having these tissues or organs rejected by human immune system.

Advanced Cell Technologies (ACT) Company has reported the successful cloning of a human embryo by removing DNA from the skin of a man's leg and inserting it into a cow's egg, which previously had its nucleus removed.

The cloned embryo was allowed to develop for 12 days before halting the experiment. Several more clonings have reportedly been done with the goal of harvesting stem cells from embryos. Stem cells are found inside embryos during the first two weeks of their development and have the potential to develop into any kind of cell in the human body. After two weeks, stem cells differentiate into more specialized tissues.

24.9   METHODS TO CLONE MAMMALS

There are basically three methods to clone mammals

1. Splitting of a cell from an embryo (twinning)
2. The Roslin technique (used to create Dolly)
3. The Honolulu technique

24.9.1   Twinning to Create Clones

Once a sperm has fertilized an egg, it soon starts dividing (cleavage). At the eight-cell embryo stage the eight cells are separated and implanted into the uteri of eight separate mothers. This produces eight clones from different mothers.

24.9.2   Roslin Technique

The udder cells of a Finn Dorset sheep were selected to provide the genetic information for the clone. The researchers allowed the cell to divide (*in vitro*) and form culture. This produced multiple copies of the same nucleus. This step only becomes more useful when the DNA is altered (for example, polly) because the changes can be studied to make sure that they have taken effect. A donor cell was taken from the culture and then starved in a mixture, which had basic nutrients to keep the cell alive. This caused the cell to begin shutting down all active genes and enter the G_0 stage. The egg cell of a Blackface ewe was then enucleated and placed next to donor cell. Within eight hours, using an electric pulse, donor and recipient cells were fused together to activate the embryo development.

The surviving embryo was allowed to grow for about six days incubating in a sheep's oviduct (the cells placed in oviducts early in their development are much more likely to survive than those incubated in the lab). Finally, the embryo was placed into the uterus of a surrogate mother ewe. That ewe then carried the clone until it was ready to give birth. Assuming nothing went wrong, an exact copy of the donor animal was born. This newborn sheep had all the characteristics of a normal newborn sheep.

There is a risk of cancer or other genetic diseases that can occur with the gradual damage to DNA over time, as in Dolly or other animals cloned with this method.

24.9.3   Honolulu Technique

Immediately after extraction, the donor nuclei were inserted into mouse egg cells. Then the egg cell was placed in a chemical culture to jump-start the cell growth. In the culture, cytochalasin B was used to stop the polar body formation. After being jump-started, the cells developed into embryos. These embryos were then transplanted into surrogate mothers and carried to term. The most successful of the cells for the process were cumulus cells. This new technique allows for further research into exactly how an egg reprograms a nucleus.

FOR ADDITIONAL READING

1. Bates, G.P., Wainwright, B.J., Williamson, R. and Brown, S.D.M. (1986). "Microdissection of and microcloning from the short arm of human chromosome 2." *Mol. Cell. Biol.* 6:3826–3830.

2. Blackman, K. (2001). "The advent of genetic engineering." *Trends Biochem. Sci.* 26:268–270.

3. Brown, T.A. (2001). *Gene cloning and DNA analysis: An introduction*, 4th edn. Blackwell Scientific Publishers, Oxford.

4. Burke, D.T., Carle, G.F. and Olson, M.V. (1987). "Cloning of large segments of exogenous DNA into yeast by means of artificial chromosome vectors." *Science.* 236:806–812.

5. Dale, J.W. (1998). *Molecular Genetics of Bacteria*, 3rd edn. Wiley, Chichester.

6. Fiedler, W., Claussen, U., Ludecke, H.J., Senger, G., Horsthemke, B., Geurts-Van-Kessel, A., Goertzen, W. and Fahsold, R. (1991). "New markers for the neurofibromatosis-2 region generated by microdissection of chromosome 22." *Genomics.* 10:786–791.

7. Green, E.D. and Olson, M.V. (1990). "Chromosomal region of the cystic fibrosis gene in yeast artificial chromosomes: a model for human genome mapping." *Science.* 240:94–98.

8. Harrison-Lavoie, K.J., John, R.M., Porteous, D.J. and Little, P.F.R. (1989). "A cosmid clone map derived from a small region of human chromosome 11." *Genomics*. 5:501–509.

9. Kim, U.J., Shizuya, H., de Jong, P.J., Birren, B. and Simon, M.I. (1992). "Stable propagation of cosmid and human DNA inserts in an F-factor-based vector." *Nucleic Acids Res*. 20:1083–1085.

10. Ludecke, H.J., Senger, G., Claussen, U. and Horsthemke, B. (1990). "Construction and characterization of band-specific DNA libraries." *Hum. Gen*. 84:512–516 (Review).

11. Ludecke, H.J., Senger, G., Claussen, U. and Horsthemke, B. (1989). "Cloning defined regions of the human genome by microdissection of banded chromosomes and enzymatic amplification." *Nature*. 338:348–350.

12. Monaco, A.P. and Larin, Z. (1994). "YACs, BACs, PACs and MACs-artificial chromosomes as research tools." *Trends Biotechnol*. 12:280–286.

13. Peijnenburg, A.A.C.M, Bron, S. and Venema, G. (1987). "Structural plasmid instability in recombination- and repair-deficient strains of *Bacillus subtilis*." *Plasmid*. 17:167–70.

14. Pierce, J.C., Sauer, B. and Sternberg, N. (1992). "A positive selection vector for cloning high molecular weight DNA by the bacteriophage P1 system: Improved cloning efficiency." *Proc. Natl. Acad. Sci.*, USA. 89:2056–2060.

15. Potter, H. (1988). "Electroporation in biology: methods, applications and instrumentation." *Anal. Biochem*. 174:361–73.

16. Pridmore, R.D. (1987). "New and versatile cloning vectors with kanamycin resistance marker gene." *Gene*. 56:309–12.

17. Primrose, S.B., Twyman, R.M. and Old, R.W. (2003). *Principles of Gene Manipulation*, 6th edn. Blackwell Scientific Publishers, Oxford.

18. Primrose, S.B. and Twyman, R.M. (2002). *Principles of Genome Analysis*, 3rd edn. Blackwell Science, Oxford.

19. Qin, S., Zhang, J., Isaacs, C.M., Nagafuchi, S., Jani Sait, S.N., Abel, K.J., Higgins, M.J., Nowak, N.J. and Shows, T.B. (1993). "A chromosome 11 YAC library." *Genomics*. 16:580–585.

20. Robinson, K., Gilbert, W. and Church, G.M. (1994). "Large scale bacterial gene discovery by similarity search." *Nature Genet*. 7:205–14.

21. Rubin, G.M. and Spradling, A. (1983). "Vectors for P element-mediated gene transfer in *Drosophila*." *Nucl. Acids Res*. 11:6341–51.

22. Shepherd, N.S., Pfronger, B.D., Coulby, J.N., Ackerman, S.L., Vaidyanathan, G., Sauer, R.H., Balkenhol, T.C. and Sternberg, N. (1994). "Preparation and screening of an arrayed human genomic library generated with the P1 cloning system." *Proc. Natl. Acad. Sci.*, USA. 91:2629–2633.

23. Shizuya, H., Birren, B., Kim, U.J., *et al.* (1992). "Cloning and stable maintenance of 300-kilobase-pair fragments of human DNA in *Escherichia coli* using an F-factor-based vector." *Proc. Natl Acad. Sci.*, USA. 89:8794-8797.

24. Silverman, G.A., Green, E.D., Young, R.L., Jockel, J.I., Domer, P.H. and Korsmeyer, S.J. (1990). "Meiotic recombination between yeast artificial chromosomes yields a single clone containing the entire BCL2 protooncogene." *Proc. Natl. Acad. Sci.*, USA. 87:9913-9917.

25. Southern, E.M. (2000). "Blotting at 24." *Trends Biochem. Sci.* 24:585-588.

26. Stallings, R.L., Torney, D.C., Hildebrand, C.E., Longmire, J.L., Deaven, L.L., Jett, J.H., Doggett, N.A. and Moyzis, R.K. (1990). "Physical mapping of human chromosomes by repetitive sequence fingerprinting." *Proc. Natl. Acad. Sci.*, USA. 87:6218-6222.

27. Sternberg, N. (1990). "Bacteriophage PI cloning system for the isolation, amplification, and recovery of DNA fragments as large as 100 kilobase pairs." *Proc. Natl Acad. Sci.*, USA. 87:103-107.

28. Sternberg, N. (1994). "The P1 cloning system: past and future." *Mammalian Genome*. 5:397-404.

29. Van Dilla, M.A. and Deaven, L.L. (1990). "Construction of gene libraries for each human chromosome." *Cytometry*. 11:208-218.

30. Vieira, J. and Messing, J. (1982). "The pUC plasmids, an MI3mp7-derived system for insertion mutagenesis and sequencing with synthetic universal primers." *Gene*. 19:249-268.

31. Wang, M., Chen, X.N., Shouse, S., Manson, J., Wu, Q., Li, R., Wrestler, J., Noya, D., Sun, Z.G., Korenberg, J. and Lai, E. (1994). "Construction and characterization of a human chromosome 2-specific BAC library." *Genomics*. 24:527-534.

32. Watson, J.D., Gilman, M., Witkowski, J. and Zoller, M. (1992). *Recombinant DNA*, 2nd edn. W.H. Freeman, New York.

33. Wigler, M., Sweet, R., Sim, G.K. *et al.* (1979). "Transformation of mammalian cells with genes from prokaryotes and eukaryotes." *Cell*. 16:777-85.

34. Wiles, M.V., Vauti, F., Otte, J. *et al.* (2000). "Establishment of a gene-trap sequence tag library to generate mutant mice from embryonic stem cells." *Nature Genet*. 24:13-14.

35. Yansich-Perron, C., Vieira, J. and Messing, J. (1985). "Improved M13 phage cloning vectors and host strains: nucleotide sequences of the M13 mp18 and pUC 19 vectors." *Gene*. 33:103-19.

36. Yu, D., Ellis, H.M., Lee, E.C., *et al.* (2000). "An efficient recombination system for chromosome engineering in *Escherichia coli*." *Proc. Natl. Acad. Sci.*, USA. 97:5978-83.

37. Yu, S.F., Von Ruden, T., Kantoff, P.W. *et al*. (1986). "Self-inactivating retroviral vectors designed for transfer of whole genes into mammalian cells." *Proc. Nat. Acad. Sci.*, USA. 83:3194–8.

38. Zhu, H. and Dean, R.A. (1999). "A novel method for increasing the transformation efficiency of *Escherichia coli*—application for bacterial artificial chromosome library construction." *Nucl. Acids Res.* 27:910–11.

MODEL QUESTIONS

1. What are the reasons for human cloning?

2. Describe various methods to clone mammals.

3. What are restriction enzymes? Mention four enzymes and their cut-site.

4. Discuss in detail the various methods of cloning and their usage.

5. What is a cDNA library?

6. Draw diagrams of pBR 322 and explain how it is used for cloning.

7. How is DNA cloned in *E. coli*?

8. Discuss different enzymes used in cloning.

9. Describe various types of cut ends produced by restriction enzymes with examples.

10. Discuss various methods of DNA isolation for cloning.

11. Using molecular techniques how are cloning results confirmed?

12. List various problems that might arise when a large segment of DNA is cloned.

13. Describe the strategy involved in cloning of somatostatin gene.

STEM CELL BIOLOGY

25

25.1   INTRODUCTION

A stem cell is a cell that has the ability to multiply independently and differentiate into a specific organ or organism, when an appropriate environment is given. Stem cells are of two types—the **embryonic stem cells** and **the adult stem cells**. The embryonic stem cells (ESC) are derived from the inner cell mass of the blastocyst, before the blastocyst gets implanted into the uterine wall. The ESC is pluripotent, that is, it has the potential to give rise to many or several types of cells. Similarly the adult stem cells (ASC) are undifferentiated cells but found in specific organ systems. They can renew themselves and become specialized cell types of that particular tissue, when required. Until now, there is no evidence that the ASC are capable of forming all cells of the body (unipotent). The term totipotent is applicable only for fertilized egg (zygote), because it has the potential to generate all the cells and tissues that make up the embryo.

The source of the ESC is blastocyst. Sources of ASC have been found in the bone marrow, bloodstream, cornea and retina of the eye, liver, skin, gastrointestinal tract and pancreas.

Table 25.1 lists various embryonic stem cells from three different germ layers and their ability to differentiate into specific tissues.

Table 25.1   Embryonic germ layers from which differentiated tissues develop

Embryonic germ layer	Differentiated tissue
Endoderm	Thymus
	Thyroid, parathyroid glands
	Larynx, trachea, lung
	Urinary bladder, vagina, urethra
	Gastrointestinal (GI) organs (liver, pancreas)
	Lining of the GI tract
	Lining of the respiratory tract
Mesoderm	Bone marrow, RBC, WBC
	Lymphocyte, macrophage
	Adrenal cortex
	Lymphatic tissue
	Skeletal, smooth and cardiac muscle
	Connective tissues (bone, cartilage)
	Urogenital system
	Heart and vessels (vascular system)
Ectoderm	Skin
	Neural tissue (neuroectoderm)
	Adrenal medulla
	Pituitary gland
	Connective tissue of the head and face
	Eyes, ears

25.2   EMBRYONIC STEM CELLS

The important features of embryonic stem cells are listed below.

- They are derived from inner cell mass of the blastocyst.
- They can multiply without undergoing differentiation.
- They exhibit and maintain diploid set of chromosomes.
- They are pluripotent and give rise to specific cell types according to cell taken from germ layers.
- They can integrate into any developing tissues.

- They can maintain germ line and give rise to germ cells.
- They are clonogenic, that is, they can give rise to colony of genetically identical cells.

25.3 ADULT STEM CELLS

Adult stem cells have the following characteristic features.

- They are derived from specific organ tissues.
- They can make identical copies for long periods.
- They can differentiate to become a specific tissue cell type.
- They show plasticity.
- They can be manipulated to improve their ability.
- They are unipotent, and can give rise to a specific cell type.
- They are clonogenic.

25.4 MARKERS

The important haematopoietic stem cell surface markers of mouse and human are listed in Table 25.2. These cell surface markers disappear during differentiation into a distinct cell lineage.

Table 25.2   Surface markers of various cells

Human	Mouse
CD 34$^+$	CD 34$^{low/-}$
CD 59$^+$	SCA-1$^+$
Thy 1$^+$	Thy 1$^{+/low}$
CD 38$^{-/\,low}$	CD 38$^+$
C- kit$^{-/\,low}$	C-kit$^+$
Lin$^-$	Lin$^-$

25.5 FOETAL CELL CULTURE

Foetal cells have the ability to migrate, proliferate, differentiate and establish functional connections with neighbouring cells. This happens in response to environmental clues and also to its intensive programme. Foetal cells proliferate more rapidly than mature cells and require lower oxygen level. Foetal cells produce high levels of angiogenic and neurotrophic factors. These factors enhance the foetal cell's ability to grow during grafting.

There are two principal sources for haematopoietic stem cells. They are umbilical cord blood or the foetus itself. Haematopoietic stem cells circulate in the foetal blood. It can be isolated from the placenta through the umbilical cord after birth. 150 ml of foetal blood contains at least 20 ml of haematopoietic stem cells (HSC). These HSC have been safely isolated and then transplanted into recipients to treat bone marrow failure.

The umbilical vein is punctured after clamping. Blood is collected by gravity into a standard blood bag. Collected blood is stored at 4°C for further processing. For a better yield, the blood is separated within 24 hours.

The umbilical cord blood is diluted to 4 : 1 with Hank's balanced salt solution. The blood bag is hung in the inverted position at 22°C. This helps in the sedimentation of RBCs. The RBCs are transferred into a second bag. The bag containing leukocyte-rich plasma is centrifuged at 1200 rpm for 10 minutes by maintaining 22°C temperature. The supernatant plasma is again removed and mixed with cryoprotectant solution containing 20% dimethyl sulphoxide (DMSO). The cells are cultured in tissue culture medium M-199. The cell samples are stored at 1°C/min. in a programmable cell freezer.

25.6   STEM CELLS AND DIABETES

Glucose plays an important role in the metabolism of the cells by providing energy (ATP). The level of glucose is regulated by the hormone insulin. Insulin is produced in the pancreas. Any imbalance in the blood glucose level or high level of glucose in the blood causes diabetes mellitus. There are two kinds of diabetes; type 1 (insulin-dependent) and type 2 (non-insulin-dependent). Both the forms have a different clinical and genetic basis.

The increase in blood glucose is either due to the lack of insulin sensitivity or the destruction islet cells of the pancreas (which normally secrete insulin). Islet cell clusters typically respond to the concentration of glucose by releasing insulin in two phases: a quick release of higher concentration of insulin and a slower release of lower concentrations of insulin. High level of sugar in the bloodstream is responsible for most of the complication of diabetes which includes retinopathy, kidney failure, heart disease, stroke and neuropathy.

Several groups of researchers are developing cell-based therapy for diabetes. They use foetal tissue as a source for islet cells. They found that insulin content was initially higher in the fresh tissue and also in purified islets. However insulin concentration decreased after some days in the whole tissue grafts but remained the same in the purified islet grafts. When cultured islet cells were implanted in the pancreas, the insulin content increased over a period of time. The cells lines are genetically engineered to express the beta islet cell gene, *PDX-1*, which in turn stimulates the expression of the insulin gene. Such cell lines can be used for transplantation to reverse diabetes.

25.7 REPAIR OF DAMAGED HEART USING STEM CELLS

Cardiomyocytes do not regenerate during adult life. When the cardiomyocytes are damaged due to ischemia or infarction, it is possible to revert the condition with the help of stem cell therapy. Using cell surface markers, the adult bone marrow cells (pluripotent) are isolated and injected into the damaged wall (Figure 25.1). These injected cells develop into new cardiomyocytes, vascular endothelium and smooth muscle cells. The newly formed myocardium can replace the damaged heart up to 65 percent. This capacity of adult stem cells is referred to as "plasticity".

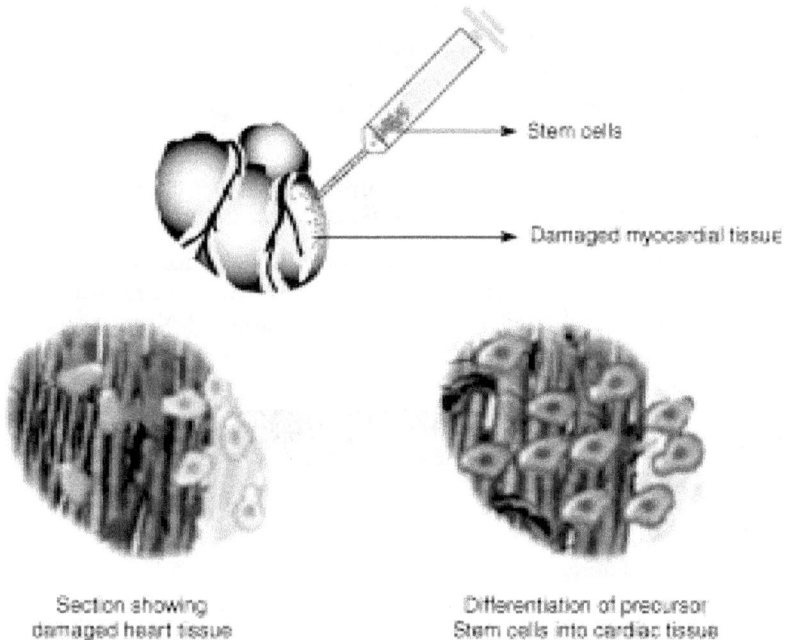

Stem cells

Damaged myocardial tissue

Section showing
damaged heart tissue

Differentiation of precursor
Stem cells into cardiac tissue

Figure 25.1   Myocardium regeneration using stem cells

FOR ADDITIONAL READING

1. Assady, S., Maor, G., Amit, M., Itskovitz-Eldor, J., Skorecki, K.L. and Tzukerman, M. (2001). "Insulin production by human embryonic stem cells." *Diabetes*. 50.

2. Beattie, G.M., Otonkoski, T., Lopez, A.D. and Hayek, A. (1997). "Functional beta-cell mass after transplantation of human fetal pancreatic cells: differentiation or proliferation?" *Diabetes*. 46:244–248.

3. Soria, B., Roche, E., Berna, G., Leon-Quinto, T., Reig, J.A. and Martin, F. (2000). "Insulin-secreting cells derived from embryonic stem cells normalize glycemia in streptozotocin-induced diabetic mice." *Diabetes*. 49:157–162.

4. Zulewski, H., Abraham, E.J., Gerlach, M.J., Daniel, P.B., Moritz, W., Muller, B., Vallejo, M., Thomas, M.K. and Habener, J.F. (2001). "Mulitpotential nestin-positive stem cells isolated from adult pancreatic islets differentiate *ex vivo* into pancreatic endocrine, exocrine, and hepatic phenotypes." *Diabetes*. 50:521–533.

5. Beltrami, A.P., Urbanek, K., Kajstura, J., Yan, S.M., Finato, N., Bussani, R., Nadal-Ginard, B., Silvestri, F., Leri, A., Beltrami, C.A. and Anversa, P. (2001). "Evidence that human cardiac myocytes divide after myocardial infarction. *N. Engl. J. Med*. 344:1750–1757.

6. Jackson, K.A., Majka, S.M., Wang, H., Pocius, J., Hartley, C.J., Majesky, M.W., Entman, M.L., Michael, L.H., Hirschi, K.K. and Goodell, M.A. (2001). "Regeneration of ischemic cardiac muscle and vascular endothelium by adult stem cells." *J. Clin. Invest*. 107:1–8.

7. Kehat, I., Kenyagin-Karsenti, D., Druckmann, M., Segev, H., Amit, M., Gepstein, A., Livine, E., Binah, O., Itskovitz-Eldor, J. and Gepstein, L. (2001). "Human embryonic stem cells can differentiate into myocytes portraying cardiomyocytic structural and functional properties." *J. Clin. Invest*. 108:407–414.

8. Kessler, P.D. and Byrne, B.J. (1999). "Myoblast cell grafting into heart muscle: cellular biology and potential applications." *Annu. Rev. Physiol*. 16:219–242.

9. Orlic, D., Kajstura, J., Chimenti, S., Jokoniuk, I., Anderson, S.M., Li, B., Pickel, J., Mckay, R., Nadal-Ginard, B., Bodine, D.M., Leri, A. and Anversa, P. (2001). "Bone marrow cells regenerate infarcted myocardium." *Nature*. 410:701–705.

MODEL QUESTIONS

1. What are stem cells?

2. What are the different types of stem cells?

3. Define totipotent and pluripotent.

4. List out the various markers of stem cells.

5. Describe the protocol of stem cell culture.

6. How are stem cells useful in insulin production (in *in vivo* as well as *ex vivo* condition)?

7. Comment on the use of cell therapy in replacing the tissues of the damaged heart.

8. What is the lifetime of replacement cells and their function?

9. Do stem cells have the pacemaking as well as conduction capabilities of native cardiac muscle cells?

GENE THERAPY

26.1   INTRODUCTION

There are more than 4000 genetic diseases that have been identified. Unfortunately our traditional pharmacological, surgical and dietary interventions provide less effective treatment for certain diseases. As a result, there has been a great deal of effort to the development of alternative approach. Gene therapy offers a promising alternative approach for the treatment of various genetic as well as acquired diseases. Gene therapy came as treatment after four groundbreaking discoveries.

- DNA is the genetic material (Hershey and Chase)
- Double helical structure of DNA (Watson and Crick)
- Genetic code (Brenner)
- Cloning of globin gene (Maniatis and his colleagues)

26.2   WHAT IS GENE THERAPY

When treating hereditary diseases using cloning (recombinant DNA technology) became a reality, it lead the way to the birth of gene therapy. Gene therapy is defined as "intracellular delivery (insertion) of genetic material or gene to cells to generate a therapeutic effect by correcting an existing genetic abnormality". Thus the cells are provided with new functions. Based on the cell type, gene therapy is grouped into two types; germ line and somatic cell gene therapy.

26.2.1   Germ-line Gene Therapy

Germ-line gene therapy involves the insertion of genes or genetic material into germ cells (ova or sperm) of an individual. This kind of therapy helps to eradicate many hereditary disorders. But it is currently considered to be ethically unacceptable.

26.2.2   Somatic Cell Gene Therapy

Somatic cell gene therapy involves the insertion of genes or genetic material into diploid cells of an individual. Some of the strategies adopted in this kind of therapy are

- *in vivo* approach
- *ex vivo* approach
- *in situ* approach

In the ***in vivo*** approach, the gene of interest is injected systematically in affected individuals. This is the least advanced strategy at present.

In the ***ex vivo*** approach, a suitable cell type is harvested from a donor and grown in tissue culture. The gene is inserted into the cell and the cells expressing transgene are amplified. The genetically altered cells are then harvested and reimplanted into an affected host.

Administering the genetic material directly to the tissues is the major approach called "***in situ*** delivery" and is the major field of clinical interest. One of the diseases which showed success using this strategy is cystic fibrosis. Also, for the treatment of tumours and myocardial infarction, this strategy is showing greater success.

However, the *in vivo* approach requires a highly efficient method for gene transfer whereas *ex vivo* is labour-intensive and cumbersome but the advantage is that it does not require a highly efficient method for gene transfer.

There are many potential risks that must be considered before going for gene therapy.

- Sometimes insertional mutagenesis can lead to cancer or tumour.
- Transfer of exogenous genetic material is possible.
- Over-expression of inserted gene can occur.
- Along with gene of interest, transfer of other genetic material is also possible.
- Contamination by other disease-causing viruses or organisms is possible.

There are three methods through which the gene functions can be altered. They are

- **Gene correction**   Defective gene or mutant gene is corrected by gene insertion.
- **Gene replacement**   When a particular gene expression is nil or abnormal, the required gene is directly inserted to produce functional product.
- **Gene augmentation**   Gene expression is increased for homeostasis in metabolism.

26.2.3   General Requirements

There are certain basic requirements for the application of gene therapy.

- Identification of genetic defect or pathogenic mediators of the disease.
- Availability of cDNA or gene needed for therapy.
- Availability of suitable vector for transferring gene of interest.
- Availability of good animal model available for preclinical studies.
- Absence of any effective treatment.

26.3   TARGET CELLS

Following are the various target cells in which the research works have been carried out.

- Fibroblasts
- Hepatocytes
- Lymphocytes
- Haematopoietic stem cells
- Vascular endothelial cells
- Tumour cells

26.4   GENE TRANSFER TECHNIQUE

Many techniques and vector systems now exist for use in gene delivery. Some of them are as follows:

- Virus-mediated
- Liposome-based or
- Direct insertion/implantation

26.4.1   Virus-mediated System

Different kinds of viral vectors are being used in gene delivery system. The most widely used viral vectors are retrovirus, adenovirus, herpesvirus and adeno-associated virus. In all these viruses, their genome is modified for incompetent replication.

Retroviral delivery   Retroviruses are RNA viruses having a protein envelop. They inject eukaryotic cells. The replication is facilitated by RNA-directed DNA polymerase (reverse transcriptase). The reverse transcribed DNA subsequently integrates into eukaryotic host chromosomes. Further to DNA integration, it is transcribed using the host machinery. Retrovial genomes contain three core genes termed *gag, pol* and *env*. These core genes are flanked by long terminal repeat (LTR) sequences. For gene therapy, the *gag, pol* and *env* coding units are replaced with a transferase of interest. The LTR and packaging *cis*-acting sequences are retained. This helps the virus to inject the cells and also to direct the expression of transgene. But the virus cannot replicate and make its own progeny. Currently, moloney murine leukemia virus (MoMuLV) is being widely used for oncogenic studies.

Adenovirus-based vector   Adenovirus vectors have emerged as vehicles of choice for *in vivo* gene therapy. It is a DNA virus. They have a transgene capacity of 7 kb to 35 kb. The advantages of this vector are the ability to produce high quantity of purified particles, having a wide range of infection (all types of tissues and specifically lungs) and being highly immunogenic.

Adeno-associated viral vector (AAV)   It is a DNA-based vector system, also known for its nonpathogenicity. This virus integrates at a specific site on chromosome 19. This vector has been shown to be a weak immunogen. The major limitation is its low transgenic capacity (≤ 4.8 kb).

Herpes simplex viral vector (HSV)   It is one of the recent viral vector systems introduced in gene therapy. It has the potential to package 150 kb of transgene particles. Some of the important limitations are immunogenicity of the particle, packaging constraints and long-term maintenance of transgene.

26.4.2   Liposome-mediated Approach

Liposomes are lipid micelles, used to transfer genes during therapy. Liposomes have positively charged lipid and a colipid for stabilization of the liposome complex. Commonly used liposome formulations are (N-[1-(2,3-dioleoloxy)] propyl-N-N-N-trimethyl ammonia chloride (DOTMA) and dioleoyl phosphatidylethanolamine (DOPE). Negatively charged DNA and positively charged lipid result in the formation of complexes. This type of gene therapy has very low efficiency.

26.4.3   Naked DNA Insertion

Vector with transgene can be directly infected into the organs. This method has been proved to be highly effective. Skin, thymus, cardiac muscle and skeletal muscle are the tissues where transgene expression is highly effective after plasmid DNA infection. Only limitation is that the age and type of species can influence the transfer efficiencies.

26.5   GENE THERAPY FOR CYSTIC FIBROSIS

Cystic fibrosis is a lethal disease, mostly found in Caucasian population. It is inherited in a recessive mode. The common clinical problems are lung damage, and respiratory failure. These problems are the result of rapid bacterial growth in the thick mucus, which accumulate in the lungs. Some of the other clinical features are perinatal obstruction of the small intestine, atrophy of the pancreatic duct and vas deferens.

Cystic fibrosis is caused due to deletion/mutation in the cystic fibrosis transmembrane conductance regulator (CFTR) gene. It expresses as CFTR protein, which is a CAMP-regulated chloride channel. The CFTR protein consists of two transmembrane regions, two nucleotide-binding domains and a regulatory region with many sites for phosphorylation. The CFTR gene lies on the long arm of chromosome 7. This gene was cloned in 1989 by Riordan *et al.* and Rommens *et al.* There is a 3 bp deletion in exon 10 of the CFTR gene, which causes CF in 70% of the cases. Presence of abnormal levels of chloride in sweat secretions is one of the earliest diagnosis in children.

The full length of cDNA of CFTR (~ 4.7 kb) is used for the insertion. The gene expression is under the control of eukaryotic viral promoters such as cytomegalovirus, simian virus 40 or Rous sarcoma virus promoters. CFTR expression is detected by immunoprecipitation or by immunohistochemistry.

The reporter genes such as β-galactosidase, luciferase gene, etc. are also used to find the expression level. The safety of the gene delivery was assessed by

- Continued expression of viral vector genes
- *Trans*-complementation of deleted functions
- Cellular and humoral immune response of the host

26.6   GENE THERAPY FOR ADENOSINE DEAMINASE DEFICIENCY

Adenosine deaminase (ADA) deficiency is a metabolic disorder. Its deficiency results in severe combined immunodeficiency (SCID) syndrome. Adenosine deaminase catalyses the deamination of adenosine to inosine. The enzyme is present in all body tissues. The accumulation of the products of adenosine such as dATP, plasma and urinary deoxyadenosine (dAdo) and low levels of S-adenosyl homocysteine hydrolase is toxic to T-lymphocytes and B lymphocytes. This results in abnormal functioning of T and B cells.

Mutation in ADA gene (1.1 kb) which is located in chromosome 20q13.11 can cause low expression or absence of the enzyme. The retroviral gene transfer into CD34$^+$ showed effective treatment in this case.

26.7   GENE THERAPY FOR CARDIOVASCULAR DISEASES

Atherosclerosis, post-angioplasty restenosis, post-bypass atherosclerosis, peripheral atherosclerotic vascular disease and vascular graft failures are common heart problems. Atherosclerosis-related diseases are characterized by the accumulation of lipid in macrophage smooth muscle cells, proliferation of smooth muscle cells, accumulation of connective tissue components and thrombus formation in arterial wall. They lead to the obstruction of blood vessels, ischemic symptoms such as pain, palpitation and sweating, thrombotic complications, myocardial infarction, stroke and peripheral vascular problems.

Blood vessels are easily accessible by current catheter-mediated techniques. Arteries can be temporarily closed by balloon catheters, and viral vectors required for gene therapy can be injected directly into the vessels. Coronary bypass grafts are also obstructed within a few years after the operation. Gene therapy strategies are directed towards the inhibition of smooth muscle cell proliferation and various growth factor activities. The most commonly suggested therapeutic approaches are as follows:

- Oligonucleotides against c-myb, cdk2 kinase and proliferating cell nuclear antigen have been shown to reduce intimal thickening in restenosis.

- Dominant negative mutations could be used to block signal transduction through growth factor receptors.

- The expression of thymidine kinase and retinoblastoma genes reduces neointimal thickening.

- Ribozymes are catalytically active RNA molecules, which can perform targeted inactivation of specific mRNA species.

26.7.1   Prevention of Restenosis

Among the many potential applications of gene therapy for CVD, prevention of restenosis has received perhaps the most attention to date. The possibility of manipulating genetic information of the vessel wall comes at a time when restenosis has continued to resist conventional medical therapies. Use of proteins such as fusion toxins derived from recombinant technology represents another form of therapy.

26.7.2   Prevention of Thrombosis

Prevention of thrombosis of bioprosthetic conduits constitutes the next suitable target for gene therapy. Increasing fibrinolytic activity within synthetic vascular grafts or stents by seeding them with genetically engineered endothelial cells that over-express the human tissue plasminogen activator gene is currently under investigation.

26.7.3   Management in the Post-infarction Period

The recent description of direct gene transfer to the vascular wall *in vivo* as well as the delivery of DNA to the myocardium by direct injection with subsequent transgene expression has provided a basis for the use of gene therapy. This helps in the management of coronary and myocardial diseases in the post-infarction period. Stimulation of the development of the collaterals around an ischemic portion of myocardium through the administration of recombinant angiogenic factors already has been reported in an animal model in the post-infarction period. This strategy can also be mediated through the transfer of genetic material. Direct gene transfer may also offer new treatment strategies for ischemic myocardium. This adjunctive therapy serves the compromised myocardium as a means of modifying the electrophysiological properties of the heart. Thus, it reduces the need for drugs with multiple toxicities.

26.8   GENE THERAPY FOR FAMILIAL HYPERCHOLESTEROLAEMIA

Patients with familial hypercholesterolaemia lack functional LDL receptor activity and develop a high plasma cholesterol level. Conventional drug therapies are inefficient in these patients. Only plasmapheresis or liver transplantation can significantly reduce the plasma cholesterol level in these cases. An alternative strategy for the treatment of familial hypercholesterolaemia is to transfer functional LDL receptors into hepatocytes. This strategy helps in lowering the plasma cholesterol level by restoring the LDL receptor activity.

FOR ADDITIONAL READING

1. Kenneth, R., Chien, *et al.* (1999). *Molecular basis of Cardiovascular Diseases.* W.B. Saunders Company, USA.

2. Robert Roberts. (1993). *Molecular basis of Cardiology.* Blackwell Scientific Publications, USA.

3. Lemoine, N.R. (1999). *Understanding gene therapy.* BIOS Scientific Publishers, UK.

4. Murphy, S.J. (1999). *Viral delivery systems for gene therapy.* BIOS Scientific Publishers, UK.

5. Miller, A.D. (1999). *Nonviral delivery systems for gene therapy.* BIOS Scientific Publishers, UK.

6. Vassaux, G. (1999). *Gene therapy for monogenic diseases.* BIOS Scientific Publishers, UK.

7. Genotype–Phenotype Consortium. C.F. (1993). "Correlation between genotype and phenotype in patients with cystic fibrosis." *New Engl. J. Med.* 329:1308–1313.

8. Riordan, J.R., Rommens, J.M., Kerem, B.S., Alon, N., Rozmahel, R., Grzelczak, Z., Zielenski, J., Lok, S., Plavsic, N., Chou, J.L., Drumm, M.L., Iannuzzi, M.C., Collins, F.S. and Tsui, L.C. (1989). "Identification of the cystic fibrosis gene: cloning and characterization of complementary DNA." *Science.* 245:1066–1073.

9. Rommens, J.M., Iannuzzi, M.C., Kerem, B.S., Drumm, M.L., Melmer, G., Dean, M., Rozmahel, R., Cole, J.L., Kennedy, D., Hidaka, N., Zsiga, M., Buchwald, M., Riordan, J.R., Tsui, L.C., Collin, S.F.S. (1989). "Identification of the cystic fibrosis gene: chromosome walking and jumping." *Science.* 245:1059–1065.

10. Brenner, M.K., Cunningham, J.M., Sorrentino, B.P. and Heslop, H.E. (1995). "Gene transfer into human hematopoietic progenitor cells." *Br. Med. Bull.* 51:167–191.

11. Hirschhorn, R. (1993). "Overview of biochemical abnormalities and molecular genetics of adenosine deaminase deficiency." *Pediatr. Res.* 33:S35–41.

12. Hock, R.A., Miller, A.D. (1986). "Retrovirus-mediated transfer and expression of drug resistance genes in human haematopoietic progenitor cells." *Nature.* 320:275–277.

13. Touchette, N. (1996). "Gene therapy: Not ready for prime time." *Nature Med.* 2:7–8.

14. Van Beusechem, V.W., Bakx, Ta., Kaptein, L.C., Bart Baumeister, J.A., Kukler, A., Braakman, E., Valerio, D. (1993). "Retrovirus-mediated gene transfer into rhesus monkey hematopoietic stem cells: the effect of viral titers on transduction efficiency." *Hum. Gene. Ther.* 4:239–247.

MODEL QUESTIONS

1. Explain in detail the various methods of gene transfer.
2. List out the ethical aspects involved in gene therapy.
3. What is the role of gene therapy in medical treatment?
4. Explain how gene therapy helps in treating cardiovascular disease.
5. Gene therapy is best as well as worst for treating any genetic disorder. Discuss with examples.
6. Is gene therapy a "terror" or "treatment"?
7. How can cystic fibrosis be treated by gene therapy?
8. Detail how gene therapy can be applied for treating adenosine deaminase deficiency.

ORIGIN OF LIFE AND GENOME EVOLUTION

27

27.1   INTRODUCTION

Accident or incident might have caused our genome to originate, but mutation and recombination provided opportunity for the genome to evolve. We can understand the genome evolution clearly by comparing between the genome of different organisms. This chapter will guide you on genome evolution since its biochemical origin to the present day. It is believed that 14 billion years ago, the universe began with the gigantic primordial fireball (called big bang). After about 4 billion years galaxies began to fragment to form the sun and its planets. Simultaneously the earth was formed and covered with water. Some 3.5 billion years ago, the land masses began to appear along with cellular life. The cellular life was preceded by self-replicating polynucleotides, which happen to be the progenitors of the first genomes.

27.2   THE ORIGIN OF GENOME

Earth atmosphere had low oxygen content until photosynthesis evolved and had similar salt composition to that of First Ocean. The most abundant atmospheric gases were probably methane and ammonia. Experiments by Miller (1953) proved that electrical discharges in a methane–ammonia mixture resulted in a range of amino acids such as alanine, glycine, valine and several others. Further to this, hydrogen cyanide and formaldehyde were also formed to prepare other amino acids as well as purines and pyrimidines. These chemical syntheses lead to the accumulation of some biomolecules in the earth's atmosphere.

Polymerization of the building blocks into biomolecules might have occurred in the oceans or on solid surface or may be in hydrothermal vents. But the important event is the synthesis of polymeric biomolecules similar to the ones found in the present living system. The global ocean could have contained as many as 10 biomolecules per litre and necessary events could have also taken place in billion years. The first biochemical systems could have been centred entirely on RNA. The major supportive evidence came in the mid 1980s, as that RNA can have catalytic activity, synthesis and copying of RNA, and transfer of amino acid when forming dipeptide (similar to the action of tRNA). We now imagine that RNA molecules could have multiplied in a slow and haphazard fashion without much accuracy. Moreover a greater accuracy in multiplication would have enabled the RNAs to increase in length without losing their sequence specificity. This could have paved the way for RNA genomes or specifically protogenome.

Other interesting events are: How could this RNA develop into a DNA? How were first cell-like structures formed? How were long-chain unbranched fatty acids formed? Many such questions were unanswered due to lack of supporting evidences. But transfer of the coding function to the more stable DNA seems almost inevitable. This happens by the replacement of uracil with its methylated derivative thymine. Similarly reduction of ribonucleotides gave deoxyribonucleotides, which could then be polymerized into copies of protogenomes by a reverse transcriptase. The first DNA genomes comprised many separate molecules. Each molecule specifies a single protein probably from a single gene. The linking of these genes would have taken either before or after DNA genome formation. In this way large chromosomes would have been made.

Computer modelling and experimental simulations show that biochemical evolution occurred many times in the oceans or atmosphere. So, there is a possibility that 'life' arose on more than one occasion. Though we discuss that present-day organisms have single origin due to remarkable similarities between molecular, biological and biochemical mechanisms, it could be possible for origin of life at different occasions! Further natural selection affects the individual organism that alters genes over evolutionary time. The genes evolve by the duplication, accumulation of mutations, gene conversion and transposition.

Morphological evolution was accompanied by genome evolution. As we move up the evolutionary tree, we see more complex genomes. One indication of this complexity is gene number, which is approximately 1000 in some bacteria and 30,000–40,000 in vertebrates. The increase may not be a gradual one but there seems to be two sudden outbursts. This happened when the number of genes increased from 5000 to 10,000 before 1.4 billion years, as seen in prokaryotes and early eukaryotes. Similarly the second expansion occurred at the end of the Cambrian period when the first vertebrate was established. These protovertebrates have at least 30,000 genes.

There are several ways in which gene duplication could occur

- duplication of entire genome
- duplication of complete or part of a chromosome
- duplication of single gene or group of genes

After gene duplication, one of the separate copies of a gene may undergo changes in sequence. The original gene may continue to function and the copy either becomes nonjunctional (pseudogene) or a gene with altered expression. This subtle altered function or expression is sometimes advantageous to the organism. Further during evolution, the misalignment between pseudogene copies and functional copies during recombination cause changes in the inactivation, which is known as gene conversion. In this way, gene conversion can create a gene with a new function.

The other possible way in which a genome can acquire new genes is to obtain them from another species. The genome of most bacteria appears to have been acquired from Archaea. However conjugation enables plasmids to move between bacteria and result in acquisition of new genes. Few other bacteria such as *Bacillus, Pseudomonas* and *Streptococcus* can take up DNA from the surrounding environment. The efficiency of DNA uptake varies according to genera. Contrastingly, in plants new genes can be acquired by polyploidization.

27.3   TRANSPOSONS AND INTRONS

Transposons have wide-ranging effects on molecular evolution. The most significant effect of transposons is to initiate recombination and lead the genome to rearrangements. In many cases, the rearrangements cause deletion of important gene, which further leads to lethality. In some cases, along with transposons a short segment of DNA may get transduced and become functional. Similary, microsatellite repeats and intron appearance can also have an impact on genome evolution.

27.4   EVOLUTIONARY GENETICS

Evolution of the species as a whole results from the differential rates of survival and reproduction of the various types of change over time. For example, the population of the next generation will contain a higher frequency of those types that most successfully survive and reproduce under the existing environmental conditions. Thus, the frequencies of various types within the species will change over time.

There is an obvious similarity between the process of evolution as Darwin described it and the process by which the plant or animal breeder improves a domestic stock. The plant breeder selects the highest-yielding plants from the current population and uses them as the parents of the next generation. If the characteristics causing the higher yield are heritable, then the next generation should produce a higher yield. Darwin chose the term natural selection to describe his model of evolution in three principles through differential rates of reproduction of different variants in the population.

- **Principle of variation** (among individuals within any population, there is variation in morphology, physiology, and behaviour)
- **Principle of heredity** (offsprings resemble their parents more than they resemble unrelated individuals)
- **Principle of selection** (some forms are more successful at surviving and reproducing than other forms in a given environment)

27.5   VARIATION AND DIVERGENCE OF POPULATIONS

In evolution, the various forces of breeding, mutation, migration, and selection are acting simultaneously in populations. Also, these forces operate together and mould the genetic composition of populations to produce both variations within local populations.

The genetic variation within and between populations is a result of the interplay of the various evolutionary forces. Generally, forces that increase or maintain variation within populations prevent the differentiation of populations from each other, whereas the divergence of populations is a result of forces that make each population homozygous. Thus, random drift (or inbreeding) produces homozygosity while causing different populations to diverge. Divergence and homozygosity are counteracted by the constant flux of mutation and the migration between localities, which introduce variation into the populations again and tend to make them more like each other.

27.5.1   Heritability of Variation

The first rule of any reconstruction or prediction of evolution is that the phenotypic variation must be heritable. All variable traits are not heritable. Certain metabolic traits (resistance to high salt concentrations in *Drosophila*) show individual variation but no heritability. Further, behavioural traits have lower heritabilities than morphological traits, especially in organisms with more complex nervous systems that exhibit immense individual flexibility in central nervous system. The evolution of a particular quantitative trait is based on its genetic variance in the population. Thus, suggestions that certain traits in the human species such as performance in IQ tests, temperament, and social organization are in the process of evolving or have evolved at particular epochs in human history and depend critically on evidence about whether there is genetic variation for these traits. Reciprocally, traits that appear to be completely invariant in a species may nevertheless evolve.

27.5.2   Observed Variation within and between Populations

About one-third of all protein-encoding loci are polymorphic, and all classes of DNA, including exons, introns, regulatory sequences, and flanking sequences, show nucleotide diversity among individuals within populations. Several of these examples also documented some differences in the genotype frequencies between populations. The relative amounts of variation within and between populations vary from species to species, depending on history and environment. In humans, some gene frequencies (e.g. skin colour or hair form) are well-differentiated between populations and major geographical groups (so-called geographical races).

27.6   PROCESS OF SPECIATION

Allopatric speciation occurs through an initial geographical and mechanical isolation of populations that prevents any gene flow between them, followed by genetic divergence of the isolated populations sufficient to make it biologically impossible for them to exchange genes in the future.

The forms of biological isolating mechanisms that arise between species include:

27.6.1   Prezygotic Isolation

This results from failure to form zygote. It may be due to the following.

- Lack of mating opportunity

  Temporal isolation   activity, fertility, or mating at different times or seasons

  Ecological isolation   restriction to different, habitats, presence in non-overlapping habitats or ecological niches

- Lack of mating compatibility

  Sexual, psychological, or behavioural incompatibility

  Mechanical isolation   the failure of genitalia or flower parts to match

  Gametic isolation   physiological incompatibility of sperm with the reproductive tract of the female in animals or of the pollen with the style in plants or a failure of successful fertilization of the egg cell or ovule

Examples of prezygotic isolating mechanisms are well known in animals. The light signals that are emitted by male fireflies which attract females differ in intensity and timing between species. In the tsetse fly, *Glossina*, mechanical incompatibilities cause severe injury and even death if males of one species mate with females of another.

27.6.2   Postzygotic Isolation

This results from failure of fertilized zygote to contribute gametes to future generation.

Hybrid inviability   hybrids either fail to develop or have a lower fitness than individuals of the parental species

Hybrid sterility   partial or complete inability of adult hybrids of either sex to produce gametes in normal numbers

Hybrid breakdown   sterility or inviability of the offspring of mating among hybrids or between hybrids and the parental species

Postzygotic isolation is more common in animals than in plants, because of incompatibilities and chromosomal variations. When the eggs of the leopard frog, *Rana pipiens*, are fertilized by sperm of the wood frog, *R. sylvatica*, the embryos do not succeed in developing. Horses and asses can easily be crossed to produce mules, but, as is well known, these hybrids are sterile.

27.7   ORIGIN OF NEW GENES

It is clear that evolution consists of more than the substitution of one allele for another at the loci of defined function. New functions have arisen that promoted new ways of making a living. Many of these new functions, for example, the development of the mammalian inner ear from a transformation of the reptilian jawbones, result from continuous transformations of shape for which we do not have to invoke totally new genes and proteins. Older metabolic functions must have necessarily been maintained while new ones were being developed which in turn means that old genes had to be preserved while new genes with new functions had to evolve. Where does the DNA for new functions come from?

27.7.1   Polyploidy

One process for the provision of new DNA is the duplication of the entire genome by polyploidization, much more common in plants than in animals.

27.7.2   Duplications

A second process for the increase in DNA is the duplication of small sections of the genome as a consequence of misreplication of DNA. At first, after a duplicated segment has arisen, there is the possibility of an increase in the production of the polypeptide, but then functional differentiation between the sequences may occur in one of two directions. In one, no functional change occurs and there is simply a duplication of polypeptide production. The general function of the original sequence is maintained in the new DNA, but there is some differentiation of the sequences by accumulated mutations so that variations on the same protein are produced, allowing a somewhat more complex molecular structure.

A classic example is the set of gene duplications and divergences that underlie the production of human haemoglobin. Adult haemoglobin is a tetramer consisting of two α polypeptide chains and two β chains, each with its bound haeme molecule. The gene encoding the α chain is on chromosome 16 and the gene for the β chain is on chromosome 11, but the two chains are about 49 percent identical in their amino acid sequences, an identity that clearly points to the common origin. However, in foetuses, until birth, about 80 percent of β chains are substituted by a related γ chain. These two polypeptide chains are 75 percent identical, and the gene for the γ chain is close to the β-chain gene on chromosome 11 and has an identical intron–exon structure. The early embryo begins with α, γ, ε and ζ chains and, after about 10 weeks, the ε and ζ are replaced by α, β, and γ. Near birth, β replaces γ and a small amount of yet a sixth globin, δ is produced.

In the evolution of haemoglobin, the duplicated DNA encodes a function closely related to that served by the original gene from which it arose. The other possibility for evolution of duplicated DNA is a complete qualitative divergence in function. Birds and mammals, like other eukaryotic organisms, have a gene encoding lysozyme, a protective enzyme that breaks down the cell wall of bacteria. This gene has been duplicated in mammals to produce a second sequence that encodes a completely different, nonenzymatic protein, α-lactalbumin.

27.7.3   Imported DNA

DNA can be inserted into chromosomes from other chromosomal locations and from other species, and genes from totally unrelated organisms can become incorporated into cells to become part of their heredity and function.

27.8   PHYLOGENETICS

Taxonomy is the practice of classifying biodiversity. In 1758, Carl Linnaeus proposed a system for classifying plants and animals (Figure 27.1). Linnaeus gave each animal two names—genus and species (e.g. *Homo sapiens*). Then he grouped genus into families, families into order, order into classes, class into phyla. This phyla is grouped into kingdoms (Animalia for animals and Plantae for plants).

Later on in 1969, Thomas Whittaker proposed a "five kingdom" system (Figure 27.2). First, he grouped organisms with a true nucleus (eukaryotes) and organisms without a nuclear membrane (prokaryotes). In the five kingdom system, the Monera (bacteria), Protista (unicellular eukaryotes) and fungi are grouped separately from plants and animals. Further to this many other scientists reorganized the classification and proposed various systems.

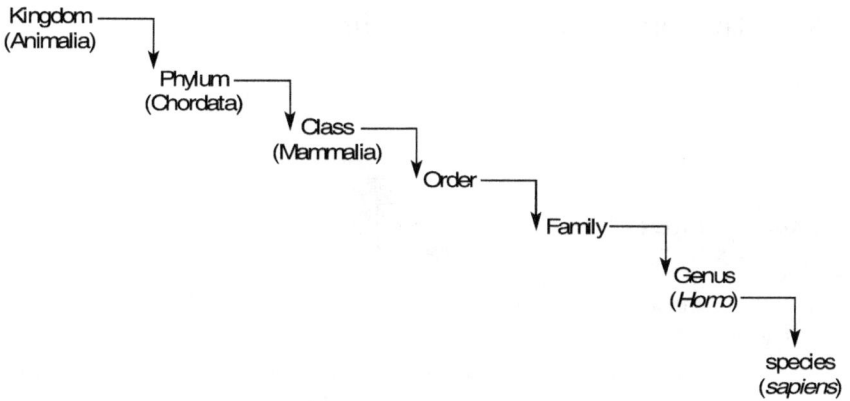

Figure 27.1   Linnaeus classification

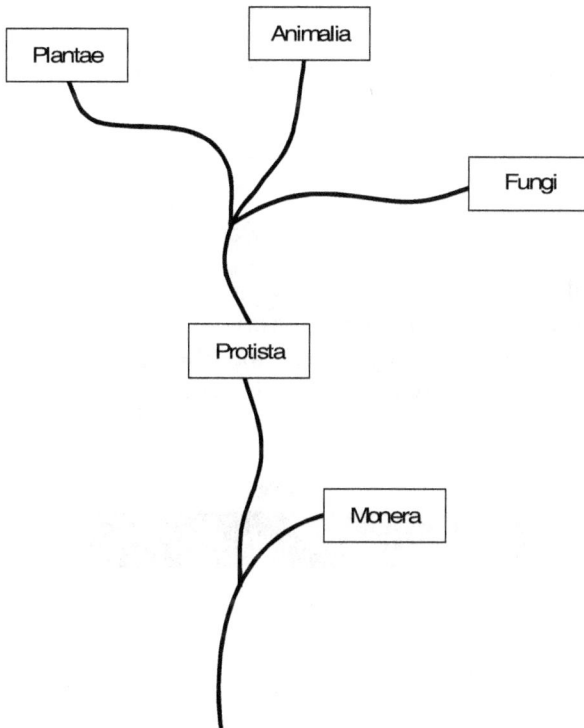

Figure 27.2   Five kingdom system

In 1950, the German taxonomist Willi Hennig developed an analysis called cladistic analysis to study the relationship of an organism in a phylogenetic tree which shows how ancestral species split into a descendant species and their relationships.

27.8.1   The Origin of Bats and Flight

Molecular phylogenetics plays an important role when there is a conflict among phylogeny and different morphological characters. For example, there was a doubt whether flight evolved twice in mammals (Figure 27.3); also

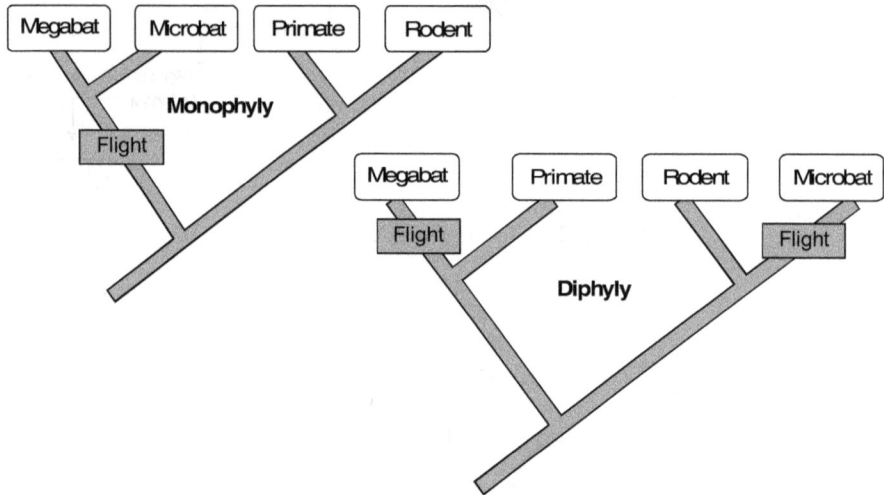

Figure 27.3   Bat phylogeny and origin of flight

whether megabats (Megachiroptera) are closer to primates or to microbats (Microchiroptera). Based on mitochondrial DNA sequences and higher proportions of the G ≡ C, they argued that the apparent **monophyly** of bats that was observed in the molecular studies is due to convergent evolution.

FOR ADDITIONAL READING

1. Bartel, D.P. and Unrau, P.J. (1999). "Constructing an RNA world." *Trends. Genet.* 12:M9–M13.

2. Brookfield, J.F.Y. (1997). "Genetic redundancy." *Adv. Genet.* 36:137–155.

3. Gibbons, A. (1998). "Which of our genes makes us human?" *Science.* 281:1432–1434.

4. McDonald, J.F. (1995). "Transposable elements: possible catalysts of organismic evolution." *Trends Ecol. Evol.* 10:123–126.

5. Orgel, L.E. (2000). "A simpler nucleic acid." *Science.* 290:1306–1307.

6. Futuyama, D.J. (1998). *Evolutionary Biology*, 3rd edn. Sinauer, Sunderland, MA.

7. Gesteland, R.F., Cech, T.R. and Atkins, J.F. (eds.) (1999). *The RNA World*, 2nd edn. Cold Spring Harbor Laboratory Press, Cold Spring Harbor.

8. Jackson, M., Strachan, T. and Dover, G.A. (1996). *Human Genome Evolution*. BIOS Scientific Publishers, Oxford.

9. Li, W.H. (1997). *Molecular Evolution*. Sinauer, Sunderland, M.A.

MODEL QUESTIONS

1. Summarize the origin of the present genome?

2. With examples, detail gene duplication.

3. Are gene duplication and domain shuffle caused by genome evolution?

4. Outline the impact of transposons, introns and microsatellites on genome evolution.

5. What are the ways in which the size of the genome increases?

6. Discuss various kinds of speciation.

7. What is cladystic analysis?

8. How does phylogeny help in tracing adaptation in life?

9. Explain Linnaeus mode of classification.

HUMAN GENOME PROJECT

28

28.1 INTRODUCTION

We have learnt from the previous chapters that the size of the genome varies according to species and their adaptation. It is interesting to know about the entire wealth of human genome but it is cumbersome.

A genome is the entire DNA in an organism or a cell. **Robert Sinsheimer**, a molecular biologist by training, made the first proposal of Human Genome Project (HGP) in 1985. While he was the Chancellor of the University of California, he organized a scientific meeting to discuss the possibility of the project. **Charles DeLisi,** Head, Division of Health and Environmental Research, Department of Energy (DOE) came to know about the HGP proposal and became an avid supporter of the project. In 1986, DeLisi convened a meeting with scientists of DNA research laboratories at Livermore and Los Alamos in USA and suggested to carry out the project with a primary goal of determining the nucleotide sequence of human genome.

Due to legal problems, the National Academy of Sciences appointed a committee and the committee suggested that both DOE and National Institute of Health (NIH) should be involved with a common advisory board. In 1987, under the leadership of **James Wyngaarden**, NIH secured $17.4 million for the project. **James D. Watson** became the first director of the new 'Office of Human Genome Research (OHGR). The OHGR appointed Norton Zinder as chairman of program Advisory committee on the Human Genome. In 1990, the office became a 'centre' called The National Center for Human Genome Research (NCHGR). In 1998, NCHGR became **National Human Genome Research Institute** and later **Francis Collins** (2003) was appointed as director. The project has been the largest and most complex of international collaborations with funding from their governments and many charitable societies across the world. Important scientific contributions also came from other collaborators, viz. Japan, France, Germany and China.

The project goals are to

- identify all the approximate 30,000 genes in human DNA
- determine 3 billion nucleotide base pairs of human DNA
- store information in databases
- develop tools for data analysis
- transfer related technologies to the private sector, and
- address the ethical, legal and social issues that may arise from the project

28.2   CONTENT OF THE GENOME

The first working draft of the entire human nuclear genome appeared in the **February 2001** issues of the journals *Nature* and *Science*. Due to rapid technological advancement, the project was completed by even April 2003 (two years ahead) and the complete high quality reference sequence was made available to researchers worldwide for practical applications. A number of genes and their association with human diseases have also been established. The content and some of the salient features of the human genome (Figure 28.1) are highlighted below and complete details on the content of chromosome are listed in Table 28.1.

- The human genome contains 3.2 billion nucleotide bases (A, C, T and G).
- The size of the genes vary greatly. The average gene consists of 3000 bases. The largest known human gene is dystrophin (2.4 million bases).
- The functions are unknown for more than 50% of discovered genes.

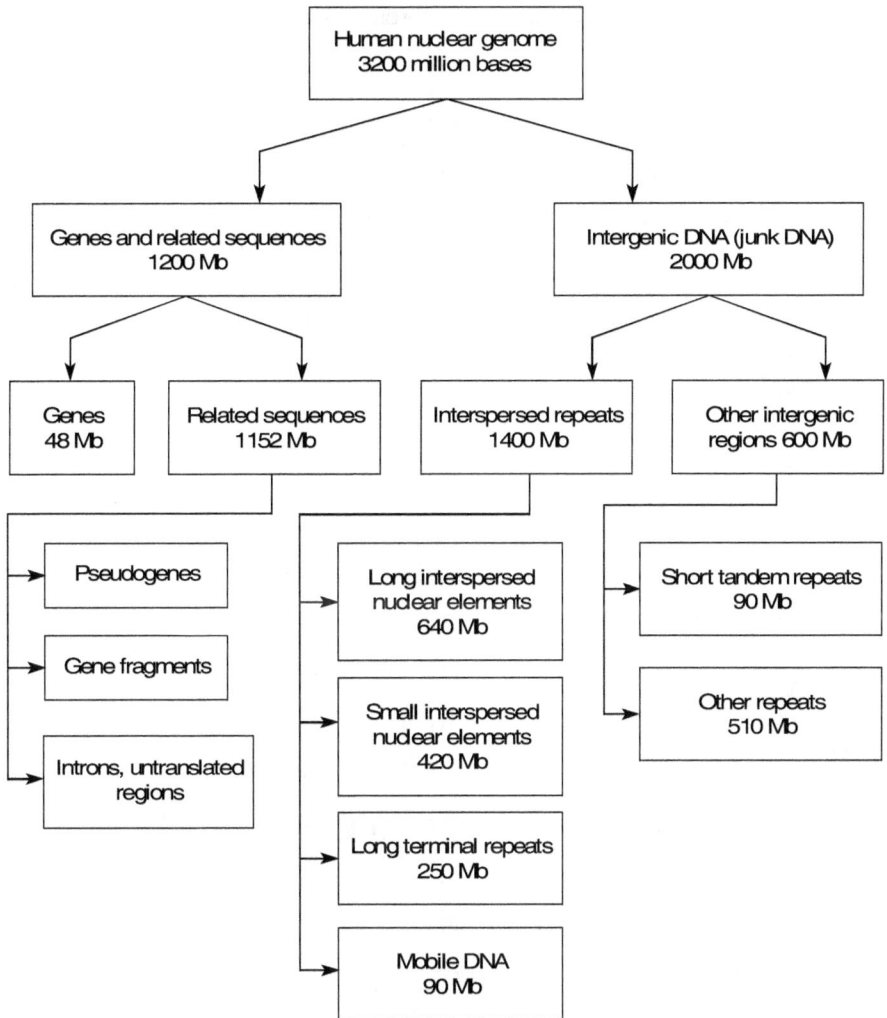

Figure 28.1   Content of the human genome based on IHGSC (April 2003)

- The sequences of human genome remain the same in 99.9% people.
- About 2% of the genome encodes instructions for the synthesis of proteins.
- Repeat sequences (do not code for proteins) make up about 50% of the genome.
- Repeat sequences are thought to maintain chromosome structure and dynamics. By rearrangement, it creates entirely new genes or modifies and reshuffles existing genes.

Table 28.1   Details on human chromosomes after the draft

Chromosome	Length in (bps)	Gene count	Known gene count	Pseudogene count
1	245,522,847	2281	1988	723
2	243,018,229	1482	1246	690
3	199,505,740	1168	1033	499
4	191,411,218	866	743	424
5	180,857,866	970	834	434
6	170,975,699	1152	1050	459
7	158,628,139	1116	916	454
8	146,274,826	794	692	364
9	138,429,268	919	778	309
10	135,413,628	862	730	317
11	134,452,384	1426	1264	359
12	132,449,811	1104	1009	396
13	114,142,980	399	318	213
14	106,368,585	733	646	260
15	100,338,915	766	589	308
16	88,827,254	957	839	216
17	78,774,742	1257	1104	305
18	76,117,153	322	267	168
19	63,811,651	1468	1337	164
20	62,435,964	631	592	180
21	49,944,323	271	243	87
22	49,554,710	552	471	129
X	154,824,264	931	766	380
Y	57,701,691	104	76	54
Mitochondrial gene	16,571	37	13	–

- About 40% of the human proteins showed similarity with fruit fly or worm proteins.
- Genes appear to be spread randomly throughout the genome with vast expanses of non-coding DNA in between.
- Chromosome 1 is the largest human chromosome and has 2968 genes, and the Y chromosome is the smallest having 231 genes.
- Candidate genes were identified for numerous diseases and disorders including breast cancer, muscle disease, deafness and blindness.
- Single nucleotide polymorphism can occur in 3 million locations.
- Every 2 kb contains a microsatellite (short tandem repeat).

28.3 THE HUMAN MITOCHONDRIAL GENOME

In 1981, Anderson *et al.* have decoded the entire sequence of human mitochondria. The circular and double-stranded genome contains 16569 base pairs and 37 genes. Among them, thirteen genes code for respiratory complex proteins and the other 24 genes represent RNA molecule for the expression of mitochondrial genome.

28.4 BENEFITS OF HUMAN GENOME RESEARCH

The 'Periodic Table of Life' developed from this project will be beneficial to everyone in many ways. James Watson and the joint NIH–DOE genome advisory panel were against patenting the genes. They are of the view that since public was paying for deciphering genome, they must decide what to do with the information. Also, scientists should have access to all gene data for the advancement of genome research programme. In 1997, NIH established **Gen Bank** and made it possible for everyone to access information through Internet. This encouraged many to refrain from taking patent on raw sequence data.

The following are the applications of the data obtained from HGP in the various fields of study.

Molecular medicine

- to develop better disease diagnosis
- detect genetic predispositions to disease
- design drugs based on molecular information and individual genetic profiles
- useful for better gene therapy

Microbial genomics

- fast to detect and treat pathogens
- develop new biofuels
- protect citizens from biological and chemical warfare
- clean up toxic waste safely and efficiently

Risk assessment

- evaluate the level of health risk in individuals who are exposed to radiation or mutagens
- detect pollutants and monitor environments

Anthropology and evolution

- to study evolution due to germ-line mutations
- to study migration of different population groups
- to study mutations on the Y chromosome to trace lineage and migration of males

DNA identification

- identify criminals whose DNA may match evidence left at crime scenes
- exonerate persons wrongly accused of crimes
- establish paternity and other family relationships
- identify endangered and protected species
- detect bacteria and other organisms that may pollute environment
- match donors with recipients in organ transplant programs
- determine pedigree for seed or livestock breeds

Agriculture and animal science

- grow crops of disease and drought resistance
- high productivity
- breed farm animals
- develop biopesticides

28.5 ETHICAL, LEGAL, AND SOCIAL IMPLICATIONS (ELSI)

As the advances in the understanding of human genetics and genomics will have important implications for individuals and society, examination of the ethical, legal, and social implications of genome research is, therefore, an integral and essential component of the HGP. The ability to generate any individual's genetic profile raises important questions of privacy, confidentiality, ownership and autonomy. How should information be protected? Who should have access to the information and under what circumstances? What rights, if any, do employers, insurers, and family members have to an individual's genetic information? Diagnosis of many genetic disorders will be possible before treatment becomes available. How will we resolve dilemmas raised by such a gap? In addition, the potential trend towards understanding diseases or traits in genetic terms may raise questions about human nature and our fundamental beliefs about ourselves and our species. Will we view an individual as merely a product of interacting genes? How will we define normalcy, abnormalcy, or disability? The recognition of the importance of such questions, and the increased dialogue surrounding them has resulted in an international commitment to and interest in addressing issues of ethical, legal, and social importance. Up to 5% of the money in some countries is being spent on the educational, ethical, legal and social importance. A body of scholarship has been generated by the ELSI program in the areas of privacy and fair use of genetic information, safe and effective integration of genetic information into clinical settings, ethical issues surrounding genetics research, and professional and public education.

The HUGO Ethics Committee was made with the following purposes:

- **to promote** discussion and understanding of social, legal and ethical issues as they relate to the conduct of, and the use of knowledge derived from, human genome research. This may encompass consideration of research directions, practices and results, and the issues of human diversity, privacy, and confidentiality, intellectual property rights, patents, and commercialization, disclosure of genetic information to third parties, the non-medical use of such information, and the medical, legal and social aspects of testing, screening, accessibility, DNA banking, and genetic research;
- **to act** as an interface between the scientific community, policy makers, educators, and the public;
- **to foster** greater public understanding of human variation and complexity;
- **to collaborate** with other international bodies in genetics, health, and society with the goal of disseminating information;
- **to deliberate** about policy issues in order to provide advice to the HUGO Council and to issue statements where appropriate;
- **to report** on its activities at least annually to the HUGO Council: and to act on any other related matter.

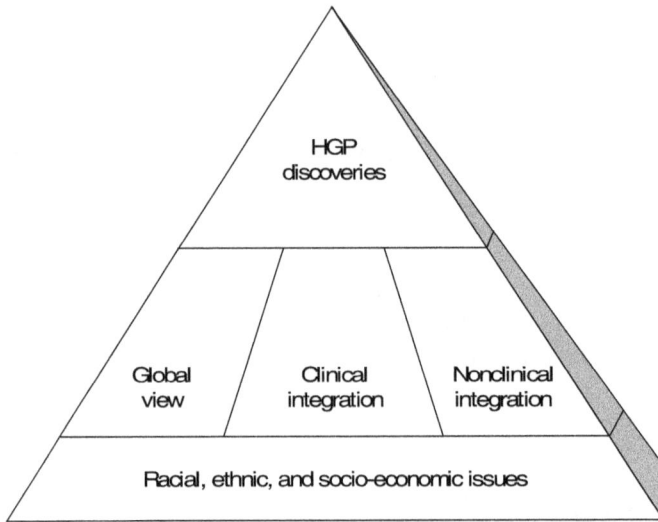

Figure 28.2   Pyramid of goals for ELSI

The new goals for ELSI research and education can be visualized as a pyramid of interrelated issues and activities (Figure 28.2). The first goal, at the top of the pyramid, deals with the issues around the completion of the first human DNA sequence and the study of human genetic variation, making concrete that the vision that advances in genome science will be an important factor contributing to the ELSI research agenda. The second and third goals focus on the integration of the information generated by these new discoveries into clinical, nonclinical, and research settings. The fourth goal examines the interaction of this information with philosophical, theological, and ethical perspectives. Finally, providing the foundation for all of these explorations is the fifth goal, examining how the understanding and use of genetic information are affected by socio-economic factors and concepts of race and ethnicity.

FOR ADDITIONAL READING

1. Anderson, S., Bankier, A.T., Barriel, A.G. *et al.* (1981). "Sequence and organization of the human mitochondrial genome." *Nature*. 290:457–474.

2. Baltimore, D. (2001). "Our genome unveiled." *Nature*. 409:814–816.

3. Bork, P. and Copley, R. (2001). "Filling in the gaps." *Nature*. 409:818–820.

4. Cantor, C.R. (1990). "Orchestrating the Human Genome Project." *Science*. 248:49–51.

5. Green, E.D. and Waterston, R.H. (1991). "The human genome project: prospects and implications for clinical medicine." *J. Am. Med. Assoc.* 266:1966–1975.

6. IHGSC (International Human Genome Sequencing Consortium). (2001). "Initial sequencing and analysis of the human genome." *Nature.* 409:860–921.

7. Kay, L.E. (1993). *The Molecular Vision of Life.* Oxford University Press, Oxford.

8. Lander, E.S. and Weinberg, R.A. (2000). "Genomics: journey to the center of biology." *Science.* 287:1777–1782.

9. Leader, P. (1990). "Can the human genome project be saved from its critics... and itself?" *Cell.* 63:1–3.

10. Li, W.H., Gu, Z., Wang, H. and Nekrutenko, A. (2001). "Evolutionary analyses of the human genome." *Nature.* 409:847–849.

11. Orel, V. (1995). *Gregor Mendel: The first Geneticist.* Oxford University Press, Oxford.

12. Pennisi, E. (2000). "And the gene number is .....?" *Science.* 288:1146–1147.

13. Rowen, L., Koop, B.F. and Hood, L. (1996). "The complete 685-kilobase DNA sequence of the human βT cell receptor locus." *Science.* 272:1755–1762.

14. SNP Group (The International SNP Map Working Group) (2001). "A map of human genome sequence variation containing 1.42 million single nucleotide polymorphisms." *Nature.* 409:928–933.

15. Stephens, J.C., Cavanaugh, M.L., Gradie, M.I., Mador, M.L. and Kidd, K.K. (1990). "Mapping the human genome: current status." *Science.* 250:237–244. (Review).

16. Turner, P.C., McLennan, A.G., Bates, A.D. and White, M.R.H. (1997). *Instant Notes in Molecular Biology.* BIOS Scientific Publishers, Oxford.

17. Venter, J.C., Adams, M.D., Myers, E.W., *et al.* (2001). "The sequence of the human genome." *Science.* 291:1304–1351.

18. Watson, J.D. (1990). "The Human Genome Project: past, present, and future." *Science.* 248:44–49 (Review).

19. Collins, F.S., Patrinos, A., Jordan, E., Chakravarti, A., Gesteland, R., Walters, L. and the members of the DOE and NIH planning groups. (1998). "New Goals for the U.S. Human Genome Project: 1998–2003." *Science.* 282:683–689.

MODEL QUESTIONS

1. Name some organisms whose genomes have been successfully sequenced.

2. How will the human genome project be of benefit to various researchers and human beings?

3. What are the contents of the human nuclear genome?

4. What is the outcome of the human genome project?

5. Mention various goals of the human genome project.

6. Why is it important to know about the human genome?

7. What are the ethical, legal and social implications of human genome project?

DNA VACCINE

29

29.1 INTRODUCTION

Vaccines are preparations which, when injected into a person, stimulate a person's immune system to respond against infection. Normally vaccines are made up of materials of harmful infectious organisms such as

- Whole organism (bacteria, virus) in dead form, (e.g. influenza virus vaccine)

- Nonharmful live organism in weakened (attenuated) form (e.g. measles, mumps and rubella virus vaccine (MMR))

- Inactive specific product or parts of an organism (e.g. tetanus vaccine)

DNA vaccines are made of modified form of DNA of infectious organisms. When the vaccine is injected into an individual, it leads to the expression of genes within the foreign DNA. This results in provoking the immune system that responds against the products of foreign DNA (DNA vaccine). This immune response is similar to the response that would be produced if the person were actually infected by the true organism. Vaccines are injected underneath our skin or into muscle tissue. New vaccines are being sprayed onto the surface of our mucosa tissue inside the nose. Other means of delivery are in the process of being developed based on immersion, gene gun and electroporation.

29.2   BASIC REQUIREMENTS

DNA vaccines usually consist of plasmid vectors of bacteria. These plasmid vectors contain transgenes inserted under the control of promoter of eukaryote; this helps in the expression of protein in mammalian cells. The basic requirements for the backbone of plasmid DNA vector are:

- eukaryotic promoter
- cloning site
- polyadenylation sequence
- selectable marker
- bacteria or mammalian cell as host

A strong promoter (generally derived from cytomegalovirus or Simian virus 40) is required for the optimal expression of genes in mammalian cells. In the cloning site, the gene of interest is attached. Inclusion of polyadenylation sequence either from bovine growth hormone (BGH) or SV40 helps in stabilization of mRNA transcript from degradation. Ampicillin- or kanamycin-resistant gene is used as a selectable marker. When mammalian cells are used, kanamycin-resistant gene is often used for better expression. Finally, *Escherichia coli* (ColE1) origin of replication is used to maintain the plasmid copy numbers in the host cell.

29.3   REGULATORY ELEMENTS

Table 29.1 describes the various promoters/enhancers or other transcriptional elements that are found in DNA vaccines. Promoters derived from viruses have provided greater gene expression *in vivo* than other eukaryotic promoters. The cytomegalovirus (CMV) enhancer-promoter has often been shown to direct the highest level of transgene expression in eukaryotic tissues when compared with other promoters.

Table 29.1   Comparison of promoters used in DNA vaccine production

Expressed antigen	Promoters/enhancers compared	*In vitro/in vivo* comparison
GFP	CMV, muscle-specific creatine kinase (CKM) promoter	Higher level expressions were driven by the CKM promoter compared to CMV
LacZ	CMV, glial fibrillary acidic protein (GFAP) promoter, neuron specific enolase promoter	GFAP is as efficient at driving *LacZ* expression as CMV
CAT	HIV-I long terminal repeat (LTR) RSV Transactivation response element (TRE)	HIV-I-LTR could be transactivated in unstimulated cells; RSV TAR could be transactivated in unstimulated cells
CAT	CMV, RSV, SV40, murine leukemia virus promoter	CMV promoter was the best promoter for muscle
CAT	CMV, SV2	CMV promoter was found to have highest transcriptional activity
Luciferase	CMV, RSV, SV40, PGK, hybrid β-actin promoter/CMV enhancer, CMV	The hybrid β-actin/CMV promoter-enhancer showed greater luciferase expression than CMV
Hepatitis B surface antigen (HBsAg)	CMV, desmin	Promoters of CTL and Th1 serum antibody responses against HbsAg in mice were similar
Bovine herpesvirus glycoprotein D(gD) (BHVgD)	RSV, CMV/IA	CMV construct produce higher neutralizing antibody titres against gD
SV40 large tumour antigen	CMV, SV40	CMV construct induced higher levels of antibody than the SV40 construct
Adenovirus E4 ORF3	CMV, RSV, SV40, UbC, EF-Iα	Constructs containing the UbC and EF-Iα promoters stimulated the most stable expression of antigen

29.4   KOZAK SEQUENCES

Flanking sequences adjacent to the AUG initiator codon with mRNA influence the recognition of eukaryotic ribosomes. This sequence is called 'kozak' consensus sequence. This sequence is required for the optimal translational efficiency of expressed mammalian genes. It has been found that efficient translation is obtained when the −3 position contains a purine base, or a guanine is positioned at +4. Prokaryotic genes and some eukaryotic genes do not possess kozak sequences. So the expression level of these genes can be increased only by the insertion of a kozak sequence.

29.5   HOW VACCINES WORK

Live vaccines always provide better protection, as they stimulate a broader part of the immune system than the dead or inactivated microorganisms. Although DNA vaccines are not living, they are more closely related to genetically modified viral vaccines than their name indicates. This is because during viral infection, mostly the viral DNA alone augments our immune reaction. This is an important aspect to be borne in mind when considering how such vaccines should be regulated. DNA vaccines and viral vaccines both possess certain quality in the way in which they work. They provide better and more lifelong protection than vaccines consisting of inactivated microorganisms. DNA vaccines and viruses both result in genetic material entering into the cells, thereby triggering the production of new proteins. Such proteins retain their natural form and are not deformed, as is often the case when microorganisms are inactivated. This leads to the activation of humoral immune response that more readily recognizes naturally existing infectious agents. Moreover, DNA vaccines and viral vaccines activate T cells that is also called as the "cellular immune response". Through immunological memory, T cells remember what is going on inside the cells of the body. The memory fragments of the cell content are transported with the major histocompatibility complex (MHC) to the cell surface and presented to the immune cells. In this way, virus-infected cells can be detected and the infection stopped by the T cells killing the infected cells.

29.6   SIMILARITIES BETWEEN
DNA VACCINES AND VIRUSES

Viruses are simple and they are dependent on the machinery of host cells to reproduce. They consist of heritable material encapsulated by proteins and in some cases a membrane. In the heritable material, viruses have genes for enzymes that are necessary for the multiplication of the virus. With certain viral infections, the virus-heritable material alone is sufficient to produce new viral particles. Hence, this type of viral infection can also be formed by the cell taking up naked viral heritable material. A DNA vaccine consisting of complete heritable material from such viruses could, therefore, trigger new viral infection. Recently, DNA vaccine for infectious poliovirus has become available. Inversely, it is conceivable that DNA vaccines could be given a viral form by encapsulating the DNA in a viral particle. Since there is the chance of an indistinct borderline between virus and DNA vaccine, it is important, when considering appropriate regulatory mechanisms, that DNA vaccines and genetically modified viral vaccines are viewed in the same context.

29.7   THE IMMUNOLOGY OF DNA VACCINES

Intramuscular injections of plasmid DNA encoding influenza nucleoprotein could protect organisms against a challenge with live influenza virus. Immunization with plasmid DNA has been shown to activate both humoral and cellular immune responses, including the generation of antigen-specific CD3+, cytotoxic T cells as well as CD4+ T helper cells.

FOR ADDITIONAL READING

1. Davis, H.L. (1997). "Plasmid DNA expression systems for the purpose of immunization." *Curr. Opin. Biotechnol.* 8:635–640.

2. Gurunathan, S., Klinman, D.N. and Seder, R.A. (2000). "DNA vaccines: Immunology, application, and optimization." *Ann. Rev. Immunol.* 18:927–974.

3. Lee, A.H., Suh, Y.S., Sung, J.H. and Sung, Y.C. (1997). "Comparison of various expression plasmids for the induction of immune response by DNA immunization." *Mol. Cells.* 7:495–501.

4. Loriat, D., Li, Z., Mancini, M., Tiollais, P., Paulin, D. and Michel, M.L. (1999). "Muscle specific expression of hepatitis B antigen: no effect on DNA raised immune responses." *Virology.* 260:74–83.

5. Xiang, Z.Q., Spitalnik, S.L., Cheng, J., Erikson, J., Wojczyk, B. and Ertal, H.C. (1995). "Immune responses to nucleic acid vaccines to rabies virus." *Virology.* 209:569–579.

6. Kodihalli, S., Goto, H., kobasa, D.L., Kawaoka, Y. and Webster, R.G. (1999). "DNA vaccine encoding haemoagglutinin provides protective immunity against H5NI influenza virus in mice." *J. Virol.* 73:2094–2098.

7. Van Drunen Little – Van den Hurk, S., Braun, R.P., Lewis, P.J., Karvonen, B.C., Baca-Estrade, M.e., Snider, M., McCartney, D., Watts, D. and Babiuk, L.A. (1998). "Intradermal immunization with a bovine herpesvirus-I DNA vaccine induces protective immunity in cattle." *J. Gen. Virol.* 79:831–839.

8. Glavin, T.A., Muller, J. and Khan, A.S. (2000). "Effect of different promoters on immune responses elicited by HIV-I *gag/env* multigenic DNA vaccine in *Macaca mulatta* and *Macaca nemestrina.*" *Vaccine.* 18:2566–2583.

9. Andre, S., Seed, B., Eberly, J., Schruat, W. and Hass, J. (1998). "Increased immune response elicited by DNA vaccination with a synthetic gp 120 sequence with optimized codon usage." *J. Virol.* 72:1497–1503.

10. Uchijima, M., Yoshida, A., Nagata, T. and Koide, Y. (1998). "Optimization of codon usage of plasmid DNA vaccine is required for the effective MHC class-I restricted T cell responses against an intracellular bacterium." *J. Immunol.* 161:5594–5599.

11. Deml, L., Bojak, A., Steck, S., Graf, M., Wild, J., Schirmbeck, R., Wolf, H. and Wagner, R. (2001). "Multiple effects of codon usage optimization on expression and immunogenicity of DNA vaccine candidate vaccines encoding the human immunodeficiency virus type I gag protein." *J. Virol.* 75:10991–11001.

12. Stratfold, R., Douce, G., Zhang-Barber, L., Fairweather, N., Eskola, J. and Dougan, G. (2001). "Influence of codon usage on the immunogenicity of a DNA vaccine against tetanus." *Vaccine.* 19:810–815.

13. Klinman, D.M., Yamschchikov, G. and Ishigatsubo, Y. (1997). "Contribution of CpG motifs to the immunogenicity of DNA vaccines." *J. Immunol.* 158:3635–3639.

14. Weeratna, R., Milan, C.L.B., Krieg, A.M. and Davis, H.L. (1998). "Reduction of antigen expression from DNA vaccines by coadministered oligo-deoxynucleotides." *Antisense Nucleic Acid Drug Devel.* 8:351–356.

15. Haddad, D., Ramprakash, J., Sedegah, M., Charoenvit, Y., Baumgartner, R., Kumar, S., Hoffman, S.L. and Weiss, W.R. (2000). "Plasmid vaccine expressing granulocyte-macrophage colony-stimulating factor attracts infiltrates inducing immature dendritic cells into injected muscles." *J. Immunol.* 165:3772–3781.

16. Lewis, P.J., Cox, G.J.M., Van Drunen Little Van den Hurk, S. and Babiuk, L.A. (1997). "Polynucleotide vaccines in animals: enhancing and modulating reponses." *Vaccine.* 15:861–864.

17. Rice, J., King, C.A., Spellerberg, M.B., Fairweather, N. and Stevenson, F.K. (1999). "Manipulation of the pathogen-derived genes to influence antigen presentation via DNA vaccines." *Vaccine.* 17:3030–3038.

18. Wu, Y., Wang, X., Csencsits, K.L., Haddad, A., Walterns, N. and Pascual, D.W.M. (2001). "Cell-targeted DNA vaccination." *Proc. Natl. Acad. Sci.*, USA. 98:9318–9323.

19. You, Z., Huang, X., Hester, J., Toh, H.C. and Chen, S.Y. (2001). "Targeting dendritic cells to enhance DNA vaccine potency." *Cancer Res.* 61:3704–3711.

MODEL QUESTIONS

1. What is a vaccine?

2. What is a DNA vaccine?

3. List various types of vaccines.

4. How are vaccines delivered to human beings?

5. What are the potential problems with a vaccine?

6. What are the basic requirements of DNA vaccine development?

7. What are the properties of DNA vaccine?

MOLECULAR TECHNIQUES

30

30.1   ELECTROPHORESIS

Electrophoresis is a technique used to separate the charged particles such as protein and nucleic acids, under the influence of an electric field in a gel matrix. The movement of the particle is proportional to the overall charge, density, size and shape. For molecules with a relatively homogeneous composition (nucleic acid), charge, density and shape are constant. So the movement is mainly dependent on size of the molecule. Smaller molecules move faster through the gel and reach farther whereas heavier molecules migrate slowly and reach less distance. Electrophoresis is carried out in a gel-like matrix, usually made from agarose (carbohydrate polymers) or poly-acrylamide (synthetic polymer). The gel is typically held between two glasses or plastic plates or a thin layer of gel covers the surface of a flat support. The nucleic acids or proteins to be separated are applied to one end of the gel, and molecules move through the pores (size of the pore also alters the velocity of the moving particles) in the matrix under the influence of an electric field.

30.2   BLOTTING TECHNIQUES

30.2.1   Southern Blotting

A powerful tool for analysing gene structure (DNA) called Southern blot was developed by E.M. Southern in 1978. This tool is used to transfer DNA from gel matrix onto the nylon membrane. The technique involves cutting genomic DNA with one or several restriction enzymes and the resultant fragments are separated by size on an agarose gel. A sheet of nitrocellulose filter or nylon membrane is placed over the gel and a flow of buffer is set up through the gel towards the nitrocellulose filter. This causes the movement of DNA fragments out of the gel and the DNA fragments get bound to the nitrocellulose filter.

A specific labelled gene probe, viz. purified RNA, a cloned cDNA, or a short synthetic oligonucleotide, is then used to hybridize the DNA molecules bound to the filter. The labelled probe will hybridize to specific molecules containing a complementary sequence. Using autoradiography, the pattern of bands indicating the number and size of the DNA fragments on the nitrocellulose filter complementary to the probe is documented.

Southern blotting method is mainly used to detect major gene deletion and rearrangements found in a variety of human diseases. This technique is also used to identify structurally related genes in the same species and homologous genes in the other species.

30.2.2   Northern Blotting

Northern blotting is a technique used to analyse the total cellular RNAs. The RNAs are separated by size on an agarose gel. The RNA molecules in the gel are transferred onto the nitrocellulose filter or nylon membrane. Northern blotting is useful to find the size of a specific mRNA and is also used to compare the size of cloned cDNAs, revealing whether the cloned cDNA is full-length. In addition, this can also indicate which tissues or cell types express a particular gene or the factors that regulate its expression.

Cells are stimulated with serum and total RNA is isolated at specific time intervals. The RNA obtained at each point is analysed by Northern blotting using cloned cDNA probes. The results show that the profile of mRNA is present in low levels in untreated cells and rapidly accumulates following serum stimulation.

30.2.3 Western Blotting

Western blotting is a technique used to analyse the proteins. It is the most specific and commonly available test used to determine HIV infection. In this technique, HIV proteins are separated by SDS-gel electrophoresis resulting in discrete bands. These proteins are then transferred from the gel onto a filter paper, and the person's serum is added. If antibodies are present, they bind to the viral proteins (primarily gp41 and p24). This can be further detected by adding antibody to human IgG labelled with either radioactivity or an enzyme such as horseradish peroxidase, which produces a visible colour change.

30.2.4 Slot and Dot Blotting

Dot and Slot blotting are performed by spotting a small sample of RNA onto nitrocellulose and then hybridizing to a specific RNA or DNA probe. The membrane is then exposed to X-ray film. It provides a way to quantitate the intensity of specific gene expression.

30.2.5 Zoo Blotting

The Southern blots of genomic DNA from a collection of organisms reveal the degree of evolutionary conservation of a gene. This blotting is called Zoo blotting. The identification of yeast genes and their relation to the *ras* oncogene in human tumour shows a remarkable example of evolutionary conservation.

30.3 DNA FINGERPRINTING

Each individual except identical twins possesses a unique genetic information that can be visualized using recombinant DNA methods. DNA fingerprinting (DNA profiling) uses hypervariable minisatellites on Southern blots to form specific bands to identify individuals/relationships. The principle of individual uniqueness and identical DNA structure within all tissues of the same body provides the basis for DNA profiling. Hence this technique of DNA fingerprinting has broad application in various disciplines which can be broadly classified into

- Human forensic sciences
- Family relationship analysis
- Diagnostic medicine

- Animal and plant science
- Wildlife poaching

30.3.1   Human Forensic Sciences

Whenever the true identity of a criminal is in doubt and there are no solid evidences against the suspect, DNA fingerprinting gives hard evidence because DNA profiles from 2 different individuals can hardly ever match. Even if the specimen is a mixture from 2 or more different individuals it can be identified.

The specimen may include

- Sperm DNA from semen recovered from a rape victim
- Bloodstains found at the site of crime
- Strands of hair at the site of crime

30.3.2   Family Relationship Analysis

Whenever true biological relationships are debated in cases of disputed parentage, illegitimate birth, missing persons or immigration, DNA fingerprinting is the best tool.

In the case of immigrants requesting status for specific close relatives, DNA profiling can be used to decide on these citizenship applicants.

30.3.3   Diagnostic Medicine

For diagnostic and therapeutic purposes this method is indispensable. The uses of DNA profiling in the clinical side are;

- Post-transplant cell population identification
- Twin zygosity analysis
- Tumour analysis
- To identify microorganisms
- Tissue culture line identification
- Pedigree analysis

Pedigree analysis is to determine the probability of recurrence of a genetic defect in a family. If inherited defects are detected, the treatment can be started early in life or the baby can be aborted if the defect is untreatable.

30.3.4   Animal and Plant Sciences

- Determination of linkage and taxonomical classification.
- Sex determination and pathogenesis.
- Classifying plants into species and subspecies is made easy although somatic stability has been noted between different tissues from same plant but not between same species as mutations are high.

30.3.5   Wildlife Poaching

A major problem in the battle against poaching is identification of seized parts and products of poached animals.

30.4   DNA FOOTPRINTING

DNA footprinting is a technique used to identify DNA sequence on which regulatory or enzymatic proteins bind. Fragments of a 5´ end-labelled, double-stranded DNA segment, partially degraded by DNase in the presence and absence of the binding protein, are visualized by electrophoresis and autoradiography. For example, the binding of *lac* repressor to *lac* operator is visualized by "footprinting".

30.5   DNA SEQUENCING

30.5.1   Maxam and Gilbert Method

In 1977, Alan Maxam and Walter Gilbert used chemical reagents for DNA sequencing (chain degradation sequencing method) to bring about base-specific cleavage of the DNA.

In the first step, the double-stranded DNA is denatured to single-stranded DNA and tagged with ^{32}P at one end. Then the molecules are treated with dimethyl sulphate, which attaches methyl group to the guanosine nucleotide. Now piperidine is added. This removes the modified nucleotide and cuts the DNA at the phosphodiester bond. The result is a set of cleaved DNA molecules. The cleaved molecules are electrophoresed in a polyacrylamide gel and the sequence is read. For other nucleotides, Table 30.1 shows different kinds of base modifiers and chemicals which cleave DNA at altered bases.

Table 30.1   Base modifier and cleavage

Base specificity	Base modification	Strand cleavage
G	Dimethyl sulphate	Piperidine
G + A	Acid	Piperidine
T + C	Hydrazine	Piperidine
C	Hydrazine + NaCl	Piperidine
A > C	NaOH	Piperidine

30.5.2   Frederick Sanger's Method

It is also called chain terminator method. In this method, the DNA to be sequenced is incubated with DNA polymerase I, a suitable primer, and the four dNTP substrates for the polymerization reaction. The reaction mixture includes a ^{32}P "tagged" to any one of the dNTPs or the primer.

A small quantity of a 2′, 3′-dideoxynucleoside triphosphate (ddNTP), which lacks the 3′-OH group of deoxynucleotides is also added to the reaction mixture. The chain growth is terminated wherever dideoxy analog is incorporated into the growing polynucleotide in place of the corresponding normal nucleotide. In this way, a series of truncated chains is generated, each of which ends with the dideoxy analog at one of the positions occupied by the corresponding base.

Recent advancements show that sequencing tasks use four different ddNTP and the reaction products are electrophoresed in parallel lanes. The lengths of the chains indicate the position of the dideoxynucleotide incorporated. This results in the sequence of the replicated strand that can be directly read from the gel. The important point here is that the sequence obtained by the chain-terminator method is complementary to the DNA strand being sequenced.

30.5.3   Automated Sequencing

Large-scale sequencing operations can be achieved by automation. The primers used in the four chain-extension reactions are linked to a different fluorescent dye. The separately reacted mixtures are combined and subjected to gel electrophoresis in a single lane. The terminal base on each fragment is identified by its characteristic fluorescence. Using fluorescence detectors, computer-controlled automated systems can idenfity ~10,000 bases per day.

30.6   POLYMERASE CHAIN REACTION

The polymerase chain reaction (PCR) technique selectively generates a large number of copies of desired DNA sequences from the original mixture of DNA sequences. This technique was devised by Kary Mullis in 1985. The format of the PCR and PCR reaction chart are shown in Figure 30.1 and Table 30.2 respectively.

In the beginning of cycle 1

94°C	Two DNA strands are separated.
60°C	Primer anneals specifically to the 3′ end of the target sequence.
60–75°C	*Taq* polymerase uses free nucleotides to synthesize DNA strands.

End cycle 1

2 long double-stranded DNA molecules are synthesized.

Begin cycle 2

94°C	Two double-stranded DNA strands are separated.
60°C	Primer anneals specifically to the 3′ end of the target sequence.
60–75°C	*Taq* polymerase uses free nucleotides to synthesize DNA strands.

End cycle 2

4 long double-stranded DNA molecules are synthesized.

Begin cycle 3

95°C	Four double-stranded DNA strands are separated.
60°C	Primer anneals specifically to the 3′ end of the target sequence.
60–75°C	*Taq* polymerase uses free nucleotides to synthesize DNA strands.

End cycle 3

6 long double-stranded DNA molecules are synthesized.

2 target molecules are synthesized (see also flow chart in the Figure 30.1).

Begin cycle 4

95°C	Double-stranded DNA is melted.

60°C   Primer anneals specifically to the 3′ end of the target sequence.

60–75°C   *Taq* polymerase uses free nucleotides to synthesize DNA strands.

End cycle 4

8 long double-stranded DNA molecules are synthesized.

8 target molecules are synthesized.

Begin cycle 5

95°C   Double-stranded DNA is melted.

60°C   Primer anneals specifically to the 3′ end of the target sequence.

60–75°C   *Taq* polymerase uses free nucleotides to synthesize DNA strands.

End cycle 5

10 long double-stranded DNA molecules are synthesized.

22 target molecules are synthesized.

By the end of 30 cycles, 60 longer DNA molecules and more than a billion (1, 073, 741, 766) target molecules are synthesized.

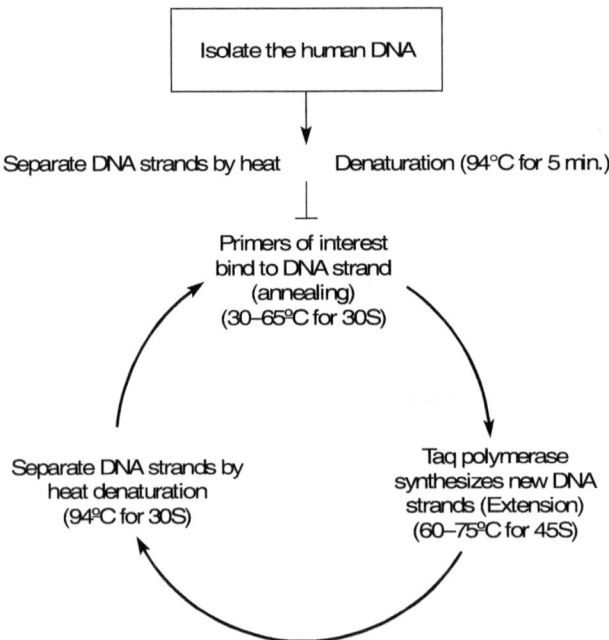

Figure 30.1   Format for PCR reaction

++ : Insertion   +– : Heterozygous   – – : Deletion

S. No.	Patient	Sign⁺	S. No.	Patient	Sign⁺	S.No.	Patient	Sign⁺
1			33			65		
2			34			66		
3			35			67		
4			36			68		
5			37			69		
6			38			70		
7			39			71		
8			40			72		
9			41			73		
10			42			74		
11			43			75		
12			44			76		
13			45			77		
14			46			78		
15			47			79		
16			48			80		
17			49			81		
18			50			82		
19			51			83		
20			52			84		
21			53			85		
22			54			86		
23			55			87		
24			56			88		
25			57			89		
26			58			90		
27			59			91		
28			60			92		
29			61			93		
30			62			94		
31			63			95		
32			64			96		

Table 30.2   PCR reaction chart

PCR reaction mixture		Cycle timer primer name:	
Primer 1 (F)	- 0.5 µM	Initial denaturation	-
Primer 2 (R)	- 0.5 µM	Denaturation	-
Taq polymerase	-	Annealing	-
MgCl₂	- 1.5	Extension	-
DNTPs	- 200 µM	Final extension	-
10x PCR buffer	-	No. of cycles	-
PCR water	-		

30.7   DNA HYBRIDIZATION

A probe is the small portion of DNA (or RNA) labelled with radioactive chemical or fluorescent dye and is used in a hybridization assay to identify other DNA or RNA sequences which are closely related to it in base sequence. Nucleic acid probes are made of single-stranded or double-stranded molecules (working probe is single strand). Nucleic acids used as probes are normally obtained by DNA cloning or chemical synthesis. There are three kinds of probes, viz. **conventional probes, RNA probes** and **oligonucleotide probes**.

30.7.1   Conventional DNA Probes

Conventional DNA probes are isolated by cell-based DNA cloning or by PCR-based DNA cloning. In the former case, the starting DNA may range in size from 0.1 kb to hundreds of kilobases in length and is double-stranded. PCR-derived DNA probes have often been less than 10 kb long and double-stranded. Conventional DNA probes are usually labelled during an *in vitro* DNA synthesis reaction.

30.7.2   RNA Probes

RNA probes can conveniently be generated from DNA, which has been cloned in a specialized plasmid vector. Such vectors normally contain a phage promoter sequence immediately adjacent to the multiple cloning sites, and specific RNA transcripts can be obtained from the cloned insert using the relevant phage RNA polymerase.

30.7.3   Oligonucleotide Probes

Oligonucleotide probes are short (typically 15–50 nucleotides) single-stranded pieces of DNA made by chemical synthesis. Oligonucleotide probes are designed with a specific sequence chosen in response to prior information about the target DNA. Oligonucleotide probes are often labelled by incorporating a ^{32}P isotope at the 5´ end.

30.8   LABELLING

DNA and RNA can conveniently be labelled by incorporation of nucleotides containing a radioisotope or fluorescent chemical. Two major types of procedures have been widely used to label DNA *in vitro*.

*   In **DNA-polymerase-based strand labelling,** the DNA is labelled by incorporating numerous labelled nucleotides during a DNA strand synthesis reaction. Normally during the DNA synthesis, at least one of the four deoxyribonucleoside triphosphates (dNTPs) contains a labelled group.

*   In **end-labelling method,** the labelled group is being added to one or a few terminal nucleotides. This method is useful for labelling single-stranded oligonucleotides and restriction mapping.

30.8.1   Labelling DNA by Nick Translation

In the nick-translation procedure, a single-strand nick is introduced using deoxyribonuclease I (DNase I) in the DNA. This procedure exposes 3´ hydroxyl termini and 5´ phosphate termini. The exposed nick can then serve as a starting point for introducing new nucleotides at the 3´ hydroxyl side of the nick using the DNA polymerase activity of *E. coli*. DNA polymerase I acts at the same time to remove existing nucleotides from the other side of the nick by the 5´ $\longrightarrow$ 3´ exonuclease activity. As a result, the nick will be moved progressively along the DNA in the 5´ $\longrightarrow$ 3´ direction. The reaction is carried out at about 15°C and the incorporation of labelled nucleotides proceeds by replacing unlabelled nucleotides with labelled ones.

30.8.2   Random-primed DNA Labelling

The random primed DNA labelling method is based on hybridization of a mixture of all possible hexanucleotides: the starting DNA is denatured and then cooled slowly so that the individual hexanucleotides can bind to suitable complementary sequences within the DNA strands. Synthesis of new complementary DNA strands is primed by the bound hexanucleotides. The synthesis is catalysed by the Klenow subunit of DNA polymerase I (which contains the polymerase activity in the absence of associated exonuclease activities) in the presence of the four dNTPs, at least one of which has a labelled group. This method produces labelled DNAs of high specific activity. Because all sequence combinations are represented in the hexanucleotide mixture, binding of primer to template DNA occurs in a random manner, and labelling is uniform across the length of the DNA.

30.8.3   End-labelling of DNA

Labelling of single-stranded oligonucleotides is usually achieved by an end-labelling reaction involving polynucleotide kinase. This enzyme can catalyse the exchange of a labelled γ-phosphate group from ATP with the 5´ terminal phosphates on single or double-stranded DNA.

30.8.4   Labelling of RNA

The preparation of labelled RNA probes (riboprobes) is most easily achieved by *in vitro* transcription of insert DNA cloned in a suitable plasmid vector. The vector is designed so that adjacent to the multiple cloning sites is a phage promoter sequence, which can be recognized by the corresponding phage RNA polymerase. By using a mix of NTPs, at least one of which is labelled, radiolabelled transcripts of high specific activity can be generated. Bacteriophage T3 and T7 promoter/RNA polymerase systems are also used commonly for generating riboprobes. Labelled sense and antisense riboprobes can be generated from any gene cloned in such vectors (the gene can be cloned in either of the two orientations) and are widely used in *in situ* hybridization.

30.9   KINDS OF LABELLING

Nucleic acids are labelled with two kinds of chemicals such as radioactive chemicals (isotopic labelling) and fluorescent chemicals (nonisotopic method).

30.9.1 Isotopic Labelling and Detection

The nucleotide probes are radiolabelled with an isotope ^{32}P, ^{35}S or ^{3}H and can be detected specifically in solution or within a solid specimen. The intensity of an autoradiographic signal is dependent on the intensity of the radiation emitted by the radioisotope and the time of exposure. Since ^{32}P emits high-energy β-particle radiation, it has been used widely in Southern blot hybridization, dot-blot hybridization, colony hybridization and plaque hybridization. Radionucleotides which provide less energetic β-particle radiation, have been preferred for DNA sequencing (^{35}S-labelled) and tissue *in situ* hybridization, and ^{3}H-labelled nucleotides for chromosome *in situ* hybridization. ^{35}S and ^{3}H have moderate and very long half-lives respectively.

^{32}P-labelled nucleotides used in DNA strand synthesis labelling reactions have the radioisotope at the α-phosphate position, because the β- and γ-phosphates from d-NTP precursors are not incorporated into the growing DNA chain. In the case of ^{35}S-labelled nucleotides, which are incorporated during the synthesis of DNA or RNA strands, the NTP or d-NTP carries an ^{35}S isotope in place of the O⁻ of the α-phosphate group. ^{3}H-labelled nucleotides carry the radioisotope at several positions. Specific detection of molecules carrying a radioisotope is most often performed by autoradiography.

30.9.2 Nonisotopic Labelling and Detection

Nonisotopic labelling can be made by the high affinity of two ligands such as **biotin** (a vitamin) and the bacterial protein **streptavidin**. Biotinylated probes can be made easily by including a suitable biotinylated nucleotide in the labelling reaction. A popular alternative is to use the plant steroid digoxigenin (obtained from Digitalis plants) as a label. A specific antibody for this compound permits detection of molecules containing it. Both digoxigenin- and biotin-labelled nucleotides are designed to have an intermediate spacer so that, once introduced into a labelled nucleic acid probe, the attached reporter groups can protrude sufficiently far from the nucleic acid backbone so as to facilitate their detection by streptavidin or the digoxigenin-specific antibody respectively.

30.10 AUTORADIOGRAPHY

Autoradiography technique is used to localize and record the radiological substances within a fractioned sample (protein or nucleic acids), to produce an image in a photographic emulsion. The fractioned sample normally is embedded within a dry gel or fixed to the surface of a dried nylon membrane/ nitrocellulose filter. The photographic emulsions consist of silver halide crystals in suspension in a clear gelatinous phase. The emission of a β-particle or a γ-ray by a radionuclide converts the Ag$^+$ ions to Ag atoms. The resulting latent image can then be converted to a visible image once the image is developed by an amplification process in which the entire silver halide crystals are reduced to give metallic silver. This results in the removal of any unexposed silver halide crystals, giving an autoradiographic image which provides a two-dimensional representation of the distribution of the radiolabel in the original sample.

Direct autoradiography involves placing the sample in intimate contact with an X-ray film, a plastic sheet with a coating of photographic emulsion. The radioactive emissions from the sample produce dark areas on the developed film. This method is best suited for the detection of weak to medium-strength β-emitting radionuclides (e.g. ^{3}H, ^{35}S). However, it is not suited for high-energy β-particles (e.g. ^{32}P). Such emissions pass through the film, resulting in the wasting of the majority of the energy. Indirect autoradiography is a modification in which the emitted energy is converted to light by a suitable chemical (scintillatory or Fluor) using either of the following methods. The sample is impregnated with a **liquid scintillator**. The energy transferred from the radioactive emissions causes the scintillator molecules to emit photons, thereby exposing the photographic emulsion. This method is largely used to improve the detection of weak emitters such as ^{3}H or ^{35}S. If, for example, the sample occurs within a gel the weak energy of the ^{3}H or ^{35}S radiolabel would result in a significant amount of the energy being absorbed by the sample, unless the gel is impregnated with a **fluor** such as PPO (2,5-diphenyl oxazole).

Intensifying screens are sheets of solid inorganic scintillatory materials which are placed behind the film in the case of samples emitting high-energy radiation, such as ^{32}P. Those emissions which pass through the photographic emulsion are absorbed by the screen and converted to light. By effectively superimposing a photographic emission upon the direct autoradiographic emission, the image is intensified.

FOR ADDITIONAL READING

1. Sambrook, J., Fritsch, E.F. and Mantiatis, T. (1989). *Molecular Cloning: A Laboratory Manual*, 2nd edn. Cold Spring Harbor Laboratory Press, New York.

2. Vosberg, H.P. (1989). "The polymerase chain reaction: an improved method for the analysis of nucleic acids." *Hum. Gen.* 83:1–15.

3. Eisenstein, B.I. (1990). "The polymerase chain reaction: a new method of using molecular genetics for medical diagnosis." *N.E.J. Med.* 322:178–183.

4. Innis, M.A., Gelfand, D.H., Sninsky, J.J. and White, T.J., eds. (1990). *PCR protocols: A Guide to Methods and Applications.* Academic Press, New York.

5. Brown, T.A. (1994). *DNA Sequencing: The Basics.* Oxford University Press, Oxford.

6. Sanger, F., Nicklen, S. and Coulson, A.R. (1977). "DNA sequencing with chain terminating inhibitors." *Proc. Natl. Acad. Sci.,* USA. 74:5463–5467.

7. Maxam, A.M. and Gilbert, W. (1977). "A new method for sequencing DNA." *Proc. Natl. Acad. Sci.,* USA. 74:560–564.

8. Mullikan, J.C. and McMurray, A.A. (1999). "Sequencing the genome, fast." *Science.* 283:1867–1868.

9. Emery, A.E.H. (1984). *An introduction to Recombinant DNA.* John Wiley and Sons, Chichester.

10. Antonarakis, S.E. (1989). "Diagnosis of genetic disorders at the DNA level." *N.E.J. Med.* 320:153–163.

11. Verma, R.S. and Babu, A. (1989). *Human Chromosomes.* Pergamon Press, New York.

12. Nakamura, Y., Leppert, M., O'Connell, P., Wolff, R., Holm, T., Culver, M., Martin, C., Fujimoto, E., Hoff, M., Kumlin, E. and White, R. (1987). "Variable number tandem repeat (VNTR) markers for human gene mapping." *Science.* 235:1616–1622.

13. Weber, J.L. and May, P.E. (1989). "Abundant class of human DNA polymorphisms which can be typed using the polymerase chain reaction." *Am. J. Hum. Genet.* 44:388–396.

14. Lander, E.S. (1989). "DNA fingerprinting on trial." *Nature.* 339:501–505.

15. Southern, E.M. (2000). "Blotting at 25." *Trends Biochem. Sci.* 25:585–588.

MODEL QUESTIONS

1. What is the principle of electrophoretic technique?

2. Describe the following
   i. Southern blotting
   ii. Northern blotting
   iii. Western blotting
   iv. Slot and Dot blotting
   v. Zoo blotting

3. Describe in detail DNA fingerprinting and its application in various disciplines.

4. Comment on DNA footprinting and its application in molecular biology.

5. Explain the polymerase chain reaction technique in detail.

6. What are the applications of autoradiography in molecular biology?

7. What is hybridization?

8. List and explain various probes used in hybridization technique.

9. What are the two major kinds of labelling techniques?

10. What is DNA sequencing? Explain various kinds of sequencing techniques.

APPENDIX

BIOINFORMATICS FOR GENETICIST

Since the Gene Bank Database established in 1988, Bioinformatics grew as a field to maintain, organize, analyse and make accessible large amounts of gene and genomic sequence information to scientific community. By mid 1990s, many varieties of databases had been developed for gene and protein sequences, comparison, mapping, expression sequence tags, evolutionary tree and for other types of molecular data. Currently there are hundreds of public online molecular databases available for scientific usage. Some of the important databases are listed below.

Description	*URL*
Sites relevant to the Human Genome Project	
CEPH (various maps)	http://www.cephb.fr/bio/ceph-genethon-map.html
dbESt (EST sequences)	http://www.ncbi.nlm.nih.gov/dbEST/
dbSNP (SNPs)	http://www.ncbi.nlm.nih.gov/SNP/
dbSTS (STS sequences)	http://www.ncbi.nlm.nih.gov/dbSTS/

(Contd.)

Description	URL
Genethon (various maps)	http://www.genethon.fr/
Radiation hybrid database	http://www.ebi.ac.uk/RHdb
Sites relevant to other eukaryotic genome projects	
Various eukaryotes	http://www.sanger.ac.uk/Projects/
Microbial eukaryotes	http://www.mips.biochem.mpg.de/
Candida albicans	http://alces.med.umn.edu/Candida.html
Dictyostelium discoideum	http://glamdring.ucsd.edu/others/dsmith/dictydb.html
Neurospora crassa	http://www.mips.biochem.mpg.de/
Plasmodium falciparum	http://ben.vub.ac.be/malaria/who.html
Plasmodium	http://www.wehi.edu.au/biology/malaria/who.html
Saccharomyces cerevisiae	http://genome-www.stanford.edu/Saccharomyces
Invertebrates	
Bombyx mori	http://www.ab.a.u-tokyo.ac.jp/sericulture/shimada.html
Caenorhabditis elegans	http://www.ddbj.nig.ac.jp/htmls/c-elegans/html/CE_INDEX.html
Drosophila melanogaster	http://flybase.bio.indiana.edu/
Mosquito	http://klab.agsci.colostate.edu/
Vertebrates	
Chicken	http://www.ri.bbsrc/ac/uk/chickmap/chickgbase/manager.html

Cow	http://locus.jouy.inra.fr/cgi-bin/bovmap/intro2.pl
Dog	http://mendel.berkeley.edu/dog.html
Mouse	http://www.informatics.jax.org/
Pufferfish	http://fugu.hgmp.mrc.ac.uk
Rat	http://ratmap.gen.gu.se/
Zebrafish	http://zfish.uoregon.edu/

Plants

Various plants	http://ars-genome.cornell.edu/
Arabidopsis thaliana	http://www.arabidopsis.org/
Rice	http://www.tigr.org/tdb/at/at.html
Beans	http://www.ba.cnr.it/Beanref/
Cotton	http://algodon.tamu.edu/
Forest trees	http://dendrome.ucdavis.edu/
Maize	http://aestivum.moulon.inra.fr/imgd/

Sites relevant to prokaryotic genome projects

| Various species | http://www.tigr.org/tdb/mdb/mdb.html |

Sequence databases

| European Bioinformatics Institute | http://www.ebi.ac.uk |
| Genbank | http://www.ncbi.nlm.bih.gov |

(Contd.)

Description	URL
DNA Database of Japan	http://www.ddbj.nig.ac.jp/
Bioinformatics resources	http://www.ncgr.org/
Codon usage data	http://www.kazusa.or.jp/codon/
Eukaryotic promoters	http://epd.isb-sib.ch/
Restriction endonucleases	http://neb.com/rebase/rebase.html
RNA editing	http://rna.ucla.edu/
Telomeres	http://www.genlink.wustl.edu/teldb/index.html
Vertebrate gene evolution	http://pbil.univ-lyon1.fr/databases/hovergen.html
Allele frequencies and DNA polymorphisms	http://alfred.med.yale.edu
All gene mutations related to Alzheimer disease	http://molgen-www.uia.ac.be/ADmutations/
Microarray gene expression data	http://www.ebi.ac.uk/arrayexpress
Genes, cytogenetics, and clinical features of cancer and prion	http://www.infobiogen.fr/services/chromcancer/
Integrated genetic and marker maps of human chromosomes	http://gai.nci.nih.gov/CHLC/
Protein–protein interactions	http://dip.doe-mbi.ucla.edu
Single-nucleotide polymorphisms	http://www.ncbi.nlm.nih.gov/SNP/
All known nucleotide and protein sequences	http://www.ddbj.nig.ac.jp
All known nucleotide and protein sequences	http://www.ebi.ac.uk/embl.html
Annotated information on eukaryotic genomes	http://www.ensembl.org/

(Contd.)

Expert protein analysis system	http://ca.expasy.org
Human genes and genomic maps	http://www.gdb.org
Human genes, markers, and phenotypes	http://www.citi2.fr/GENATLAS/
All known nucleotide and protein sequences	http://www.ncbi.nlm.nih.gov/
Integrated database of human genes, maps, proteins, and diseases	http://bioinfo.weizmann.ac.il/cards/
Medical genetics information resource	http://www.geneclinics.org/
Nomenclature for human genes	http://www.gene.ucl.ac.uk/cgi-bin/nomenclature/searchgenes.pl
mRNA expression levels of human genes in normal tissues	http://hugeindex.org
Human genome genetic and physical map data	http://www.inforbiogen.fr/services/Hugemap
Links to locus-specific mutation databases	http://www.hgmd.org
Metabolic and regulatory pathways	http://www.genome.ad.jp/kegg
Human mitochondrial genome	http://www.gen.emory.edu/mitomap.html
Catalogue of human genetic and genomic disorders	http://www.ncbi.nlm.nih.gov/Omim
Biologically significant protein patterns and profiles	http://www.expasy.org/prosite
Comprehensive, annotated, nonredundant protein sequences	http://pir.georgetown.edu

Description	URL
Abstracts of journal articles	http://www.ncbi.nlm.nih.gov/PubMed/
Non-redundant sequences from genomes and proteins	http://www.ncbi.nlm.nih.gov/LocusLink/refseq.html
Single-nucleotide polymorphism	http://snp.cshl.org
Raw and normalized data from microarray experiments	http://genome-www.stanford.edu/microarray
Protein sequences	http://www.expasy.ch/sprot
Genome assemblies and annotation	http://genome.ucsc.edu/
Gene Finding and Analysis	
Ensembl at EBI/Sanger Centre	http://www.ensembl.org/
Human Genome Browser at UCSC	http://genome.ucsc.edu
Map viewer at NCBI	http://www.ncbi.nlm.nih.gov/cgi-bin/Entrez/map_search
Protein Atlas of the genome	http://www.confirmant.com/
SWISS-2DPAGE database	http://ca.expasy.org/ch2d/
Ensembl 4.28.1 announcement	http://www.ensembl.org/Dev/Lists/announce/msg00070.html
NCBI gene model builder	http://www.ncbi.nlm.nih.gov/PMMGifs/Genomes/ Model MarkerHelp.html
InterPro at EBI	http://www.ebi.ac.uk/interpro/
Proteome analysis at EBI	http://www.ebi.ac.uk/proteome/
RefSeq at NCBI	http://www.ncbi.nlm.nih.gov/LocusLink/refseq.html

(Contd.)

International Protein index	http://www.ebi.ac.uk/IPI/IPIhelp.html
Derwent sequence patient databases	http://www/derwent.com/geneseq/index.html
BLAST at NCBI	http://www.ncbi.nlm.nih.gov/BLAST/
BLAT at UCSC	http://genome.ucsc.edu/cgi-bin/hgBlat?command=start)
DAS-distributed annotation	http://biodas.org/
Exofish at Genoscope	http://www.genoscope.cns.fr/externe/tetraodon/
Fgenesh at Sanger Institute	http://genomic.sanger.ac.uk/gf/Help/fgenesh.html
Expasy translation tool	http://ca.expasy.org/tools/dna.html
CAP3 nucleotide assembly tool	http://bio.ifom-firc.it/ASSEMBLY/assemble.html
GeneWise at Sanger Institute	http://www.sanger.ac.uk/Software/Wise2/
Genscan at MIT	http://genes.mit.ed/GENSCAN.html
SSAHA at Sanger Institute	http://www.sanger.ac.uk/Software/analysis/SSAHA/

Mutagenesis Centres and Databases

ENU Mutagenesis Programme (Harwell)	http://www.mgu.har.mrc.ac.uk/mutabase/
RIKEN Mouse Functional Genomics Group	http://www.gsc.riken.go.jp/Mouse/
GSF ENU Mouse Mutagenesis Screen project	http://www/gsf.de/isg/groups/enu-mouse.html
Neuroscience Mutagenesis lab at the Jackson	http://www.jax.org/nmf/
Neurogenomics centre at northwestern university	http://genome.northwestern.edu/
Tennessee Mouse Genome Consortium	http://www.tnmouse.org/

Description	URL
McLaughlin Research Institute	http://www.montana.edu/wwwmri/enump.html
Baylor college of Medicine	http://www.mouse-genome.bcm.tmc.edu/ENU/MutagensisProj.asp
Medical Genome Centre (Australia)	http://jcsmr.anu.edu.au/group_paegs/mgc/CancerGenLab.html
The Mouse Heart, Lung, Blood, and Sleep Disorders Centre (JAX)	http://www.jax.org/hlbs/index.html
SNP detection tools	
Sequencher	http://www.genecodes.com/features.htm
Polyphred	http://droog.mbt.washington.edu/PolyPhred.html
POLYBAYES	http://www.genome.wustl.edu/gsc/polybayes/
Repeat masking tools	
RepeatMasker	www.genome.washington.edu/uwgc/analysistools.repeatmask.html
MaskerAid	http://sapiens.wustl.edu/maskeraid/
Primer design tools	
Primer3	http://www-genome.wi.mit.edu/cgi-bin/primer3_www.cgi
TSC primer db	ftp://snp.cshl.org/pub/SNP/.
Primer design tips	http://www.alkani.com/primers/refdsgn.htm.

(Contd.)

Tools for sequence extraction and manipulation

SNPper   http://bio.chip.org:8080/bio/.

UCSC HGB   http://genome.ucsc.edu/index.html

Sequence manipulation and translation

Sequence manipulation site   http://www/bioinformatics.org/sms/

Amino acid properties

Properties of amino acids   http://www.russell.embl-heidelberg.de/aas/

Secondary structure prediction

TMPRED   http://www.ch.embriet.org/software/TMPRED/TMPRED_form.html

SOSUI   http://sosui.proteome.bio.tuat.ac.jp/sosuiframe0.html

TMHMM   http://www.cbs.dtu/services/TMHMM/

PREDICTPROTEIN   http://www.embl-heidelberg.de/predictprotein/

GPCRdb 7TM plots (Snake plots for most 7TMs)   http://www.gpcr.org/7tm/seq/snakes.html

Tertiary structure prediction and visualization

Swiss-Model   http://expasy/hcuge.ch/swissmod/SWISS-MODEL.html

SCOP   http://scop.mrc-lmb.cam.ac.uk/scop/

Identification of functional motifs

INTERPRO   http://www.ebi.ac.uk/interpro/scan.html

Description	URL
PROSITE	http://www.ebi.ac.uk/searches/prosite.html
PFAM	http://www.sanger.ac.uk/Software/Pfam/
NetPhos (serine, threonine and tyrosine phosphorylation)	http://www.cbs.dtu.dk/services/NetPhos/
NetOGlyc (O-glycosylation)	http://www.cbs.dtu.dk/services/NetOGlyc/
NetNGlyc (N-glycosylation)	http://www/ebs.dtu.dk/services/NetNGlyc/
SIGNALP (signal peptide prediction)	http://wwwebs.dut.dk/services/SignalP/
Swissprot (functional annotation)	http://www/expasy.ch/cgi-bin/sprot-search-ful
RNA secondary structure prediction	
Homepage of M.Zucker	http://bioinfo.math.rpi.edu/~zukerm/
Pattern definition	
CoreSearch (ftp)	ariane.gsf.de/pub/unix/coresearch_1.2.tar.Z
CONSENSUS simple.html	http://bioweb.pasteur.fr/seqanal/interfaces/consensus-
S/MAR detection	
MARFinder	http://www.ncgr.org/MAR-search/
SMARTest	http://www.genomatix.de

UTR analysis

UTR database — http://bighost-area.ba.cnr.it/BIG/BioWWW/#UTRdb

Genomatix tools

SMARTest — http://www.genomatix.de/free_services/

PromoterInspector

MatInspector professional

GEMS Launcher

ELDorado

Sequence tools

Software and Databases for the Statistical Analysis

Tag extraction, tag identification, statistical comparison	www.sagenet.org	
Tag identification, statistical, xProfiler, Virtual Northern	www.ncbi.nlm.nih.gov/SAGE/	
Tag extraction, tag identification, statistical comparison, management	www.cmbi.kun.nl/usage/	
Tag extraction, statistical comparison, data management	ehm@umich.edu	
Detection of sequencing errors in SAGE libraries	georg.feger@serono.com	
Statistical comparison	http://www.igs-server.cnrs-mrs.fr/~audic/significance.html	

(Contd.)

Description	URL
Statistical comparison	j.m.ruijter@amc.uva.nl or www.cmbi.kun.nl/usage/
Statistical comparison	Michael.man@pfizer.com
Full Text Journal Access	
PubMed	http://www.ncbi.nlm.nih.gov/entrez
Scirus	http://ww.scirus.com/
HighWire	http://highwire.stanford.edu/
Biomednet	http://www.bmn.com
ScienceDirect (Elsevier)	http://www.sciencedirect.com
IDEAL	http://www.ideallibrary.com/
Nature Publishing Group	http://www.nature.com/nature/
Wiley InterScience	http://www.interscience.wiley.com/
Microarray	
Information and Educational site on Microarray	
	arrays.sdsc.edu/
	arrays.rockefeller.edu/xenopus
	bioinf.man.ac.uk/microarray/
	bioinformatics.duke.edu/camda/
	discover.nci.nih.gov/nature2000/

ee.tamu.edu/~camdi/subpages/cdna.html

genome-www.stanford.edu/nci60/

info.med.yale.edu/microarray/

mach1.nci.nih.gov/

quantgen.med.yale.edu/

www.bio.davidson.edu/biology/courses.genomics/chip/chip.html

www.cse.ucsc.edu/research/compbio/genex/genex.html

www.ebi.ac.uk/microarray/

www.epa.gov/nheerl/epamac/links.html

www.mcb.arizona.edu/wardlab/microarray.html

www.mged.org/

www.microarrays.org/

www.nhgri.nih.gov/DIR/LCG/15K/HTML/

www.nigms.nih.gov/funding /microarray.html

www.sghms.ac.uk/depts/medmicro/bugs/

www.tigr.org/tdb/microarray

www-stat.stanford.edu/~tibs/lab/

(Contd.)

Description	URL
Commercial /dealers site	
	www.aat-array.com
	www.affymetrix.com
	www.axon.com/GN_Genomics.html
	biodiscovery.com/
	www.cartesiantech.com/
	www.corning.com/cmt/
	www.d-trends.com/
	www.genemachines.com
	www.incyte.com
	www.mediacy.com/arraypro.htm
	www.microarrays.com/
	www.perkinelmer.com
	www.premierbiosoft.com/dnamicroarray/dnamicroarray.html
	www.scanalytics.com/product/hts/microarray.html
	stratagene.com/
	www.bio-rad.com
Complete facilities of microarray	www.emory.edu/WHSC/YERKES/VRC/genomicsstats.html

sgio2.biotec.psu.edu/

www.array.saci.org/

www.bcm.tmc.edu/microarray/

www.ccgpm.org

www.uhnres.utoronto.ca/services/microarray/

www.umich.edu/~caparray/

linkage.rockefeller.edu/wli/microarray/

www.biologie.ens.fr/en/genetiqu/puces/links.html

www.bsi.vt.edu/ralscher/gridit

www.gene-chips.com/

www.sciencesmag.org/feature/e-market/benchtop/micro.shl

Internet Search Engines

http://www.google.com

http://www.altavista.com/

http://www.alltheweb.com/

http://www.northernlight.com/

http://www.excite.com/

http://hotbot.com/

http://www.lycos.com/

http://www.metacrawler.com/

http://www.scirus.com/

GLOSSARY

a Alanine (amino acid), adenine (nucleotide).

abortive transduction The failure of a transducing DNA that has to be incorporated into the recipient chromosome.

acentric fragment A piece of chromosomal segment without a centromere.

acridine-orange A mutagen that tends to cause frameshift mutations.

acrocentric chromosome A Chromosome whose centromere lies very near one end.

activator A protein that, when bound to an operator or an enhancer, activates transcription.

activator AC element A transposable element in corn with short inverted terminal repeat sequences.

active site The site of an enzyme where the actual enzymatic function is performed.

adaptation In the evolutionary sense, some heritable feature of an individuals's phenotype that improves its chances of survival and reproduction in the existing environment.

adaptive peak A high point on an adaptive landscape selection that tends to drive the genotype composition of the population towards a genotype corresponding to the adaptive peak.

additive genetic variance Genetic variance associated with the average effects of substituting one allele for another.

adenine A purine base that pairs with thymine in the DNA double helix or with uracil in RNA.

adenine methylase An enzyme that methylates adenine before replication, to distinguish old strands from the newly synthesized strand.

adenosine The nucleoside containing adenine as its base.

adjacent segregation In a reciprocal translocation, the passage of a translocated and a normal chromosome to each of the poles.

aflatoxin B1 (AFB1) A mutagen, that cleaves the bond between guanine and deoxyribose and produce an apurinic site.

albino A pigmentless "white" phenotype, determined by a mutation in a gene coding for a pigment-synthesizing enzyme.

alkylating agent It is a chemical that can add alkyl groups to another molecule; many mutagens act through alkylation.

allele   An alternative form of a gene, also called as allelomorph.

allele frequency   A measure of the commonness of an allele in a population; the proportion of all alleles of that gene in the population.

allelic exclusion   A process by which one immunoglobulin light chain and one heavy chain gene are transcribed in any one cell; other genes are repressed.

allopatric speciation   The formation of new species from populations that are geographically isolated from each other.

allopolyploidy   Polyploidy produced by the hybridization of two species.

allosteric protein   A protein whose shape is changed when it binds a ligand. Due to this, the protein's ability to react to a second ligand is altered.

allosteric site   A change in the conformation of the protein that modifies the activity of its active site.

allosteric transition   A change from one conformation of a protein to another.

allotype   Mutant of a nonvariant part of an immunoglobulin gene that follows simple Mendelian inheritance.

allozygosity   Two alleles are alike but unrelated in homozygosity.

allozymes   Forms of an enzyme that differ in electrophoretic mobility but controlled by alleles of the same locus.

alternate segregation   A separation of centromeres during meiosis in a reciprocal translocation heterozygote such that balanced gametes are produced.

alternative splicing   Splicing out introns in eukaryotic pre-mRNAs, resulting in one gene producing several different mRNA and protein products.

altruism   A form of behaviour in which an individual risks lowering its fitness for the benefit of another.

***alu* family**   Intermediately repetitive DNA sequence that is dispersed in the human genome. The sequence is more or less 300 bp long. The name *Alu* comes from the restriction endonuclease that cleaves it.

amber codon   Codon UAG, a nonsense codon.

amber suppressor   A mutant allele encoding a tRNA whose anticodon is altered, so that the tRNA inserts an amino acid at an amber codon in translation.

Ames test   A widely used test to detect possible chemical carcinogens, based on mutagenicity in the bacterium *Salmonella*.

aminoacyl site   Site on the ribosome, occupied by an aminoacyl-tRNA, prior to peptide bond formation.

aminoacyl-tNRA   synthetases   Enzymes that attach amino acids to their proper tRNAs.

amniocentesis   A technique to test the genotype of an embryo or foetus *in utero* using amniotic fluid with less risk to the mother or foetus.

amphidiploid   An allopolyploid, a polyploidy formed from the union of two separate chromosome sets and their subsequent doubling.

anagenesis The evolutionary process in which one species evolves into another without splitting the phylogenetic tree.

analogous structures Structures of different species that are functionally similar but that are developed from different embryonic structures as, for example, the wings of birds and the wings of insects.

anaphase The stage of cell division (mitosis and meiosis) in which sister chromatids or homologous chromosomes are separated by spindle fibres.

androgen A family of steroid hormones that promote male development, such as testosterone and 5-hydroxytestosterone.

aneuploids Individuals or cells exhibiting aneuploidy.

aneuploidy The condition of the cell that has additions or deletions of whole chromosomes.

animal breeding Practical application of genetic analysis for development of lines of domestic animals.

annealing Spontaneous alignment of two single DNA strands to form a duplex.

antibody A protein produced by B-lymphocyte that protects the individual against antigens or infectious agents.

anticoding strand The DNA strand that forms the template for both the transcribed mRNA and the coding strand.

anticodon The three-base sequence on tRNA, which is complementary to an mRNA codon.

antigen A foreign substance capable of triggering an immune response in an organism.

antimutator mutations DNA polymerase mutations that reduce the overall mutation rate of a cell or an organism.

anti-oncogene A gene that represses malignant growth or tumour and the absence of which results in malignancy.

antiparallel strands DNA strands that run in opposite directions with respect to their 3′ and 5′ ends.

antisense RNA RNA product of *mic* (mRNA-interfering complementary RNA) genes that regulates another gene by base pairing, and thus blocking its mRNA.

AP endonuclease An enzyme that removes apurinic sites in DNA so that repair synthesis can replace them with appropriate complementary strands.

apoptosis The cellular pathways responsible for cell death and subsequent removal of the remains of the dead cell may be induced by intracellular damage or by signals from neighbouring cells.

AP site Site resulting from the loss of a purine or pyrimidine residue from the DNA.

apurinic site DNA site that has lost purine residue.

apyrimidinic site DNA site that has lost pyrimidine residue.

ascospore A sexual spore from certain fungus species in which spores are found in a sac called an ascus.

attached X A pair of *Drosophila* X chromosomes joined at one end and inherited as a single unit.

attenuator region A control region at the promoter end of repressible amino acid operons that exerts transcriptional control based on the translation of a small leader peptide gene.

autogamy   Nuclear reorganization in a single paramecium cell similar to the changes that occur during conjugation.

autonomous controlling element   A controlling element that seems to have both regulator and receptor functions combined in a single unit, which enters a gene and causes an unstable mutation.

autonomous replication sequence (ARS)   A segment of DNA needed for the initiation of its replication; generally a site recognized and bound by proteins of the replication system.

autophosphorylation   A protein kinase, which phosphorylates specific amino acid residues on itself.

autopolyploidy   Polyploidy in which all the chromosomes come from the same species.

autoradiogram   A pattern of dark spots in a developed photographic film or emulsion, in the technique of autoradiography.

autoradiography   A technique in which radioactive molecules make their location known by exposing photographic plates.

autosomes   The non-sex chromosomes or somatic chromosomes.

autotrophs   Organisms that utilize carbon dioxide as a carbon source.

autozygosity   Homozygosity in which the two alleles are identical by descent.

bacillus   A rod-shaped bacterium.

backcross   The cross of an individual with one of its parents.

back mutation   A change in a nucleotide pair in a mutant gene that restores the original sequence and hence the original phenotype (reverse mutation).

bacteriophages   Bacteria infecting viruses.

balanced polymorphism   Stable genetic polymorphism maintained by natural selection.

balancer   A chromosome with multiple inversions, used to retain favourable allele combinations in the uninverted homologue.

balbiani rings   Larger polytene chromosome or chromosome puffs.

Barr body   Heterochromatic body found in the nuclei of normal females but absent in the nuclei of normal males.

base analog   A chemical which mimics a DNA base, and because of the mimicry, the analogue may act as a mutagen.

base correction   The production of a properly paired nucleotide pair from a sequence of hybrid DNA that contains a wrong pair.

base excision pair   In this pathway, subtle base-pair distortions are repaired by creation of apurinic sites followed by repair synthesis.

B DNA   The right-handed, double-helical DNA (Watson and Crick Model).

bead theory   A disproved hypothesis according to which genes are arranged on the chromosome like beads on a necklace indivisible into smaller units of mutation and recombination.

behaviour mutation   A mutation affecting behaviour of the mutant individual.

binary fission The simple cell division in single-celled organisms.

binomial distribution A statistical distribution having two modes.

biochemical genetics The study of the relationship between genes and enzymes (the role of genes in controlling the steps in biochemical pathways).

bivalents Structures formed during prophase of meiosis I, consisting of the synapsed homologous chromosomes.

blastocyst A very early stage in embryonic development when the embryo consists of a hollow ball of cells with a blastocoel (fluid-filled internal compartment).

blastoderm In an embryo, there is a single layer of nuclei or cells surrounding the central yolk.

blastomeres Cells that are making up the blastula.

blastopore Embryonic opening of the future gut.

blastula The first developmental stage of a developing embryo.

blending inheritance The characteristics of an individual result from the smooth blending of fluid-like influences from its parents.

blunt-end ligation The ligating or attaching of blunt-ended pieces of DNA by T4 DNA ligase.

brachydactyly A human phenotype of unusually short digits, generally inherited as an autosomal dominant.

branch migration The process in which a crossover point between two duplexes slides along the duplexes.

breakage and reunion The general mode of occurrence of recombination. DNA duplexes are broken and reunited in a cross-wide fashion according to the Holliday model.

breakage fusion-bridge cycle The damage that a dicentric chromosome undergoes in each cell cycle.

broad heritability The proportion of total phenotypic variance at the population level that is contributed by genetic variance.

bubbles The nucleic acid configuration during replication in eukaryotes, or the shape of heteroduplex DNA at the site of a deletion or insertion.

bud A daughter cell formed by mitosis in yeast, one daughter cell retains the cell wall of the parent, and the other forms a new cell wall.

CAAT box An invariant DNA sequence at upstream about −70 in many eukaryotic promoters.

canalized characters Characters that are stable in development despite environmental and genetic disturbances.

cap A sequence of methyl group added to the 5′ end of eukaryotic mRNA.

carcinogen A substance that causes cancer; cancer-causing mutagen.

caspase Enzyme of a proteases family that takes part in initiating and carrying out the cell death (apoptosis).

catabolite repression The inactivation of an operon by the presence of large amounts of the metabolic end product.

catalyst A substance that increases the rate of a chemical reaction without itself undergoing permanent changes.

cDNA library A library composed of complementary DNA, not necessarily representing all mRNAs.

cell cycle Represents growth, replication of the genetic material, and nuclear and cytoplasmic division.

cell division The process by which two cells are formed from single cell.

cell fate The ultimate differentiated state of a cell.

cell lineage A set of cells derived from a common ancestral cell by mitotic divisions.

centimorgan A chromosome mapping unit.

central dogma The fundamental thesis that the information can be transferred only from DNA to DNA, from DNA to RNA, and from RNA to protein.

centrioles Cylindrical organelles, found in eukaryotes (except higher plants) that organize the formation of the spindle.

centromere A specialized region of DNA on eukaryotic chromosome where kinetochore proteins bind.

Chargaff's rule Chargaff's observation that in the base composition of DNA, the quantity of purine is equal to the quantity of pyrimidine.

charon phages Phage lambda derivatives used as vehicles in DNA cloning.

chiasmata An x-shaped configuration seen in tetrads during the latter stages of prophase I of meiosis.

chimera An organism derived from more than one zygote.

chimeric plasmid Hybrid or genetically mixed plasmid used in DNA cloning.

chi-site Sequence of DNA at which the RecBCD protein cleaves one of the strands during recombination.

chloroplast Chlorophyll-containing organelle in plants which is the site for photosynthesis.

chromatids The subunits of a chromosome prior to anaphase of meiosis or mitosis.

chromatin The nucleoprotein material of the eukaryotic chromosome.

chromomeres Dark regions of chromatin condensation seen in eukaryotic chromosomes during meiosis or mitosis.

chromosome The genetic material in viruses and cells. A circle of DNA in prokaryotes; a DNA or an RNA molecule in viruses; a linear nucleoprotein complex in eukaryotes.

chromosome jumping A technique of isolating clones from a genomic library that are not contiguous but skip a region between known points on the chromosome This is done usually to bypass regions that are difficult or impossible to "walk" through or regions known not to be of interest.

chromosome puffs Diffuse, uncoiled regions in polytene chromosomes where transcription is actively taking place.

chromosome walking A technique for studying segments of DNA, larger than can be individually cloned by using overlapping probes.

CIB method Muller's technique to screen fruit flies for lethal recessive X chromosome mutations. The CIB chromosome carries a recessive lethal (l), a dominant marker (B), and an inversion (crossover suppressor, C).

cis The geometric configurations of atoms or mutants on the same chromosome.

cis-dominant Mutants that control the functioning of genes on that same piece of DNA.

cis-trans **complementation test** A mating test to determine whether two mutants on opposite chromosomes will complement each other; a test for allelism.

cistron Term coined by Benzer for the smallest genetic unit (gene) that exhibits the *cis-trans* position effect.

cladogenesis The evolutionary process whereby one species splits into two or more species.

clone A group of cells arising from a single ancestor.

coding strand The DNA strand with the same sequence as the transcribed mRNA.

codominance The relationship of alleles such that the phenotype of the heterozygote shows the individual expression of each allele.

codon The three letters of RNA or DNA nucleotides that specify either an amino acid or termination of translation.

codon preference The preference of one or a few codons of amino acids with several codons and their disproportionate use.

coefficient of relationship The proportion of alleles held in common by two related individuals.

complementary DNA (cDNA) DNA synthesized by reverse transcriptase using RNA as a template.

concordance The amount of similarity in phenotypic expression between individuals.

conjugation The union of two bacterial cells, during which chromosomal material is transferred from the donor to the recipient cell.

consanguineous Inbreeding or incestuous mating.

consensus sequence A sequence of the common nucleotides found in many different DNA or RNA samples of homologous regions.

conservative replication Mode of DNA replication in which an intact double helix acts as a template for a new double helix.

copper fist Configuration of a DNA binding protein that resembles a fist closed around a penny (copper ions). The knuckles of the fist of the yeast ACE1 protein interact with the promoter of the metallothionein gene, enhancing its transcription.

copy-choice hypothesis Hypothesis that states that recombination resulted from the switching of the DNA-replicating enzyme from one homologue to the other.

co-repressor The metabolite which when bound to the repressor forms a functional unit that can bind to its operator and block transcription.

correlation coefficient A statistic that gives a measure of how closely two variables are related.

cosmid A hybrid plasmid that contains *cos* sites at each end. *Cos* sites are recognized during head filling of lambda phages. Cosmids are useful for cloning large segments of foreign DNA.

co-transduction The simultaneous transduction of two or more genes.

cot values Co, the original concentration of denatured, single-stranded DNA and t, time in seconds, giving a useful index of renaturation. Cot ½ values are the midpoint values on cot curves.

coupling An allelic arrangement in which mutants are on the same chromosome and wild-type alleles on the homologue.

criss-cross pattern of inheritance The phenotypic pattern of inheritance controlled by X-linked recessive alleles.

critical chi-square A chi-square value for a given degree of freedom and probability level to which an experimental chi-square is to be compared.

crossing over A process in which homologous chromosomes exchange materials by a breakage and reunion.

cyclic AMP AMP (adenosine monophosphate) used frequently as a second messenger in eukaryotes and as a catabolite repressor in prokaryotes.

cytogenetics The cytological approach to genetics, mainly consisting of microscopic studies of chromosomes.

cytokinesis The cytoplasmic division of a cell into two daughter cells during cell division.

cytoplasm The material between the nuclear and cell membranes; includes fluid, organelles, and various membranes.

cytoplasmic inheritance Extra-chromosomal inheritance that is controlled by non-nuclear genomes such as plasmids and mitochondrial DNA.

cytosine A pyrimidine base that pairs with guanine.

cytoskeleton The protein cable systems and associated proteins that together form the architecture (mesh-like network) of a eukaryotic cell.

cytosol The fluid part of the cytoplasm, outside the organelles.

cytotoxic T lymphocytes A group of T cells that are responsible for attacking host cells that have been infected with an invading bacterium or virus (antigens).

dauermodification The persistence for several generations of an environmentally induced trait.

daughter cell Two identical cells formed by division of a cell.

daughter chromatids Two identical chromatids formed by the replication of one chromosome.

degenerate code The genetic code is so called when there are many codons to specify a single amino acid.

degrees of freedom An estimate of the number of independent categories in a particular statistical test.

deletion chromosome A chromosome with part deleted either in the p-arm or q-arm.

depauperate fauna A fauna, especially common on islands, lacking many species found in similar habitats.

determinant A spatially localized molecule that causes cells to adopt a definite fate or set of related fates.

determination The process of commitment of cells to particular fates.

diakinesis The final stage of prophase I of meiosis when chiasmata terminalize.

dicentric chromosome A single chromosome with two centromeres.

dicytotene A prolonged diplonema of primary oocytes that can last many years.

dideoxy method A DNA sequencing method that uses chain-terminating (dideoxy) nucleotides.

dihybrid An organism heterozygous at two loci.

diploid A cell having two chromosome sets or an individual organism having two chromosome sets in each of its cells.

diplonema (diplotene) The stage of prophase of meiosis I in which chromatids appear to repel each other.

disassortative mating The mating of two individuals with dissimilar phenotypes.

discontinuous replication In DNA lagging strand, the replication in short $5'$ to $3'$ segments using the $5'$ to $3'$ strand as a template while going backward, away from the replication fork.

discontinuous variation Variation having distinct classes of phenotypes for a particular character.

D-loop The initial structure formed during DNA replication in chloroplast and mitochondrial chromosomes where the origin of replication is different on the two strands. Also called the displacement loop.

DNA–DNA hybridization When DNA from the same or different sources is heated and then cooled, complementary strands will re-form at homologous regions. This technique is useful for determining sequence similarities and degrees of repetitiveness among DNAs.

DNA glycosylases Endonucleases that initiate excision repair at the sites of mutated or improper bases in DNA.

DNA gyrase A topoisomerase that relieves supercoiling in DNA by creating a transient break in the double helix.

DNA ligase An enzyme that closes nicks (break) in a single-strand of double-stranded DNA by creating an ester bond between adjacent $3'$-OH and $5'$-PO$_4$ ends on the same strand.

DNA polymerase Enzymes that polymerize DNA nucleotides using single-stranded DNA as a template.

DNA–RNA hybridization When a mixture of DNA and RNA is heated and then cooled, RNA can hybridize (form a double helix) with DNA that has a complementary nucleotide sequence.

dominance variance Genetic variance at a single locus attributable to dominance of one allele over another.

dominant An allele that expresses itself even in the heterozygous condition. Also the trait (character) controlled by that allele.

donor organism An organism that provides DNA for use in recombinant DNA technology or DNA-mediated transformation.

dot blotting A blotting technique of DNA autoradiographs, which reveal dots rather than bands on a gel, indicative of a cloned gene.

double digest The product formed when two different restriction endonucleases act on the same segment of DNA.

double helix The normal structural configuration of DNA, which consist of two helices (two strands) rotating about the same axis.

double infection Infection of a bacterium with two genetically different phages.

double reduction It is a condition in polyploids, where a heterozygous individual produces homozygous gametes.

double sex An allele that converts fruit fly *(Drosophila)* males and females into developmental intersexes.

Down's syndrome An abnormal human phenotype due to non-disjunction. It causes mental retardation, atrioventricular canal defect, etc. This happens mainly due to a trisomy of chromosome 21. It is more common in babies born to older mothers.

downstream Also called positive strand where structural genes are found. A convention on DNA related to the position and direction of transcription by RNA polymerase ($5'\rightarrow 3'$).

duplicate genes Two identical allele pairs in one diploid individual.

duplication More than one copy of a particular chromosomal segment in a chromosome set.

dyad Two sister chromatids attached to the same centromere.

dynamic mutation An unstable expanded repeat that changes size between parent and child.

ectopic integration The insertion of a gene at a site other than its usual locus in a transgenic animal.

electrophoresis The separation of molecular entities by electric current in gel.

electroporation A technique for transfecting cells or genetic materials by the application of a high-voltage electric pulse.

elongation factors Proteins (EF-Ts, EF-Tu, EF-G) that are required for the elongation and translocation, during translation in prokaryotes (EF1 and EF2 in eukaryotes).

embryonic polarity The production of axes of asymmetry in a developing embryo.

endocrine system The organs in the body that secrete hormones into the circulatory system.

endomitosis Chromosomal replication without nuclear or cellular division that results in cells with many copies of the same chromosome (polyploidy).

endonuclease Enzyme that makes nicks (breaks) in the backbone of a polynucleotide. They hydrolyse inner phosphodiester bonds.

endopolyploidy An increase in the number of chromosome sets caused by replication without cell division.

endosperm Triploid tissue formed from the fusion of two haploid female nuclei and one haploid male nucleus.

enforced outbreeding Deliberate avoidance of mating between relatives.

enhanceosome The macromolecular assembly responsible for interaction between enhancer elements and the promoter regions of genes.

enhancer A eukaryotic DNA sequence that increases transcription of a region even if the enhancer is distant from the region being transcribed.

enucleate cell A cell having no nucleus.

environmental variance The variance due to environmental variation.

enzyme A protein that functions as a catalyst.

epiblast The layer of cells in the pregastrulation embryo which will give rise to all three germ layers of the embryo proper, plus the extraembryonic ectoderm and mesoderm.

episomes Term coined by Jacob and Wollman for genetic materials. These genetic materials can either exist independently in a cell or become integrated into the host chromosome.

epistasis Masking action of allele of one gene by allele combinations of another gene.

epitope A part of an antigen with which a particular antibody reacts or binds.

ethidium A dye that can intercalate into DNA double helices when the helix is under torsional stress.

euchromatin Regions of eukaryotic chromosomes that are diffuse during interphase. Presumably the actively transcribing DNA of the chromosomes.

eugenics The science of controlled human breeding, in an attempt to improve future generations.

eukaryote An organism having eukaryotic cells.

euploidy The condition of a cell or organism that has one or more complete sets of chromosomes.

excision repair A process whereby cells remove part of a damaged DNA strand and replace it through DNA synthesis using the undamaged strand as a template.

exconjugant Each of the two cells that are separated after conjugation.

exogenote The DNA that a bacterial cell has taken up through one of its sexual processes.

exon shuffling The hypothesis forwarded by Walter Gillbert that exons code for functional units of a protein. The evolution of new genes proceeded by recombination or exclusion of exons.

exonucleases Enzymes that digest nucleotides from the ends of polynucleotide molecules. They hydrolyse phosphodiester bonds of terminal nucleotides.

expression vector A hybrid vector (plasmid) that expresses its cloned genes.

expressivity The degree of expression of a genetically controlled trait (character).

eyes of replication The configuration of replicating DNA in eukaryotic chromosomes.

F1 First filial generation.

factorial The product of all integers from the specified number down to one.

familial trait A trait shared by members of a family.

family selection A breeding technique of selecting a pair on the basis of the average performance of their progeny.

fate map A map of an embryo showing areas that are destined to develop into specific adult tissues and organs.

fecundity selection The forces acting to cause one genotype to be more fertile than another genotype.

feedback inhibition A post-translation control mechanism in which the end product inhibits the activity of the enzyme of this pathway.

fertility factor The plasmid that allows a prokaryote to engage in conjugation with, and pass DNA into, an F-cell.

fitness, W The relative reproductive success of a genotype as measured by survival, fecundity, or other life history parameters.

fluctuation test An experiment by Luria and Delbruck that compared the variance in number of mutations among small cultures with subsamples of a large culture to determine the mechanism of inherited change in bacteria.

fluorescence *in situ* hybridization *In situ* hybridization using a probe coupled with a fluorescent molecule.

footprinting A technique to determine the length of nucleic acid in contact with a protein. While in contact, the free DNA is digested. The remaining DNA is then isolated and characterized.

forward mutation The insertion or deletion of a nucleotide pair or pairs, causing a disruption of the translational reading frame.

founder effect Genetic drift observed in a population founded by a small, nonrepresentative sample of a larger population.

fragile site A chromosomal region that has a tendency to break.

fragile–X syndrome The most common form of inherited mental retardation named for its association with an X chromosome with a tip that breaks or appears uncondensed. Inheritance involves imprinting.

frameshift A mutation in which there is an addition or deletion of nucleotides that causes the codon reading frame to shift.

frequency-dependent selection Selection that depends on frequency-dependent fitness.

functional alleles Mutants that fail to complement each other in a *cis–trans* complementation test.

functional genomics Studying the patterns of transcript and protein expression and molecular interactions at a genome level.

fundamental number The number of chromosome arms in a somatic cell of a particular species.

gamete A germ cell having a haploid chromosome complement. Gametes from organisms of opposite sexes fuse to form zygotes.

gametic selection The forces acting to cause differential reproductive success of one allele over another in a heterozygote.

gastrulation The first process of movements and infoldings of the single-layered cell sheet in early animal embryos.

G-bands Eukaryotic chromosomal bands produced by treatment with Giemsa stain.

gene Inherited determinant of the phenotype.

gene amplification A process by which a particular gene or exon increases in number.

gene cloning Production of large numbers of a piece of DNA after it is inserted into a vector and taken up by a cell. Cloning occurs as the vector replicates.

gene conversion In Ascomycetes (fungi), 2 : 2 ratio of alleles is expected after meiosis, but 3 : 1 ratios observed at times. The gene conversion is by repair of heteroduplex DNA through the Holliday model of recombination.

gene disruption Inactivation of gene by the integration of a specially engineered DNA fragment that is introduced.

gene dose The number of copies of a particular gene present in the genome.

gene family A group of genes that arise by duplication of a single ancestral gene. The genes in the family may or may not have diverged from each other.

gene flow The movement of genes from one population to another by way of interbreeding.

geneic balance theory The theory of Bridges that the sex of a fruit fly is determined by the relative number of X chromosomes and autosomal sets.

gene locus Specific site on a chromosome where a gene is located.

gene pool All of the alleles available among the reproductive members of a population.

generalized transduction The ability of certain phages to transduce any gene in the bacterial chromosome.

gene replacement The insertion of a genetically engineered transgene in place of a resident gene, by a double crossover method.

gene trap A method of selecting transgene insertions that have occurred into a gene.

genetic code The linear sequence of nucleotides that specify the amino acids

during the process of translation at the ribosome.

genetic engineering Popular term for recombinant DNA technology.

genetic load The relative decrease in the mean fitness of a population due to the presence of genotypes that have less than the highest fitness.

genetic polymorphism The occurrence together in the same population of more than one allele at the same locus, with the least frequent allele occurring more frequently that can be accounted for by mutation.

genetic variance The phenotypic variance associated with the average difference in phenotype among different genotypes.

genome The entire genetic complement of a prokaryote or virus; the haploid genetic complement of a eukaryote.

genomic library A set of cloned fragments making up the entire genome of an organism or species.

genophore The chromosome (genetic material) of prokaryotes and viruses.

genotype The genes that an organism possesses.

germ cells Sperm cell and egg cell. Also called reproductive cells.

germinal mutation Mutations occurring in the cells that are destined to develop into gametes.

germ-line theory A theory to account for the high degree of antibody variability found. The theory suggests that every B lymphocyte has all the genes for every type of immunoglobulin but transcribes only one.

giemsa stain   A stain that stains the phosphate groups of DNA.

gynandromorphs   Mosaic individuals of both the male and the female phenotype.

haemoglobin   Haem–iron and globin—the oxygen transporting blood protein.

haemophilia   A human disease in which the blood fails to clot, caused by a mutation in a christmas factor gene encoding a clotting protein, inherited as an X-linked recessive phenotype.

haplodiploidy   The sex-determining mechanism found in some insect groups among which males are haploid and females are diploid.

haploid   The state of having one copy of each chromosome per nucleus or cell.

haplosufficient   Description of a gene that, in a diploid cell, can promote wild type function in single copy.

haplotype   A series of alleles found at linked loci on a single chromosome.

harlequin chromosomes   Sister chromatids that stain differently, so one band appears dark and the other band appears light.

HAT medium   A selection medium for hybrid cell lines; contains hypoxanthine, aminopterin, and thymidine. HPRT$^+$ TK$^+$ cell lines can survive in this medium.

heat-shock proteins   Proteins that appear in a cell after the cell has been subjected to elevated temperatures.

helicase   A protein that unwinds DNA at replicating Y-junctions.

helix-turn-helix motif   Configuration found in DNA-binding proteins consisting of a recognition helix and a stabilizing helix separated by a short turn.

hemizygous   The condition of loci on the X chromosome of the heterogametic sex of a diploid species.

heredity   The biological similarity of offspring and parents by inheritance.

heritability   The degree to which the variance in the distribution of a phenotype is due to genetic cause.

hermaphrodite   An individual with both male and female genitalia.

heterochromatin   Chromatin that remains tightly coiled and darkly stained throughout the cell cycle.

heterodimer   A protein with two non-identical polypeptide subunits.

heteroduplex analysis   Duplex DNA formed by strands from different sources which will have loops and bubbles in regions where the two DNAs differ. This heterogeneous DNA is referred to as a heteroduplex.

heteroduplex mapping   The use of heteroduplex analysis to determine the location of various inserts, deletions or heterogeneities in a piece of DNA.

heterogametic   The sex with heteromorphic sex chromosomes. During meiosis, it produces different kinds of gametes with regard to these sex chromosomes.

heterogeneous nuclear mRNA (hnRNA)   The original RNA transcripts found in eukaryotic nuclei before post-transcriptional modifications.

heterokaryon   A cell that contains two or more nuclei from different origins.

heteromorphic chromosomes   Chromosomes of which the members of a homologous pair are not morphologically identical (e.g. sex chromosome).

heteromultimer A protein consisting of more than two non-identical polypeptide subunits.

heteroplasmy The existence within an organism of genetic heterogeneity within the populations of mitochondria or chloroplasts.

heterotrophs Organisms that require an organic form of carbon as a carbon source.

heterozygote A diploid with different alleles at a particular locus.

heterozygote advantage A selection model in which heterozygotes have the highest fitness.

heterozygous gene pair A gene pair having different alleles in the two chromosome sets of the diploid cell.

hfr High frequency of recombination. Bacteria that has F factor in its chromosome and can transfer the chromosome during conjugation.

histocompatibility antigens Antigens that determine the acceptance or rejection of a tissue graft.

histocompatibility genes The genes that encode the histocompatibility antigens.

holandric trait Trait controlled by a locus found only on the Y chromosome. Involves father-to-son transmission.

holoenzyme The complete enzyme including all subunits (sigma factor). Often used with reference to RNA and DNA polymerases.

homeodomain The sixty amino acid proteins translated from the homeo box.

homeobox A consensus sequence of about 180 base pairs discovered in homeotic genes in *Drosophila*.

homeotic gene A gene that controls the fate of segments along the anterior–posterior axis of higher animals.

homeotic mutants Mutants in which a given cell develops along a pathway normally followed by a different cell type.

homodimer A protein consisting of two identical polypeptide subunits.

homogametic The sex with homomorphic sex chromosomes; it produces only one kind of gamete.

homologous chromosomes Members of a pair of essentially identical chromosomes that synapse during meiosis.

homologous recombination Breakage and reunion between homologous lengths of DNA mediated by RecA and RecBCD.

homomultimer A protein consisting of two or more identical polypeptide subunits.

homoplasmy The existence within an organism of only one type of plastid, usually refers to genetic identity of mitochondria or chloroplasts.

homozygote A diploid with identical alleles at a locus.

homozygous gene pair A gene pair having identical alleles in both copies.

hormones Chemicals (protein) that are secreted by one type of cell and act on a second type of cell.

H-Y antigen The histocompatibility Y-antigen, a protein found on the cell surfaces of male.

hybrid   Offspring of unlike parents.

hybrid DNA   DNA whose two strands have different origins.

hybridize   To form a hybrid by performing a cross, to anneal nucleic acid strands from different sources.

hybridoma   A cell resulting from the fusion of a spleen cell and multiple myeloma cell.

hybrid screening   Radioisotope technique used to determine whether a hybrid plasmid contains a gene of interest or DNA region.

hybrid vehicle   An episome or plasmid containing an inserted piece of foreign DNA.

hyperploid   Aneuploid containing a small number of extra chromosomes.

hypervariable loci   Loci with many alleles; variation is due to members of tandem repeats.

hypoploid   Aneuploid with a small number of chromosomes missing.

hypostatic gene   A gene whose expression is masked by an epistatic gene.

idiogram   A photograph of the chromosomes of a cell arranged in an orderly fashion.

idiotypic variation   Variation in the variable parts of immunoglobulin genes.

immune system   The animal cells and tissue that recognize and attack an antigen within the body.

inbreeding   The mating of genetically related individuals.

inbreeding coefficient, F   The probability of autozygosity.

inbreeding depression   A depression of vigour due to inbreeding.

incomplete   dominance   The situation in which both alleles of the heterozygote influence the phenotype. The phenotype is usually intermediate between the two homozygous forms.

independent assortment   Mendel's second rule, describing the independent segregation of alleles of different loci.

induced mutation   A mutation that arises after mutagen treatment.

inducer   An environmental agent that triggers transcription from an operon.

initiation   factors   (IF1,   IF2, IF3)   Proteins required for the proper initiation of translation.

initiator proteins   Proteins that recognize the origin of replication on a replicon and take part in primosome construction.

inosine   A rare base that is important at the wobble position of tRNA anticodons.

insertional translocation   The insertion of a segment from one chromosome into another non-homologous one.

insertion mutagenesis   Change in gene action due to an insertion event that either changes a gene directly or disrupts control mechanisms.

insertion sequences (IS)   Small and simple transposons.

interchromosomal recombination   Recombination   resulting   from independent assortment.

intergenic suppression   A mutation at a second locus that apparently restores the wild-type phenotype to a mutant at a first locus.

internal resolution site A region of replicative transposable elements necessary for co-integrate recombination.

interphase The metabolically active, nondividing stage of the cell cycle.

interrupted mating A mapping technique in which bacterial conjugation is disrupted after specified time intervals.

intervening sequences Sequences of DNA within a gene that are transcribed but later removed (splicing) before translation.

intra-allelic complementation The restoration of activity of an enzyme made of subunits in a heterozygote of two mutants that, when homozygous, do not have activity caused by interaction of the subunits in the protein.

intrachromosomal recombination Recombination resulting from crossing over between two gene pairs.

intragenic suppression A second change within a mutant gene that results in an apparent restoration of the original phenotype.

in vitro A biochemical work done in the test tube rather than in living systems.

in vivo In a living cell or organism.

ionizing radiation Radiation such as X-rays, that causes atoms to release electrons and become ions.

isoaccepting tRNAs The various types of tRNA molecules that carry a specific amino acid.

isochromosome A chromosome with two genetically and morphologically identical arms.

isoschizomers Restriction endonucleases that recognize the same target sequence and cleave it the same way.

isotope One of several forms of an atom having the same atomic number but different atomic mass.

junctional diversity Variability in immunoglobulins caused by variation in the exact crossover point during V-J, V-D, and D-J joining.

kappa particles Particles that give a paramecium, the killer phenotype.

karyokinesis The process of nuclear division.

karyotype The chromosome complement of a cell.

kin selection The mode of natural selection that acts on an individual's inclusive fitness.

kinetochores The chromosomal attachment points for the spindle fibres, located within the centromeres.

Klinefelter's syndrome An abnormal human male (sterile male) phenotype due to an extra X chromosome (XXY).

***lac* operon** The inducible operon including three loci involved in the uptake and breakdown of lactose.

lagging strand Strand of DNA being replicated.

lampbrush chromosomes Chromosomes of amphibian oocytes having loops suggestive of a lampbrush.

leader peptide gene A small gene within the attenuator control region of repressible amino acid operons.

leader sequence The sequence at the 5′ end of an mRNA that is not translated into protein.

leader transcript The mRNA transcribed by the attenuator region of repressible amino acid operons. The translation of a short leader peptide gene.

leading strand   Strand of DNA being replicated continuously.

leaky mutation   A mutation that confers a mutant phenotype but still retains a low but detectable level of wild type function.

leptonema   The first stage of prophase I of meiosis in which chromosomes become distinct.

lethal-equivalent alleles   Alleles whose total effect is that of lethality.

lethal gene   A gene whose expression results in the death of an organism.

leucine zipper   Configuration of DNA-binding protein in which leucine residues on two helices interdigitate, to stabilize the protein.

leukemia   Cancer of the bone marrow resulting in excess production of leukocytes.

level of significance   The probability value in statistics used to reject the null hypothesis.

ligase   An enzyme that can rejoin a broken phosphodiester bond in a nucleic acid.

linkage   The association of loci on the same chromosomes.

linkage   disequilibrium   The condition among alleles at different loci such that allelic combinations in a gamete do not occur.

linkage equilibrium   The condition among alleles at different loci such that any allelic combination in a gamete occurs as the product of the frequencies of each allele at its own locus.

linkage groups   Associations of loci on the same chromosome. There are as many linkage groups as there are homologous pairs of chromosomes.

linkage number   The number of times one helix strand coils around the other.

locus   The position of a gene on a chromosome.

lod score   Determining the most likely recombination frequency between two loci from pedigree data.

lyon hypothesis   The hypothesis that suggests that the Barr body is an inactivated X chromosome (in squamous epithelial cells).

lysate   The contents released from a lysed cell.

lysis   The breaking open of a cell by the destruction of its wall or membrane.

lysogenic   The state of a bacterial cell that has an integrated phage (prophage) in its chromosome.

macromolecule   A large polymer such as DNA, a protein, or a polysaccharide.

mapping   The study of the position of genes on chromosomes.

map unit   The distance equal to 1% recombination between two loci.

marker   A locus or allele whose phenotype provides information about a chromosome or chromosomal segment during genetic analysis.

marker retention   A technique used to test the degree of linkage between two mitochondrial mutations in yeast.

mate-killer   A phenotype of *Paramecium* induced by intracellular bacteria like mu particles.

maternal effect The effect of the maternal parent's genotype on the phenotype of her offspring.

maternally expressed gene A gene that contributes to the phenotype of an offspring on the basis of its expression in the mother.

mating type In many species of microorganisms, individuals can be divided into two types. Mating can take place only between individuals of opposite mating types.

medium Any material on which experimental cultures are grown.

megabase One million nucleotide pairs.

meiospore Cell that is one of the products of meiosis in plants.

melting Denaturation of DNA.

Mendelian ratio A ratio of progeny phenotypes corresponding to the operation of Mendel's law.

merozygote A bacterial cell having a second copy of a particular chromosomal region in the form of an exogenote.

metabolism The chemical reactions that take place in a living cell.

metacentric chromosome A chromosome whose centromere is located in the middle.

metafemale A fruit fly with an X/A ratio greater than unity.

metaphase The stage of cell division in which spindle fibres are attached to kinetochores. The chromosomes are in the equatorial plane of the cell.

metastasis The migration of cancerous cells to other parts of the body.

microfilaments The smallest diameter cable system of the cytoskeleton. Microfilament cables are composed of actin polymers.

microsatellite DNA A type of repetitive DNA based on very short repeats such as dinucleotides.

microtubule organizing centre The part of the microtubule cytoskeleton in which all the minus ends of the microtubules are clustered, ordinarily, this is near the centre of the cell.

minisatellite DNA A type of repetitive DNA sequence based on short repeat sequences with a unique common core, used for DNA fingerprinting.

mismatch repair A repair system for repairing damage to DNA that has already been replicated.

mitotic crossover A crossover resulting from the pairing of homologues in a mitotic diploid.

modifier gene A gene that affects the phenotypic expression of another gene.

molecular chaperone A protein that aids in the folding of a second protein. The chaperone prevents proteins from forming structures that would be inactive.

molecular evolutionary clock A measurement of evolutionary time in nucleotide substitutions per year.

molecular imprinting The phenomenon in which there is differential expression of a gene depending on whether it was maternally or paternally inherited.

monoclonal antibody The antibody from a clone of cells producing the same antibody. An individual with multiple myeloma usually produces monoclonal antibodies.

monohybrids Offspring of parents that differ in only one genetic characteristic, usually implies heterozygosity at a single locus under study.

morphogen A molecule that can induce the acquisition of different cell fates.

morphological mutation A mutation affecting some aspect of the appearance of an individual.

mosaicism The condition of being a mosaic.

mRNA-Messenger RNA The basic function of the nucleotide sequence of mRNA is to determine the amino acid sequence in proteins.

multifactorial A character that is determined by some unspecified combination of genetic and environmental factors.

multihybrid An organism heterozygous at numerous loci.

multimer A protein consisting of two or more subunits.

multiple allelism The existence of several known alleles of a gene.

multiple hit hypothesis The proposal that a single cell must receive a series of mutational events to become malignant or cancerous.

mu particles Bacteria like particles found in the cytoplasm of *Paramecium* that have the mate-killer phenotype.

mutability The ability to change.

mutagen An agent that is capable of increasing the mutation rate.

mutant allele An allele differing from the allele found in the standard or wild type.

mutants Alternative phenotypes to the wild-type, the phenotypes produced by alternative alleles.

mutation The process by which a gene or chromosome changes structurally, the end result of this process.

mutational load Genetic load caused by mutation.

mutational specificity The constellation of mutational damage that characterizes a particular mutagen.

mutation breeding Use of mutagens to develop variants that can increase agricultural yield.

mutation event The actual occurrence of a mutation in time and space.

mutation rate The proportion of mutants per cell division in bacteria or single-celled organisms or the proportion of mutations per gamete in higher organisms.

mutator mutations Mutations of DNA polymerase that increase the overall mutation rate of a cell or of an organism.

muton A term coined by Benzer for the smallest mutable site within a cistron. This smallest part of a gene that can be involved in a mutation event, is now known to be a nucleotide pair.

natural selection The process in nature whereby one genotype leaves more offspring than another genotype because of superior life history attributes such as survival or fecundity.

nearest-neighbour analysis A technique of transferring radioactive atoms between adjacent nucleotides in DNA used to demonstrate that the two strands of DNA run in opposite directions.

negative assortative mating Preferential mating between phenotypically unlike patterns.

negative interference The phenomenon whereby a crossover in a particular region enhances the occurrence of other apparent crossovers in the same region of the chromosome.

neomorphic mutation A mutation with phenotypic effects due to the production of a novel gene product or novel pattern of gene expression.

neutral DNA sequence variation Variation in DNA sequence that is not under natural selection.

neutral gene hypothesis The hypothesis that most genetic variation in natural populations is not maintained by selection.

neutral mutations Mutations that have no effect on the survivorship and reproductive rate of the organisms that carry them.

neutral petite A petite that produces all wild-type progeny when crossed with wild type.

nicking Nuclease action to sever the sugar-phosphate backbone in one DNA strand at one specific site.

nitrocellulose filter A type of filter used to hold DNA for hybridization.

non-disjunction The failure of a pair of homologous chromosomes to separate properly during meiosis.

nonhistone proteins The proteins remaining in chromatin after the histones are removed. The scaffold structure is made of nonhistone proteins.

nonsense codon One of the mRNA sequences (UAA, UAG, UGA) that signals the termination of translation.

nonsense mutations Mutations that change a codon for an amino acid into a nonsense codon.

nonsense suppressor A mutation that produces an altered tRNA that will insert an amino acid in translation in response to a nonsense codon.

normal distribution Any of a family of bell-shaped frequency curves whose relative position and shape are defined on the basis of the mean and standard deviation.

northern blotting Transfer of electrophoretically separated RNA molecules from a gel onto an absorbent sheet, which is then immersed in a labelled probe that will bind to the RNA of interest.

nuclear transplantation The technique of placing a nucleus from another source into an enucleated cell.

nuclease One of the several classes of enzymes that degrade nucleic acid.

nucleoid A DNA mass within a chloroplast or mitochondrion.

nucleolar organizer The chromosomal region around which the nucleolus forms; site of ribosome construction.

nucleoprotein The substance of eukaryotic chromosomes consisting of proteins and nucleic acids.

nucleoside A sugar with base compound, that is a nucleotide precursor.

nucleosomes Arrangements of DNA and histones forming regular spherical structures in eukaryotic chromatin.

nucleotide Subunits that polymerize into nucleic acids (DNA or RNA), each nucleotide consists of a nitrogenous base, a sugar, and one or more phosphate groups.

nucleotide pair substitution The replacement of a specific nucleotide pair by a different pair, often mutagenic.

null allele A mutant allele that produces no product.

null hypothesis The statistical hypothesis that states that there are no differences between observed and expected data.

nullisomic A diploid cell missing both copies of the same chromosome.

null mutation A mutation that results in complete absence of function for the gene.

okazaki fragment In DNA replication, a small segment of DNA synthesized as part of the lagging strand.

oligonucleotide A short segment of synthetic DNA.

oncogene Genes capable of transforming a cell. They are found in the active state in retroviruses and transformed cells and in the inactive state in non-transformed cells in which they are called proto-oncogenes.

one-gene-one-enzyme hypothesis Hypothesis of Beadle and Tatum that states that one gene controls the production of one enzyme.

oogenesis The process of ovum formation in female animals.

operator Is the site on DNA at which a repressor protein **binds** to prevent transcription.

operon A sequence of adjacent genes under the transcriptional control of the same operator.

organelle A subcellular structure having a specialized function.

origin of replication The point at which DNA replication is initiated.

outbreeding The mating of genetically unrelated individuals.

overdominance A phenotypic relation in which the phenotypic expression of the heterozygote is greater than that of homozygote.

oxidatively damaged base One origin of spontaneous DNA damage.

P1 Parental generation.

PAC (P1-based artificial chromosome) A derivative of phage P1 engineered as a cloning vector for carrying large inserts.

pachynema The stage of prophase I of meiosis in which chromatids are first distinctly visible.

palindrome A sequence that reads the same regardless of which direction one starts from; the sites of recognition of type II restriction endonucleases.

paracentric inversion A chromosomal inversion that does not include the centromere.

paralogous genes Two genes in the same species that have evolved by gene duplication.

passenger DNA Foreign DNA incorporated into a plasmid.

pathogen An organism that causes diseases in another organism.

pattern formation The developmental processes resulting in the complex shape and structure of higher organisms.

pedigree A representation of the ancestry of an individual or family; a family tree.

penetrance The normal appearance in the phenotype of genetically controlled traits.

peptide bond A bond joining two amino acids.

peptidyl site The site on the ribosome occupied by the peptidyl tRNA just before peptide bond formation.

peptidyl transferase The enzymatic centre responsible for peptide bond formation during translation at the ribosome.

pericentric inversion A chromosomal inversion that includes the centromere.

phenocopy A phenotype that is not genetically controlled but looks like genetically controlled phenotypes.

phenome The description of the phenotypes elicited by mutations in all of the genes in the genome, singly and in combination.

phenotype The observable attributes of an organism.

phenotypic sex determination Sex determination by non-genetic means.

pheromone A chemical signal, analogous to a hormone, that passes information between individuals.

phosphodiester bond A diester bond linking two nucleotides together (between phosphoric acid and sugars) to form the nucleotide polymers DNA and RNA.

photoreactivation The process whereby dimerized thymines in DNA are restored by an enzyme (deoxyribodipyridimine photolyase) that requires light energy.

phyletic evolution The formation of new related species as a result of the splitting of previously existing species.

phyletic gradualism The process of gradual evolutionary change over time.

phylogenetic tree A diagram showing evolutionary lineages of organisms.

physical map The ordered and oriented map of cloned DNA fragments on the genome.

plasmid An autonomous, self-replicating genetic particle, usually of double-stranded DNA.

pleiotropic mutation A mutation that has effects on several different characters.

pleiotropy The phenomenon whereby a single mutant affects several apparently unrelated aspects of the phenotype.

ploidy The number of chromosome sets.

point mutations Small mutations that consist of a replacement, addition or deletion of one or a few bases.

poky mutations Mutations of *Neurospora* that produce a petite phenotype.

polarity Meaning "directionality" and referring either to an effect seen in only one direction from a point of origin or to the fact that linear entities (such as single strand of DNA) have ends that differ from each other.

polarity gene   A mitochondrial gene with alleles that are preferentially found in daughter mitochondria after a recombinational event between mitochondria.

polar mutant   An organism with a mutation, usually within an operon, that prevents the expression of genes distal to itself.

poly(A) tail   A sequence of adenosine nucleotides added to the 3′ end of eukaryotic mRNAs.

polycistronic mRNA   An mRNA that codes for more than one protein.

polydactyly   More than five fingers or toes or both inherited as an autosomal dominant phenotype.

polymerase   chain   reaction (PCR)   A method to rapidly amplify DNA segments in cycles of denaturation, primer addition, and replication.

polymerize   To form a complex compound by linking together many smaller elements.

polymorphism   The occurrence in a population of several phenotypic forms associated with alleles of one gene or homologous of one chromosome.

polynucleotide phosphorylase   An enzyme that can polymerize disphosphate nucleotides without the need for a primer, the function of this enzyme *in vivo* is probably in its reverse role as an RNA exonuclease.

polypeptide   A chain of linked amino acids, a protein.

polyploid   A cell having three or more chromosome sets or an organism composed of such cells.

polysaccharide   A   biological polymer composed of sugar subunits.

polysome   The configuration of several ribosomes simultaneously translating the same mRNA. Shortened form of the term polyribosome.

polytene   chromosome   Large chromosome consisting of many chromatids formed by rounds of endomitosis followed by synapsis.

population   A group of organisms of the same species relatively isolated from other groups of the same species.

positional cloning   Cloning a gene knowing only its chromosomal location.

position effect   An alteration of phenotype caused by a change in the relative arrangement of the genetic material.

post-transcriptional modifications Changes in eukaryotic mRNA made after transcription has been completed. These changes include additions of caps and tails and removal of introns.

Pribnow box   Relatively invariant sequence of six nucleotides in prokaryotic promoters centred at the position −10 with the consensus sequence of TATAAT.

primary structure of a protein The sequence of amino acids in the polypeptide chain.

primase   An enzyme that creates an mRNA primer for Okazaki fragment initiation.

primer   A length of double-stranded DNA that continues as a single-stranded template in the 3′ to 5′ direction during DNA replication.

primordial germ cells Cells in the embryo and foetus which will ultimately give rise to germ-line cells.

primosome A complex of two proteins, a primase and helicase, that initiates RNA primers on the lagging DNA strand during DNA replication.

promoter A region of DNA to which RNA polymerase binds in order to initiate transcription.

promoter proximal element The series of transcription-factor-binding sites located near the core promoter.

prophage A temperate phage integrated into the host chromosome.

prophase The initial stage of mitosis or meiosis in which chromosomes become visible and the spindle apparatus forms.

proplastid Mutant plastids that do not grow and develop into chloroplasts.

propositus (proposita) The person through whom a pedigree was discovered.

proto-oncogene The nonactivated form of a cellular oncogene in an untransformed cell.

pseudoalleles Alleles that are functionally but not structurally allelic within gene families, pseudoalleles are alleles that are not expressed.

pseudoautosomal gene A gene that occurs on both sex-determining heteromorphic chromosomes.

pseudodominance The phenomenon in which a recessive allele shows itself in the phenotype when only one copy of the allele is present, as in hemizygous alleles or in deletion heterozygotes.

pseudogene A mutationally inactive gene for which no functional counterpart exists in wild-type populations.

Punnett square A diagrammatic representation of a particular cross used to predict the progeny of the cross.

purines Nitrogenous bases of which guanine and adenine are found in DNA and RNA.

pyrimidines Nitrogenous bases of which thymine is found in DNA, uracil in RNA, and cytosine in both.

quantitative inheritance The mechanism of genetic control of traits showing continuous variation.

quantitative trait locus A gene affecting the phenotypic variation in continuously varying traits such as height, weight and so forth.

quaternary structure of a protein The multimeric constitution of the protein.

random genetic drift Changes in allelic frequency due to sampling error.

random mating The mating of individuals in a population such that the union of individuals with the trait under study occurs according to the product rule of probability.

random strand analysis Mapping studies in organisms that do not keep together all the products of meiosis.

reading frame The codon sequence that is determined by reading nucleotides in groups of three from some specific start codon.

reannealing Realignment of two single DNA strands to re-form a DNA double helix that had been denatured.

recessive   An allele that does not express itself in the heterozygous condition.

recessive allele   An allele whose phenotypic effect is not expressed in a heterozygote.

reciprocal altruism   An apparently altruistic behaviour done with the understanding that the recipient will reciprocate at some future date.

reciprocal crosses   A pair of crosses of the type genotype A female × genotype B male and B female × A male.

reciprocal translocation   A translocation in which part of one chromosome is exchanged with a part of a separate non-homologous chromosome.

recombinant DNA technology   Technique of gene cloning. Recombinant DNA refers to the hybrid of foreign and vector DNA.

recombinant frequency   The proportion of recombinant cells or individuals.

recombinants   Offspring with allelic arrangements made up of a combination of the original parental alleles. The term is used in mapping studies.

recombinational repair   The repair of a DNA lesion through a process similar to recombination that uses recombination enzymes.

recon   The smallest recombinable unit within a gene. The term was coined by Benzer.

rec system   Several loci-controlling genes (*recA*, *recB*, *recC*, and others) involved in post-replicative DNA repair.

regression coefficient   The slope of the straight line that most closely relates two correlated variables.

regulatory gene   A gene primarily involved in control of the production of another gene's product.

relaxed mutant   A mutant that does not exhibit the stringent response under amino acid starvation.

release factors (RF1 and RF2)   Proteins in prokaryotes responsible for termination of translation and release of the newly synthesized polypeptide .

repetitive DNA   DNA made up of copies of the same nucleotide sequence.

replication   The process of copying, DNA synthesis.

replication fork   The point at which the two strands of DNA are separated to allow replication of each strand.

replisome   The DNA-replicating structure at the Y-junction consisting of two DNA polymerase III enzymes and a primosome.

reporter gene   A gene whose phenotypic expression is easy to monitor, used to study tissue-specific promoter and enhancer activities in transgenes.

repressible system   A coordinated group of enzymes involved in an anabolic pathway is repressible if excess quantities of the end product of the pathway lead to the termination of transcription of the genes.

repressor   The protein product of a regulator gene that acts to control transcription of inducible and repressible operons.

repressor protein   A molecule that binds to the operator and prevents transcription of an operon.

reproductive success The relative production of offspring by a particular genotype.

repulsion Allelic arrangement in which each homologous chromosome has mutant and wild-type alleles.

response element Sequence usually located a short distance upstream of a promoter that makes gene expression responsive to some chemical in the cellular environment.

restricted digest The results of the action of a restriction endonuclease on a DNA sample.

restriction endonucleases Endonucleases that recognize certain DNA sequences which they cleave. They are thought to protect cells from viral infection and are useful in recombinant DNA work.

restriction fragment length polymorphism (RFLP) Variations in banding patterns of electrophoresed restriction digests.

restriction map A physical map of a piece of DNA showing recognition sites of specific restriction endonucleases separated by lengths marked in numbers of bases.

restrictive temperature A temperature at which temperature-sensitive mutants display the mutant phenotype.

retrotransposon A class of genetic elements that includes retroviruses and transposons that have an intermediate RNA stage.

retrovirus An RNA virus that replicates by first being converted into double-stranded DNA.

reverse transcriptase An enzyme that uses RNA as a template to synthesize DNA.

rho-dependent terminator A DNA sequence signalling the termination of transcription; termination requires the presence of the rho protein.

rho-independent terminator A DNA sequence signalling the termination of transcription; the rho protein is not required for termination.

rho protein A protein that is involved in the termination of transcription.

ribosomes Organelles at which translation takes place, they are made up of two subunits consisting of RNA and proteins.

ribozyme Catalytic or autocatalytic RNA.

RNA editing The insertion of uridines into mRNAs after transcription is completed.

RNA polymerase The enzyme that polymerizes RNA by using DNA as a template. It can also act as a primase, initiating Okazaki fragments during DNA replication.

RNA replicase A polymerase enzyme that catalyses the self-replication of single-stranded RNA.

Robertsonian fusion Fusion of two acrocentric chromosomes at the centromere.

rolling circle replication A model of DNA replication that accounts for a circular DNA molecule producing linear daughter double helices.

R plasmids Plasmids that carry genes that control resistance to various drugs.

rRNA Ribosomal RNA. RNA components of the subunits of the ribosomes.

sampling distribution The distribution of frequencies with which various possible events could occur or a probability distribution defined by a particular mathematical expression.

satellite DNA Highly repetitive eukaryotic DNA primarily located around centromeres. Satellite DNA usually has a different buoyant density than the rest of the cell's DNA.

screening technique A technique to determine the genotype or phenotype of an organism.

secondary sexual characteristics The sex-associated phenotypes of somatic tissues in sexually dimorphic animals.

secondary structure of a protein A spiral or zig-zag arrangement of the polypeptide chain.

sedimentation The sinking of a molecule under the opposing forces of gravitation and buoyancy.

segregation analysis The statistical methodology for inferring modes of inheritance.

segregation load Genetic load caused when a population is segregating less fit homozygotes because of heterozygote advantage.

segregation ratio The proportion of offspring who inherit a given gene or character from a parent.

segregation, rule of Mendel's first principle describing how genes are passed from one generation to the next.

selection coefficients The sum of forces acting to lower the relative reproductive success of genotype.

selection-mutation equilibrium An equilibrium in allelic frequency resulting from the balance between selection against an allele and mutation recreating this allele.

selective neutrality A situation in which different alleles of a certain gene confer equal fitness.

self To fertilize eggs with sperms from the same individual.

selfish DNA A segment of the genome with no apparent function although it can control its own copy number.

semiconservative replication The mode by which DNA replicates, each strand acts as a template for a new double helix.

sequence similarity The level of relationship of two nucleotides or amino acid sequences to one another.

sequence-tagged site A type of small repetitive DNA sequence found throughout a eukaryotic genome.

sex chromosomes Heteromorphic chromosomes whose distribution in a zygote determines the sex of the organism.

sex-controlled traits Traits that appear more often in one sex than in another.

sex determination The genetic or environmental process by which the sex of an individual is established.

sex-determining region Y (SRY) The sex switch, or testis-determining factor in human beings, located on the Y chromosome.

sexduction Sexual transmission of donor *E. coli* chromosomal genes of the fertility factor.

sex-lethal A gene in drosophila, located on the X chromosome, that is a sex switch, directing development towards femaleness when it is in the "on" state.

sex-limited traits Traits expressed in only one sex, which may be controlled by sex-linked or autosomal loci.

sex linkage The location of a gene on a sex chromosome.

sex-linked The inheritance pattern of loci located on the sex chromosomes (usually the X chromosomes in XY species).

sex-ratio phenotype A trait in *Drosophila* whereby females produce mostly, if not only, daughters.

sex switch A gene in mammals, normally found on the Y chromosomes, that directs the indeterminate gonads to develop as testes.

sexual selection The forces, determined by mate choice, acting to cause one genotype to mate more frequently than another genotype.

Shine–Dalgarno hypothesis A proposal that prokaryotic mRNA is aligned at the ribosome by complementarity between the mRNA upstream from the initiation codon and the 3′ end of the 16S rRNA.

shotgun technique Cloning a large number of different DNA fragments as a prelude to selecting one particular clone type for intensive study.

siblings (sibs) Brothers and sisters.

signal hypothesis The major mechanism whereby proteins that must insert into or across a membrane are synthesized by a membrane-bound ribosome. The initial few amino acids synthesized are termed as signal peptides.

signal sequence The N-terminal sequence of a secreted protein, which is required for transport through the cell membrane.

silencer element A *cis*-regulatory sequence that can reduce levels of transcription from an adjacent promoter.

silent mutation A mutation that has no effect on the function of a gene product.

single-strand binding proteins Proteins that attach to single-stranded DNA, usually near the replicating Y-junction to stabilize.

single copy DNA DNA sequences present in only one copy per haploid genome.

single nucleotide polymorphism A nucleotide pair difference at a given location in the genomes of two more naturally occurring individuals.

sister chromatid exchange An event similar to crossing over that can take place between sister chromatids at mitosis or at meiosis. It was detected in harlequin chromosomes.

site-specific recombination A crossover event, such as the integration of phage lambda, that requires homology of only a very short region and uses an enzyme specific for that recombination.

snRNPs Small nuclear ribonucleo-proteins. They are components of the spliceosome, the intron-removing apparatus in eukaryotic nuclei.

somatic cell genetics Asexual genetics, including the study of somatic mutation, assortment.

somatic doubling A disruption of the mitotic process that produces a cell with twice the normal chromosome number.

somatic hypermutation The occurrence of a high level of mutation in the variable regions of immunoglobulin genes.

somatic mutation theory Suggests that mutation of a basic immunoglobulin gene accounts for all the different types of immunologbulins produced by B lymphocytes.

SOS box The region of the promoters of various genes that is recognized by the LexA repressor. Release of repression results in the induction of the SOS response.

SOS response Repair systems (RecA, Uvr) induced by the presence of single-stranded DNA that usually occurs from post-replicative gaps. Single-stranded DNA is involved in the inactivation of the LexA repressor, thereby inducing the response.

Southern blotting A method, first devised by E.M. Southern, used to transfer DNA fragments from an agarose gel to a nitrocellulose gel.

specialized transduction Form of transduction based on faulty looping out by a temperate phage, only neighbouring loci to the attachment site can be transduced.

speciation A process whereby, over time one species evolves into a different species or whereby one species diverges to become two or more species.

S phase The interphase of the cell cycle in which DNA synthesis occurs.

spindle The microtubule apparatus that controls chromosome movement during mitosis and meiosis.

spliceosome Protein–RNA complex that removes introns in eukaryotic nuclear RNAs.

splicing The removal of introns and joining together of exons in RNA.

spontaneous mutation A mutation occurring in the absence of exposure to mutagens.

SSCP or SSCA Single-stranded conformation polymorphism or analysis for point mutation screening.

stacking The packing of the flattish nitrogen bases at the centre of the DNA double helix.

stringent factor A protein that catalyses the formation of an unusual nucleotide during the stringent response under amino acid starvation.

stringent response A translational control mechanism of prokaryotes that repress tRNA and rRNA synthesis during amino acid starvation.

structural alleles Mutant alleles that are altered at identical base pairs.

structural gene A gene encoding the amino acid sequence of a protein.

structural genomics The large-scale analysis of three-dimensional protein structure.

submetacentric chromosome A chromosome whose centromere lies between its middle and its end but closer to the middle.

subtelocentric chromosome A chromosome whose centromere lies between its middle and its end but closer to the end.

subvital gene A gene that causes the death of some proportion of the individuals that express it.

supercoiling Negative or positive coiling of double-stranded DNA that differs from the relaxed state.

supergenes Several loci, which usually control related aspects of the phenotype, in close physical association.

superinfection Phage infection of a cell that already harbours a prophage.

suppressor gene A gene that, when mutated, apparently restores the wild-type phenotype to a mutant of another locus.

survivor factor A ligand that, when bound to a receptor, blocks the activation of the apoptosis pathway.

Svedberg unit A unit of sedimentation during centrifugation.

synapsis The point-by-point pairing of homologous chromosomes during zygotene or in certain dipteran tissues that undergo endomitosis.

synaptonemal complex A protein-aceous complex that apparently mediates synapsis during zygotene stage and then disintegrates.

synteny test A test that determines whether two loci belong to the same linkage group by observing concordance in hybrid cell lines.

synthetic medium A chemically defined substrate upon which microorganisms are grown.

target theory A theory that predicts response curves based on the number of events required to cause the phenomenon used to determine whether point mutations are single events.

TATA box An invariant DNA sequence at about -25 in the promoter region of eukaryotic genes, analogous to the pribnow box in prokaryotes.

tautomeric shift Reversible shifts of proton position in a molecule.

telocentric chromosome A chromosome whose centromere lies at one of its ends.

telophase The terminal stage of mitosis or meiosis in which chromosomes uncoil, the spindle breaks down, and cytokinesis usually occurs.

temperate phage A phage that can enter into lysogeny with its host.

terminator sequence A sequence in DNA that signals the termination of transcription to RNA polymerase.

test-cross The cross of an organism with a homozygous recessive organism.

testis-determining factor (TDF) General term for the gene determining maleness in human beings.

tetranucleotide hypothesis The hypothesis, based on incorrect information, that DNA could not be the genetic material because its structure was too simple.

theta structure An intermediate structure formed during the replication of a circular DNA molecule.

three-point cross A cross involving three loci.

totipotent The state of a cell that can give rise to any and all adult cell types, as compared with a differentiated cell whose fate is determined.

trans The term literally means "across" and refers usually to the geometric configuration of mutant alleles across from each other on a homologous pair of chromosomes.

trans-acting Referring to mutations of, for example, a repressor gene that acts through a diffusible protein product. The normal mode of action of most recessive mutations.

transcription The process whereby RNA is synthesized from a DNA template.

transcription factors Eukaryotic proteins that aid RNA polymerase to recognize promoters analogous to prokaryotic sigma factors.

transducing particle A defective phage, carrying part of the host's genome.

transduction The movement of genes from a bacterial donor to a bacterial recipient with a phage as the vector.

transfection The process by which exogenous DNA in solution is introduced into cultured cells.

transformational evolution An evolutionary process that results from successive changes in the properties of individuals during their own lifetime.

transgene A gene that has been modified by externally applied recombinant DNA techniques and reintroduced into the genome by germ-line transformation.

transient diploid The stage of the life cycle of predominantly haploid fungi (and algae) during which meiosis occurs.

transition A type of nucleotide pair substitution involving the replacement of a purine with another purine or of a pyrimidine with another pyrimidine.

translation The ribosome-mediated production of a polypeptide whose amino acid sequence is derived from the codon sequence of an mRNA molecule.

translocation The relocation of a chromosomal segment in a different position in the genome.

transmission genetics The study of the mechanisms involved in the passage of a gene from one generation to the next.

transposable element A general term for any genetic unit that can insert into a chromosome, exit, and relocate, and includes insertion sequenes, transposons, some phages, and controlling elements.

transposase The enzyme encoded by transposable elements that undergo conservative transposition.

transposition The process by which mobile genetic elements move from one location in the genome to another.

transposon A mobile piece of DNA that is flanked by terminal repeat sequences and typically bears genes coding for transposition functions.

transversion A type of nucleotide pair substitution in which a pyrimidine replaces a purine or vice versa.

trinucleotide repeat disease The expansion of a 3-bp repeat from a relatively low number of copies to a high number of copies that is responsible for a genetic disease, e.g. Huntington's disease.

triplet The three nucleotide pairs that compose a codon.

triploid A cell having three chromosome sets.

trisomic Basically a diploid with an extra chromosome of one type, producing a chromosome number of form 2n + 1.

trisomy Having three copies of a particular chromosome.

tumour-suppressor gene A gene encoding a protein that suppresses tumour formation. The wild-type alleles of tumour suppressor genes are thought to function as negative regulators of cell proliferation.

tumour virus A virus that is capable of inducing a cancer.

Turner's syndrome An abnormal human female phenotype produced by the presence of only one X chromosome (XO).

twin spot A pair of mutant sectors within wild-type tissue, produced by a mitotic crossover in an individual of appropriate heterozygous genotype.

two-hybrid system A pair of *Saccharomyces cerevisiae* vectors used for detecting protein–protein interaction.

unequal crossover A crossover between homologues that are not perfectly aligned.

uniparental inheritance The transmission of certain phenotypes from one parental type to all the progeny, such inheritance is generally produced by organelle genes.

unstable mutation A mutation that has high frequency of reversion; a mutation caused by the insertion of a controlling element whose subsequent exit produces a reversion.

uracil A pyrimidine base that appears in RNA in the place of thymine found in DNA.

URF Unassigned reading frame. An open reading frame (ORF) whose function has not yet been determined.

uridine The nucleoside having uracil as its base.

variable A property that may have different values in various cases.

variable expression A variable extent or intensity of phenotypic signs among people with a given genotype.

variable number tandem repeat (VNTR) A chromosomal locus at which a particular repetitive sequence is present in different numbers.

variable region A region in an immunoglobulin molecule that shows many sequence differences between antibodies of different specificities.

variance A measure of the variation around the central class of a distribution, the average squared deviation of the observations from their mean value.

variation The differences among parents and their offspring or among individuals in a population.

variational evolution An evolutionary process that results from changes over time in the frequencies of different types that are present in the population as a result of different rates of survival or reproduction of the different types.

vector A nucleic acid that is able to replicate and maintain itself within a host cell, and that can be used to confer similar properties on any sequence covalently linked to it.

viability The probability that a fertilized egg will survive and develop into an adult organism.

viral transforming gene A gene within a viral genome that can induce abnormal proliferation of cells in culture and, similarly, can induce tumours in infected whole animals.

virulent phage A phage that cannot become a prophage; infection by such a phage always leads to lysis of the host cell.

western blot Membrane carrying an imprint of proteins separated by electrophoresis and can be probed with a labelled antibody to detect a specific protein.

whole genome shotgun (WGS) sequencing The sequencing of ends of clones without regard to any information about the location of the clones.

wild type The genotype of phenotype that is found in nature or in the standard laboratory stock for a given organism.

wobble The ability of certain bases at the third position of an anticodon in tRNA to form hydrogen bonds with several different possible codons.

X : A ratio The ratio between the X-chromosome and the number of sets of autosomes.

X and Y linkage The inheritance pattern of genes found on both the X and T chromosomes.

X chromosome inactivation The process by which the genes of an X chromosome in a mammal can be completely repressed as part of the dosage compensation mechanism.

Xeroderma pigmentosum (XP) A human genetic disease syndrome due to mutation in one of several genes encoding products contributing to excision repair.

X hyperactivation In *Drosophila*, the process by which the structural genes of the male X chromosome are transcribed at the same rate as the two X chromosomes of the female combined.

X linkage The inheritance pattern of genes found on the X chromosome but not on the Y.

X-ray crystallography A technique for deducing molecular structure by aiming a beam of X-rays at a crystal of the test compound and measuring the scatter of rays.

yeast artificial chromosome (YAC) A cloning vector system in *Saccharomyces cerevisiae* employing yeast centromere and replication sequences.

yeast two-hybrid system yeast two-hybrid system An important system for identifying and purifying proteins that bind to a protein of interest.

Y linkage The inheritance pattern of genes found on the Y chromosome but not on the X.

zinc finger A polypeptide motif stabilized by binding of a zinc atom. This helps protein to bind specifically to a DNA sequence commonly found in transcription factors.

zygote The cell formed by the fusion of an egg and a sperm, the unique diploid cell that will divide mitotically to create a differentiated diploid organism.

zygotically acting gene A gene whose product is expressed only in the zygote but not in the maternal contribution to the oocyte.

zygotic induction The sudden release of lysogenic phage from an Hfr chromosome when the prophage enters the F2 cell, and the subsequent lysis of the recipient cell.

INDEX

www.ingramcontent.com/pod-product-compliance
Lightning Source LLC
Chambersburg PA
CBHW031626210326
41599CB00021B/3317